LOOKING AT NATURE

LOOKING

AT NATURE

ELSIE PROCTOR

ADAM & CHARLES BLACK
LONDON

SECOND EDITION 1969
REPRINTED 1971, 1974
ISBN 0 7136 0970 2
FIRST PUBLISHED IN ONE VOLUME 1965
BY A. AND C. BLACK LTD
4, 5 AND 6 SOHO SQUARE LONDON W1V 6AD
THE BOOK IS ALSO AVAILABLE IN FOUR PARTS

ACKNOWLEDGMENTS

The illustrations in this book are by Leonora Box, E. M. Coxhead, Clare Instrell, Bernard West and the author. The colour plates are by Bernard West and C. W. Bacon. Grateful acknowledgment is made to the Radio Times Hulton Picture Library for the photographs on pages 22, 80, 294 and 295; George Thompson for the photograph on page 81; G. A. Clarke and the Royal Meteorological Society for the photographs on pages 136 and 137; Rittener for the photograph of Skomer North Haven, W. J. Sladen for the photograph of guillemots and D. A. J. Hunford for the photograph of Skomer Voles, all on page 315; and Geoffrey Kinns for the photographs on page 321.

MADE IN GREAT BRITAIN
PRINTED BY MORRISON AND GIBB LTD., LONDON AND EDINBURGH

INTRODUCTION

This book combines Nature Study and Natural Science. It contains over a thousand illustrations, many in full colour.

It begins by telling you what to look for and where to look season by season throughout the year. It describes ways of observing, collecting, experimenting and recording in a truly scientific manner. It helps you to find out many interesting facts about plants and animals; the beginnings of life on earth; rocks, soil, air and weather; the discoveries of great natural scientists and the ways in which man is now trying to protect our wild creatures.

The book is in four parts:

Part 1 gives a seasonal account of Nature.

Part 2 describes wild creatures and plants in their own homes or habitats.

Part 3 describes many experiments and things you can make or do for yourself.

Part 4 describes the bodies of plants and animals.

At various points in the book, you will find summaries, questions and answers, lists of words with their meanings and the titles of books which you will find useful and interesting.

CONTENTS

A complete list of the contents of each part will be found at the beginning of each of the four parts. On the next six pages you will find a summary of the contents of the complete book arranged under headings. The index beginning on page 352 will tell you where you can find information on any particular subject in this book.

PLANTS

ANIMALS

WEATHER STUDY

EARTH, AIR AND WATER

WAYS OF RECORDING

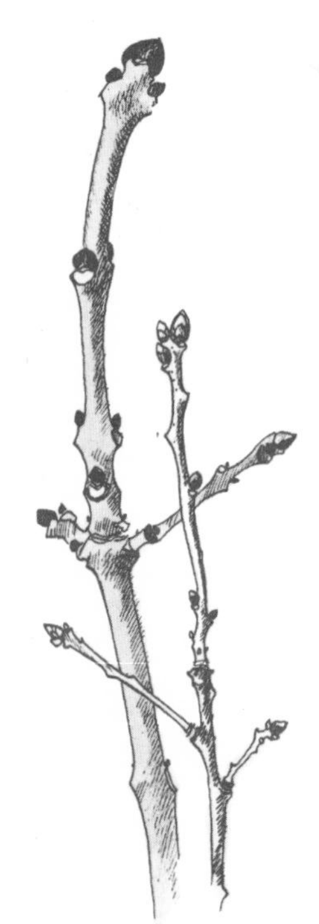

EXPERIMENTING

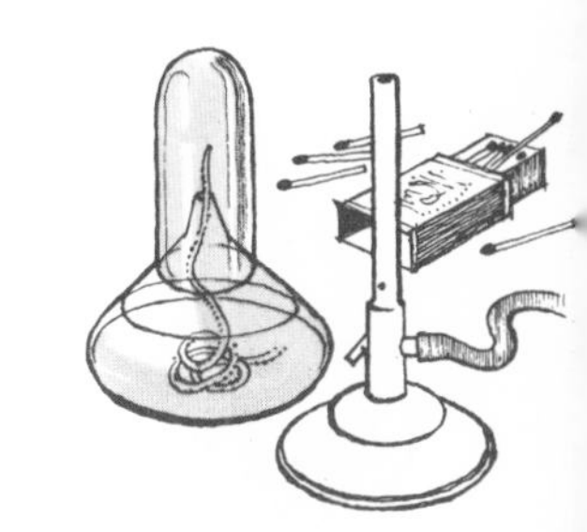

MAKING THINGS FOR YOURSELF

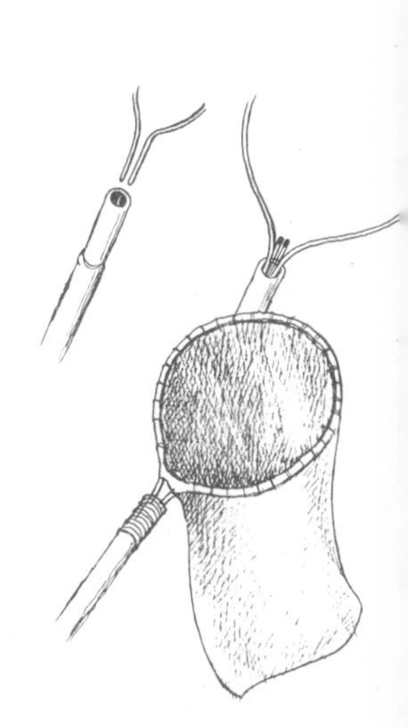

THE WORK OF NATURAL SCIENTISTS

LOOKING BACK AND LOOKING UP

COLOUR PLATES

PART 1

NATURE AWAKE AND ASLEEP

CONTENTS OF PART 1

CONTENTS OF PART 1—*continued*

THE YEAR

There are four seasons in the year, *spring*, *summer*, *autumn* and *winter*.

It is *autumn* when you go back to school after your summer holidays. At this season, plants and animals are preparing for the cold winter.

When *winter* comes, trees are bare, many plants look dead, some birds have left us and many animals are asleep.

In *spring*, all the birds come back, the plants come to life and the sleeping animals awake.

The *summer* sunshine brings out the flowers and ripens the fruit and so the year goes on until your summer holidays are here again.

AUTUMN

If you go for a walk in the woods, or in the park in your town, you will notice that the trees look different from the way they looked in summer. They were green in summer. Now they look yellow or brown. It is the leaves which are changing. Some are yellow; some are red; some are russet-brown. Many leaves are falling from the trees. You can see them on the ground, in woods, roads and gardens.

Do you collect autumn leaves? It is a good time to collect leaves, for the trees no longer need them. These leaves are dead. They are hard and brittle and not green and soft as they were in summer. Look at the leaves which you have collected. You will see that the leaves of every tree have their own special pattern or shape. Some are compound leaves and have other little leaves—or "leaflets". Here are two leaves which have leaflets:

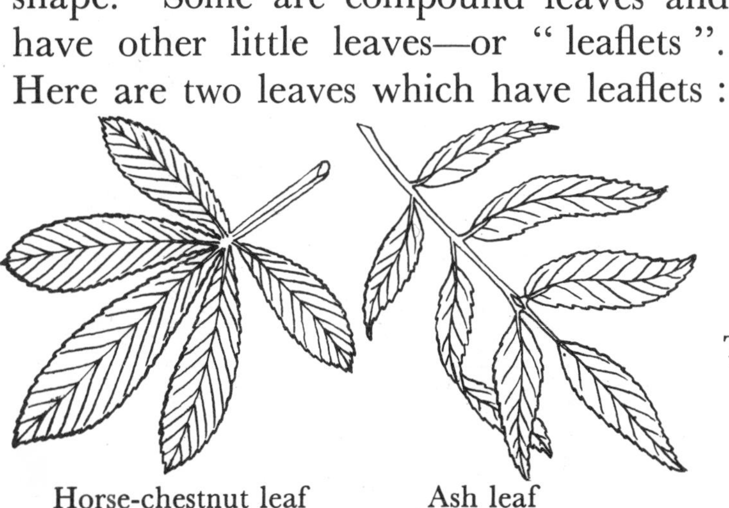

Horse-chestnut leaf   Ash leaf

The trees shown above are:
(1) Sycamore
(2) Lombardy poplar
(3) Silver birch
(4) Oak

Look at your autumn leaves. You have seen that the leaves of the ash and horse-chestnut have leaflets or little leaves on one stalk. Now look at some leaves which are *not* divided up in this way, although they may have *points* or *lobes* on them.

The oak leaf has a wavy outline with lobes like this.

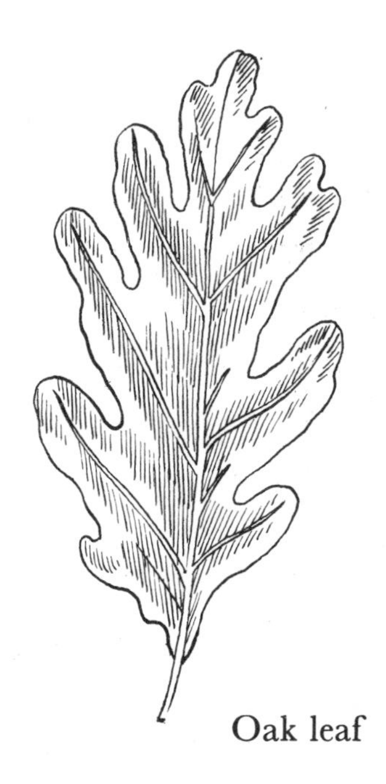

Oak leaf

Sycamore leaf

Sycamore leaves have points like this one on the left.

Maple leaves and plane leaves have points like this, too.

Some leaves have no lobes or points (except at the end). Their leaf-margin is in one flowing line. We call this *entire*.

Here is a lime leaf. It has only one point at the end of the leaf. The leaf is heart-shaped. Can you find any other *un*divided leaves?

Lime leaf

These plants do not drop their leaves in autumn, at least, not *all* their leaves. Their branches are never bare. Fir trees (except for larch) and all evergreens are like this. If you can remember your Christmas tree, you know what a fir tree is like.

This is what your Christmas tree looked like when it was growing in the woods. It has a long, straight trunk and below is one of the twigs from it.

Spruce fir

The Christmas tree is a spruce fir. All fir trees have leaves like needles. These are tough and can stand the winter very well, so the tree does not have to drop them.

Other evergreen trees and shrubs have tough leaves too, but they are not needle leaves. Have you a privet hedge? Privet is an evergreen—so is laurel, which has a large leaf like this one.

Laurel leaf

Privet

BIRDS WHICH ARRIVE IN AUTUMN

Birds are restless in autumn. Some are arriving from lands which are colder than ours. Some are leaving us to go to warmer countries.

It is colder in Norway and Sweden than it is here. Two birds come to us from these countries. They are rather like thrushes. One is called the *redwing*. The other is called the *fieldfare*.

On the right is a redwing. Do you notice that it has stripes around its eyes and it has red peeping from under its wing ?

Redwing

Fieldfare

Here is a fieldfare. It has a grey head and a grey rump. Its back is a rusty brown colour.

You can see that these two birds are very like thrushes. They both have speckled breasts. They are called " winter visitors "—because they stay with us from autumn to spring and then go back to build nests in their summer homes in Norway and Sweden.

MORE ARRIVALS IN AUTUMN (geese)

In autumn we have some visitors from very far north, where it is very cold. Some of these are large birds ; they are geese. They fly in a " V " shape as you can see in the drawing below.

Have you ever seen geese flying like this ?

Do you see their long necks sticking out ?

Here is a picture of a goose which comes to us from cold northern lands such as Greenland and Iceland.

The Brent Goose

It is called the Brent Goose. It feeds on a green weed which it finds in the shallow waters round our coasts and at the mouths of rivers. It goes back to its own country in spring.

Some ducks and other water-birds fly here to spend the winter with us too.

BIRDS WHICH LEAVE US IN AUTUMN
(migrants)

These birds with long, forked tails are swallows. Swallows leave us in autumn and fly southwards.

They travel great distances over sea and land and many of them reach South Africa. They will come back to us again in spring.

There are two other birds which look very like swallows. These are swifts and house-martins.

Here are three birds sitting on a telegraph wire. One is a swallow, one is a house-martin and one is a swift.

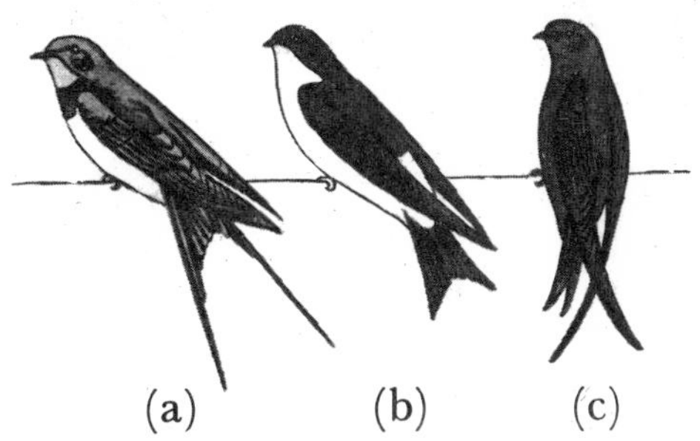

(a) is the *swallow*. It has the longest forked tail. It also has a dark throat.

(b) is the *house-martin*. It has a shorter forked tail and it has not a dark throat.

(c) is a *swift*. It also has a short forked tail and is all black.

Cuckoo

The cuckoo is another bird which leaves us before winter. It leaves us in August.

AUTUMN FRUITS

You have already noticed the colours of the autumn leaves. There is much colour in autumn. Have you seen all the brightly-coloured berries in the hedgerows? Do you collect "hips" and "haws"? These are the fruits of the wild rose and the hawthorn.

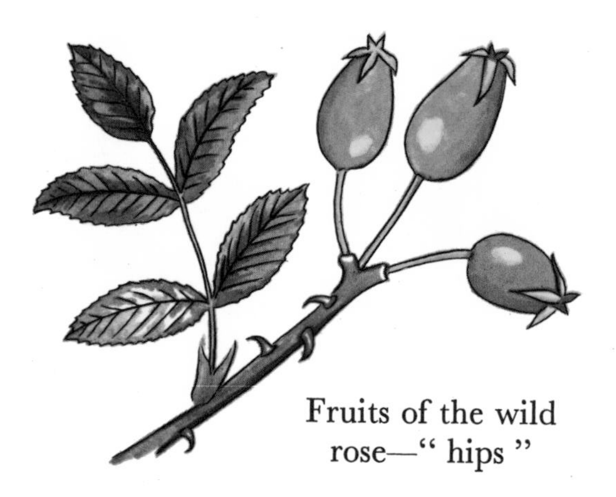

Fruits of the wild rose—"hips"

Birds like to eat the fleshy parts of these fruits. They do not like the seeds and wipe these off their beaks. Sometimes they swallow them, but these seeds have hard walls round them so they pass through the bird's body without harm. If they reach good soil, they can grow into new plants.

Hawthorn fruits—"haws"

Here is another wild fruit. It is the blackberry. Blackberry fruits are nice to eat. Have you ever gathered them to make jam? Every fruit is made up of many little fruits, each one like a tiny plum which has a "stone" with a seed inside it.

Blackberries

MORE WILD FRUITS

Here are some nuts.

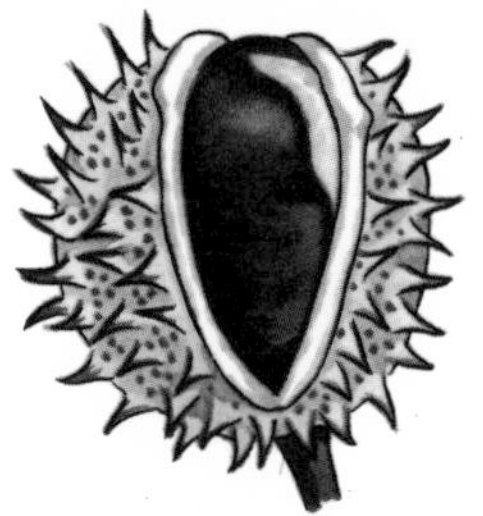

Horse-chestnut

Sweet chestnut

Beech nut

Here are some berries.

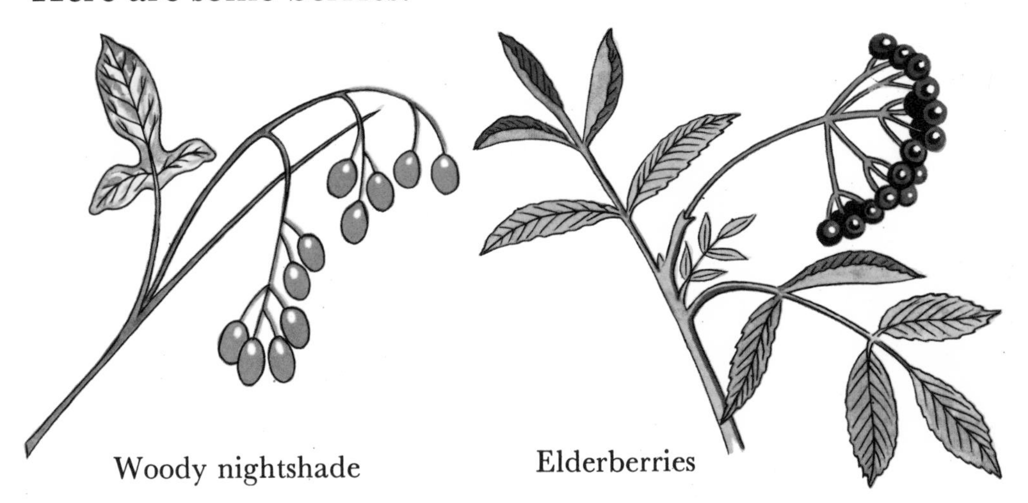

Woody nightshade

Elderberries

Here are some fruits with wings.

Ash " keys "

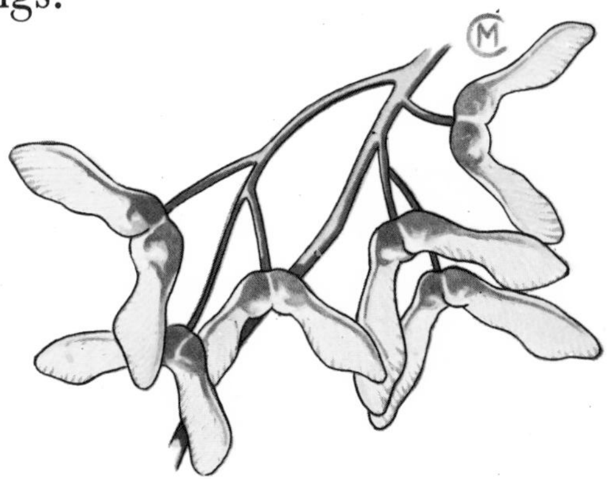

Twin fruits of sycamore

WINTER

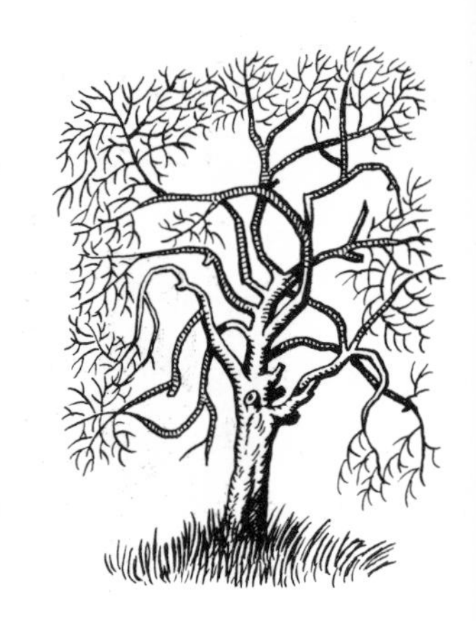

Have you ever wondered why
in winter . . .
very few flowers are seen,
many plants die,
very few birds sing,
very few insects are seen,
many animals go to sleep ?

Winter is a hard time for most living things. It is not so hard for you because you wear warm clothing, you have enough food and you live in a warm, weather-proof house. Animals live out-of-doors and have to search for their food. Most plants have tender leaves which would freeze.

Many animals cannot keep warm. Those with fur and feathers keep warmest. Which one of these animals will *not* keep warm in winter ?

Cats and mice have fur.

Snakes have cold scales.

Birds have feathers.

ANIMALS WHICH SLEEP IN WINTER

Many animals cannot find enough food in winter. Those animals which feed on grubs or insects, or on vegetable food, would find it hard to get enough food—so—they go to sleep.

The hedgehog feeds on insects and little grubs. Food is scarce as winter approaches so he covers himself with leaves and dry grass and sleeps right through the winter. He is rather weak when he wakes up and must find food as soon as possible to build up his strength.

Hedgehog asleep

This is a bat. He feeds on insects but cannot find any in winter. He sleeps from November until April. He has wings, but not like those of a bird. His wings are made of skin stretched between his limbs and body.

Bat

Snails feed on leaves and young shoots and their food dies down in winter, so they make a thin cover over the mouth of their shells and hide away in a crack or crevice until spring. Can you see this covering in the picture?

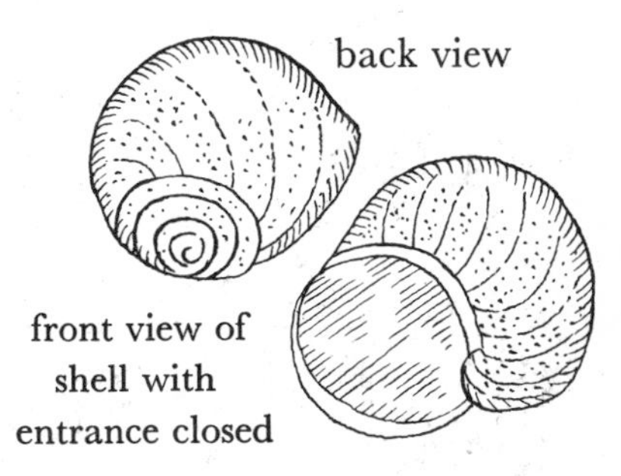

Garden snail

NATURE IN WINTER

Asleep—a dormouse curled up in its nest in a tree hollow.

Awake—a squirrel eating one of its store of nuts.

Below are more sleeping things :

(1) Plants which have died down to flat rosettes close to the ground, such as the daisy, dandelion and plantain.
(2) A hedgehog under a bank of leaves.
(3) Earthworms curled up in figures of 8 in the lowest parts of their burrows. Notice that the worms have closed the ends of the burrows with twigs.
(4) A grass snake asleep among some rotting leaves.
(5) A toad and garden snails in rock crevices.

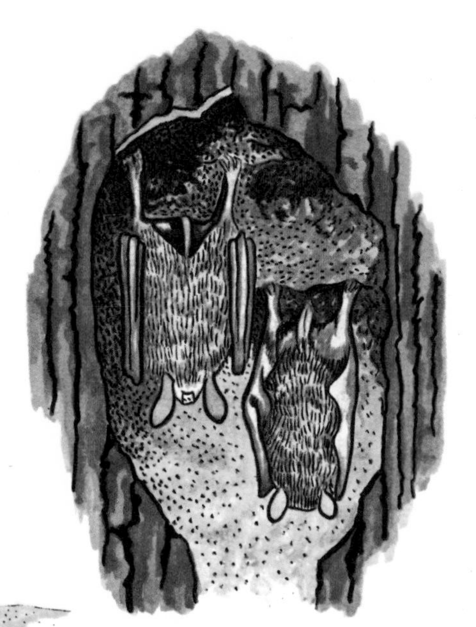

Asleep—two bats hanging upside down inside a hollow tree.

Awake—the robin sings, one of the few birds to be heard in winter.

BIRDS IN OUR GARDENS

Although some birds fly away from us in winter and many animals go to sleep and most insects die, we still have some companions through the cold wintry days. Many birds come to our gardens for food. Do you feed them? They are very hungry. Here are some that come to my feeding-table. Can you see the strings of nuts for the tits?

Can you recognise these birds?

(1) Great Tit—(black cap)
(2) Blue Tit—(blue cap)
(3) Starling—(greenish-black feathers)
(4) Blackbird—(cock is all black and has yellow beak)
(5) House sparrow—(cock has black bib)
(6) Robin—(red breast)

Do plants have to keep warm in winter? No, not in the same way that animals do, but they have to be protected against frost. As you saw in autumn, many trees drop their tender leaves which might be frost-bitten.

Winter bud of horse-chestnut

Where the leaves drop off, a little scar is left on the stem. You can see these scars in the picture of the horse-chestnut twig on the right.

All the winter buds of trees are well protected. In the horse-chestnut the outer scales of the bud are covered with a sticky varnish. This is weather-proof and keeps dampness or frost from harming the little leaves and flowers in the bud.

Here are two more twigs with winter buds :

(1) is an oak twig.

(2) is an ash twig.

Find some winter twigs yourself.

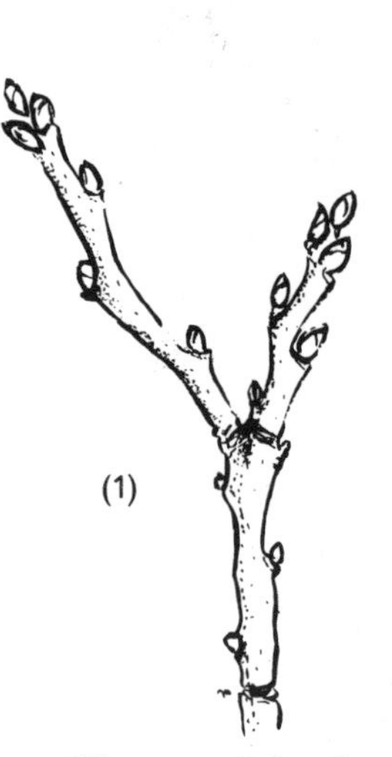

Clustered buds (oak)

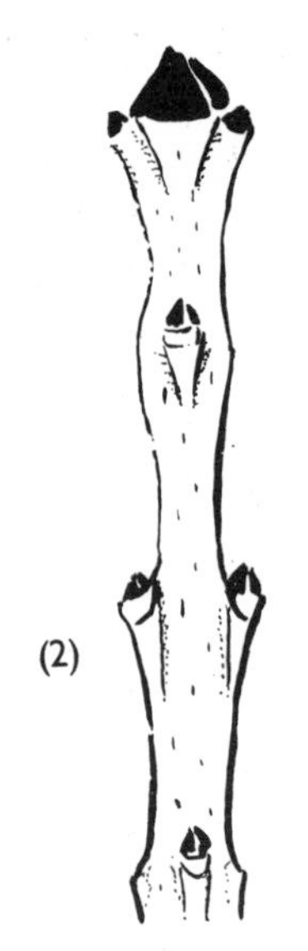

Black sooty buds (ash)

TREE SHAPES

The oak tree in winter

Winter is a good time for recognising trees by their outline. You can see how the tree is built, for there are no leaves to hide its big branches, or its trunk. The bare branches and twigs stand out well against the winter sky. Look at these two trees and compare their shape.

1. The oak is a broad and massive tree. It often has a knobbly trunk and gnarled branches.

2. The shape of the horse-chestnut is like a dome. (Look at the picture.) You will see, too, that the ends of all the twigs bend upwards.

The horse-chestnut tree in winter

LOOKING AT THE SKY IN DECEMBER

December is a good time to look at stars. The stars can be seen well in clear, frosty weather. (It is dark a long time before you have to go to bed.) Let us go out one evening and do some star-gazing. First of all we will look for the Great Plough.

It looks rather like this.

There are not really any lines between the stars. I have drawn these so that you can see the shape they make. Does it look like a plough to you? Some people think that it looks like a saucepan with a long handle.

The Great Plough

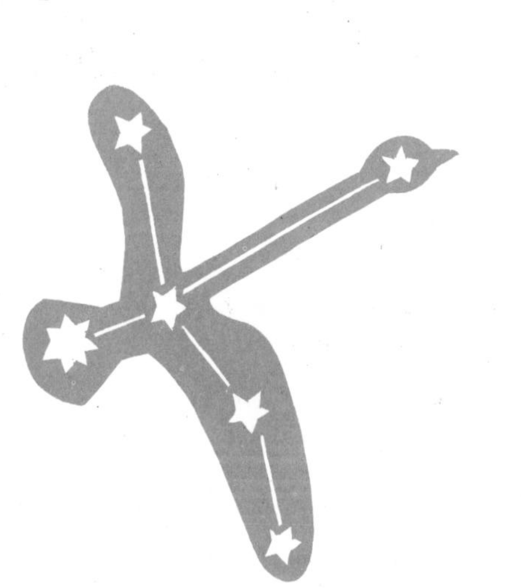

The Swan (or Cygnus)

At some distance from the Plough is a star group called "The Swan" or "Cygnus". ("Cygnus" means "Swan".) Swans fly like geese with necks stretched well out. If you look back at page 10 you will see some flying geese.

I have drawn round this star group to show you how like a flying swan it is.

The Greeks gave these star groups their names and they had legends (stories) about them. Try to find these two groups in the winter sky.

PLANTS WE SEE AT CHRISTMAS

Christmas rose

You know one Christmas plant very well—the Christmas tree or spruce fir. You have read about it on page 8.

Here is a plant which blooms at Christmas : the white *Christmas Rose.* It has dark green leaves.

" *The Holly and the Ivy* " is the title of a Christmas carol. The holly has prickly leaves and bright red berries. The ivy has a head of purple-green fruits. You will notice that the leaves of the fruiting branch of the ivy are a special shape. Look at the *ordinary* leaf of the ivy.

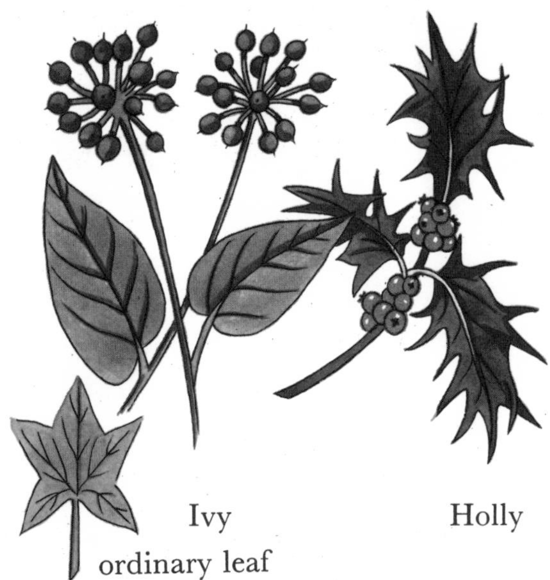

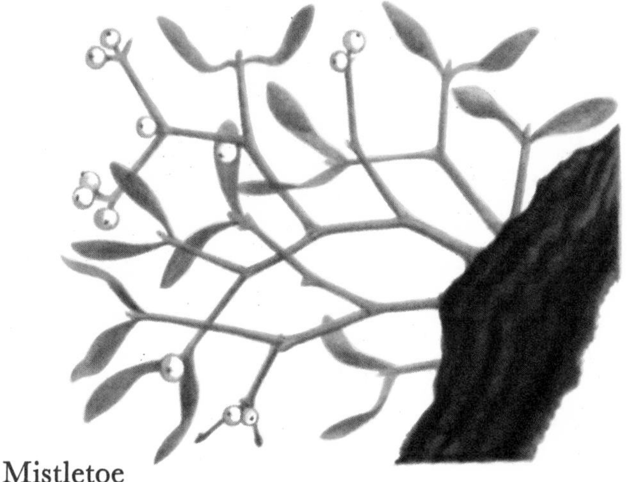

Mistletoe

Here is some *mistletoe.* It is growing out of the branch of an *apple* tree. Birds eat the berries and wipe the seeds on to a branch from their beaks. The seeds grow on the branch and a new plant begins.

(Radio Times Hulton Picture Library.)

This picture shows snow crystals as you would see them under a microscope. You will notice the great variety in the shape of these crystals. You may also notice that there is something similar about them.

Count the spokes or radiating lines which come from the centre of each crystal. How many are there ?

Snow forms when the water droplets in the air freeze. Snow-flakes are light and delicate and as they fall on the ground, much air is trapped between them. The snow forms a blanket on the ground and keeps the roots of the plants warm, although it is cold itself.

TRACKING ANIMALS IN THE SNOW

If it is snowing, put on your rubber boots, and let us go out tracking.

The first tracks we see are those of a bird (1). The bird was hopping, because the prints of both feet are together. It looks as though the bird was being stalked by a cat (2). You do not see claw marks. The cat draws its claws in.

There are some other tracks behind those of the cat. They are the footprints of a dog (3). Do you think that the dog was stalking the cat? You can see the claw-prints of the dog. He cannot draw his claws in as the cat can.

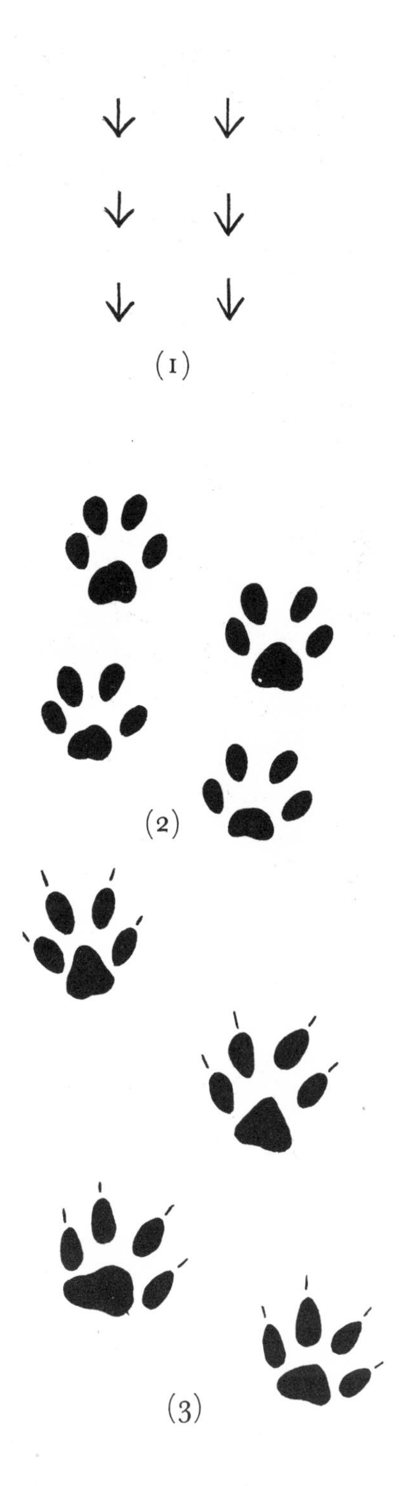

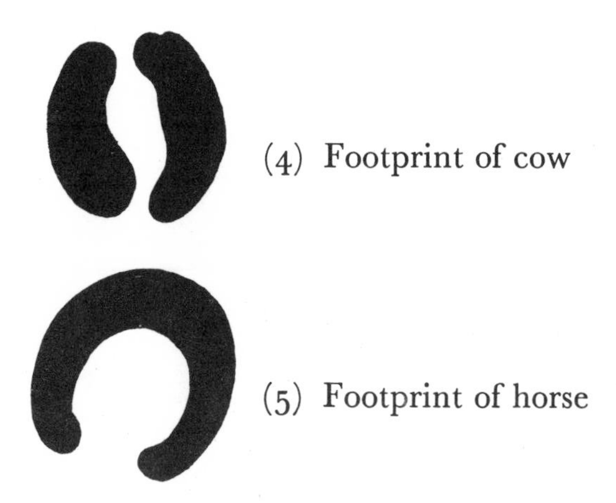

(4) Footprint of cow

(5) Footprint of horse

When does spring begin? What signs do you see? One of the signs is the first snowdrop (1). The yellow aconite (2) and the catkins on the hazel bushes (3) are other signs.

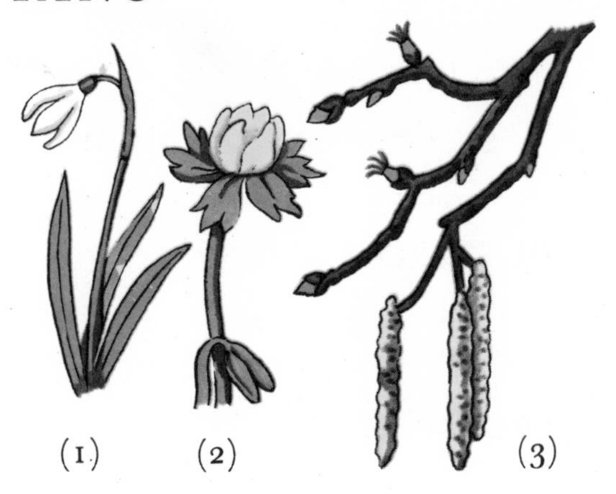

The birds give us warning of spring too. In the branches of the elm tree on the left, you see a pair of rooks. They are probably thinking of building a nest there. They like elm trees and they start looking for their nesting sites very early (even in January).

Rooks in an elm tree

Birds begin to sing. The thrush is one of the earliest to begin practising his song.

Remember to write down all the *first* sights and sounds in springtime.

A thrush singing

CATKINS

There are different kinds of catkins.

Here are some alder catkins. The alder has little cones too. Can you see these in the picture?

Alder

The elm does not have hanging catkins but fuzzy little flowers like the ones below. They have red stamens.

Elm

The hazel catkins on the right have yellow pollen which falls on the red stigmas of the female flowers. Can you see these in the picture?

Hazel

" Pussy " willow (or goat willow)

The yellow, pollen covered catkins of goat willow are shown on the left. These are on different trees from the green flowers (right). Pollen blows on to the green flowers which then produce seeds.

Female catkins of goat willow

Under the soil, seeds are coming to life and putting out roots and shoots.

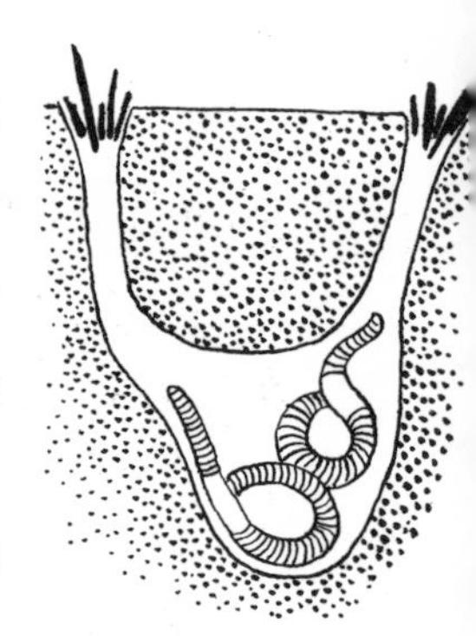

Earthworms are uncurling in their burrows. They will have to move those twigs before they can get out. Do you remember why they put the twigs in? (Page 16.)

Frog spawn

Frog spawn is appearing in ponds. You can see the little tadpoles in the spawn "bubbles". Some tadpoles have heads and tails.

Toad spawn

The mother-toad leaves the garden and goes down to the pond. She lays a string of eggs—toad spawn.

The hedgehog wakes up and comes out of his hiding-place, his prickles covered with grass, straw and dead leaves.

Hedgehog

Brimstone butterfly

Snails

This is our first spring butterfly. Its name is Brimstone. It is yellow with one orange spot on each wing.

Sleeping snails wake up. They put out their four *tentacles*. The two long tentacles have eyes.

PLANTING SEEDS

These children decided to plant some seeds and to have a flower border. Here they are carefully moving some of last year's plants ; they will take these to another part of the garden.

Below, they are making little trenches in which to sow their seeds.

They make three trenches and put in three different kinds of seeds.

These are the kinds of seeds the children put in their flower border.

In the front part of their border, they put some large wrinkled seeds like this. They had an orange-coloured flower on the packet.

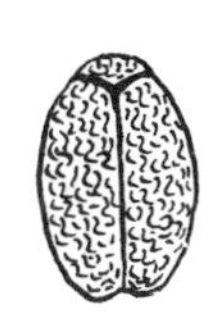

Behind these, they put a row of tiny little seeds which had a blue flower on the packet.

In the last row near the wall they planted a row of large seeds like this which had a big yellow flower on their packet.

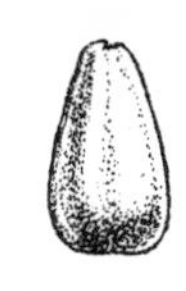

If you would like to know what happened to these three rows of seeds, turn to page 53.

EGGS AND BABY BIRDS

Later in spring, you will not hear the birds singing so much, for they are busy rearing their families. The parent birds build a nest and the mother bird lays eggs in it.

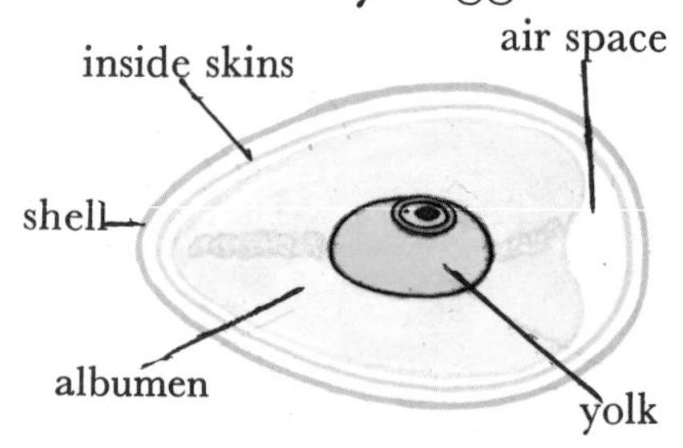

You can see inside this bird's egg. Notice the red spot.

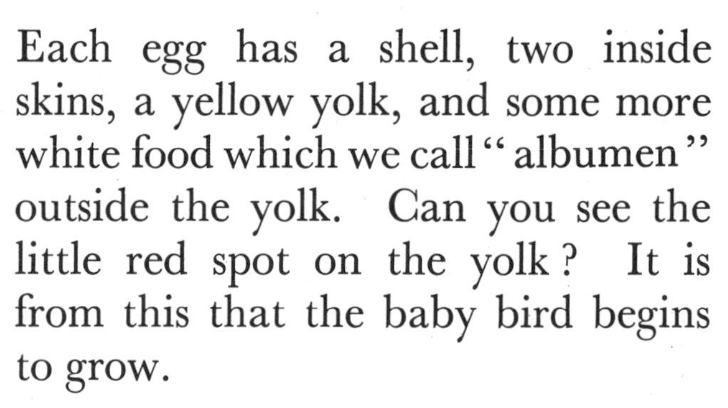

Each egg has a shell, two inside skins, a yellow yolk, and some more white food which we call "albumen" outside the yolk. Can you see the little red spot on the yolk? It is from this that the baby bird begins to grow.

When the baby birds hatch out, the parents are kept very busy feeding them. Here is a baby thrush, with its mouth wide open, waiting for some food. This baby bird has left the nest. Once the baby birds have left the nest, they never go back and their parents follow them about and feed them until they can find food for themselves.

Baby thrush

Trying to fly

Above is a baby bird trying out its little wings. It is not easy—learning to fly.

Ducklings on the pond can swim and feed themselves soon after they are hatched.

THE DAWN CHORUS

When birds awaken—as soon as it is light, or even before—they sing their first song of the day. It begins softly at first but grows stronger as the light grows stronger and soon, all the birds are singing together. We call this wonderful bird-song, " the dawn chorus ".

The birds shown in this picture are :

(1) Woodlark
(2) Woodcock (long beak)
(3) Skylark (sings as it flies)
(4) Redstart (red tail)
(5) Thrush (speckled breast)
(6) Robin (red breast)
(7) Chaffinch (white streaks on wings)
(8) Cock pheasant (beautifully coloured—makes a "barking" noise)
(9) Wren (cocked tail)

MORE ABOUT GROWING SEEDS

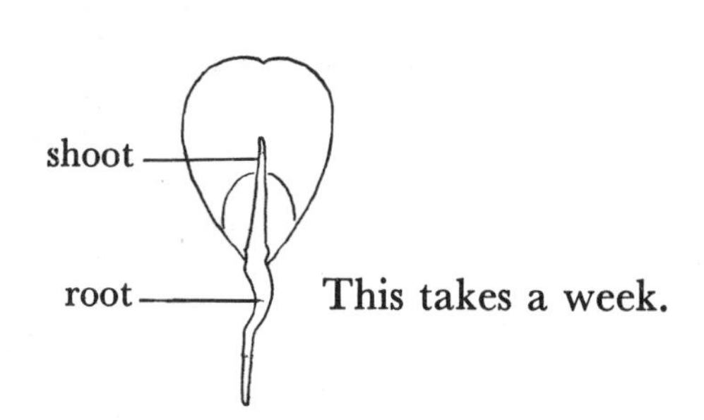

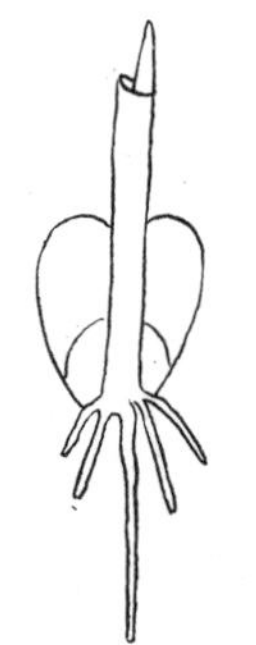

This takes about 11 days.

1. THE MAIZE SEED

Here is a seed of maize beginning to grow. It puts out a root and then a little shoot.

After a while, more roots grow and the shoot pushes upwards. The shoot has a sheath to protect it.

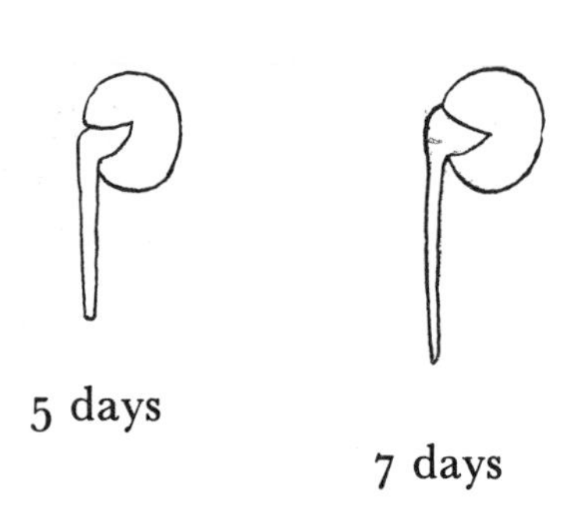

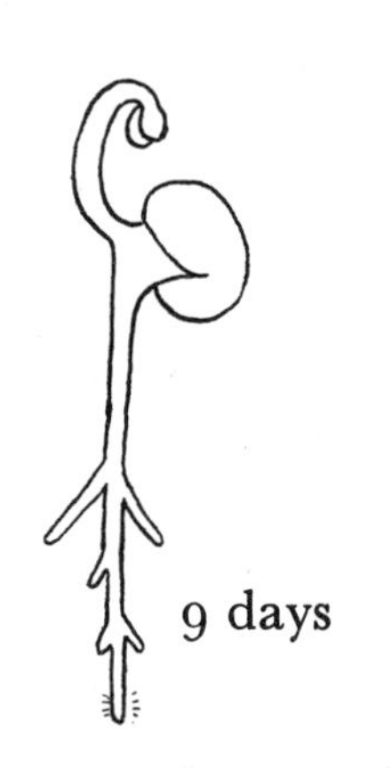

2. THE PEA SEED

First, the pea seed puts out a little root.

Next this root grows longer.

Then a shoot pulls out. Some small roots grow from the main one.

Measure the lengths of your growing seeds with a ruler.

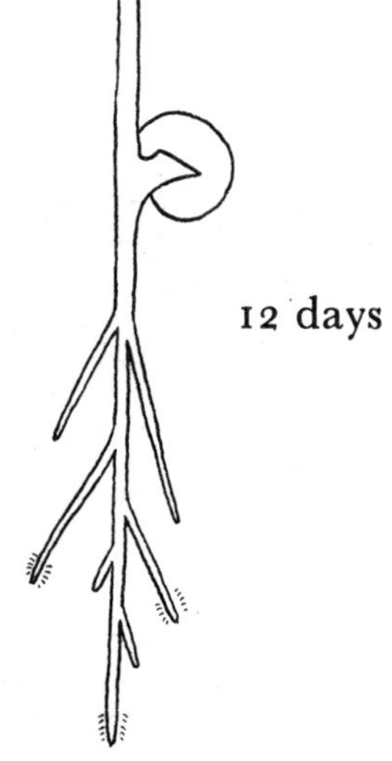

The drawings show how long it took my pea seed to reach its various stages. Now—time yours! and measure it.

(1)

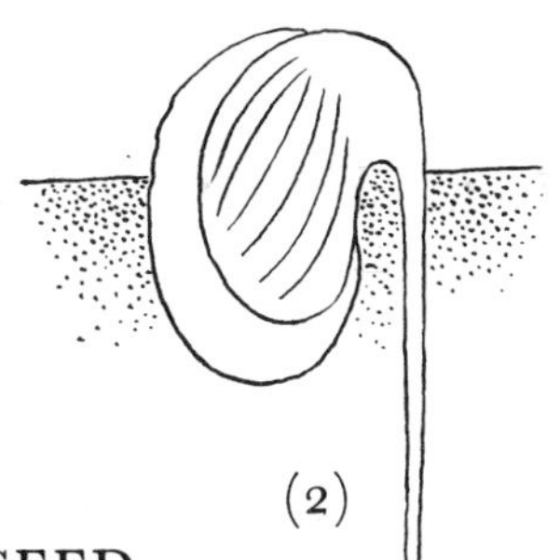

(2)

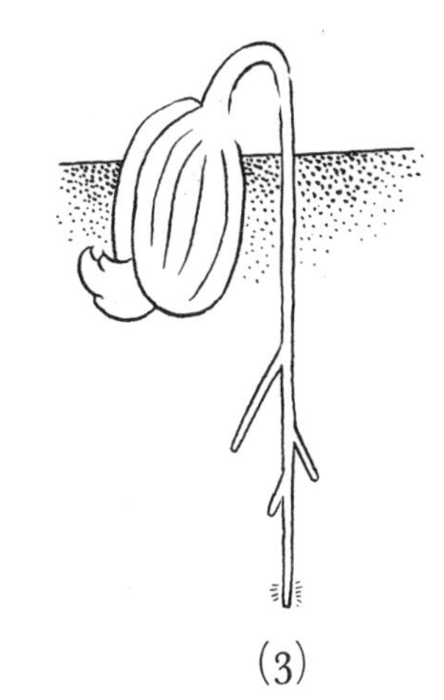

(3)

3. THE FRENCH BEAN SEED

(1) Looking inside a french bean we can see the little root within a bulge in the skin.

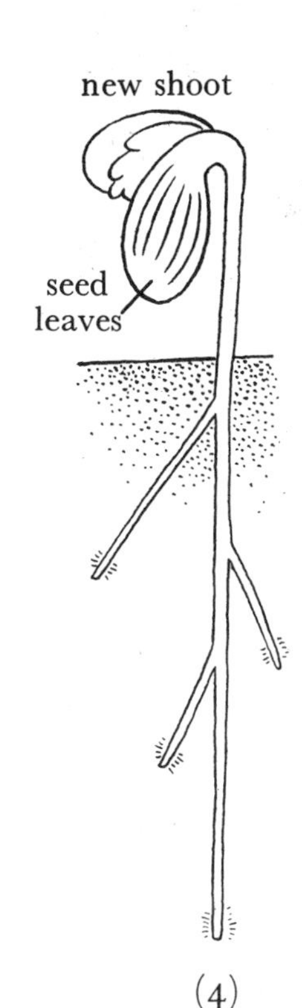

(4)

(2) The skin breaks and the root grows down into the soil.

(3) The root is becoming longer and the "seed-leaves" are pulling out of the ground.

(4) The seed leaves are not the real leaves. They take the place of the real leaves, for a time. You can see the new shoot with leaves on it, between the two seed leaves.

The old skin has gone, the seed leaves are opening. Something is pulling out between the seed leaves—what is it? It is the new shoot.

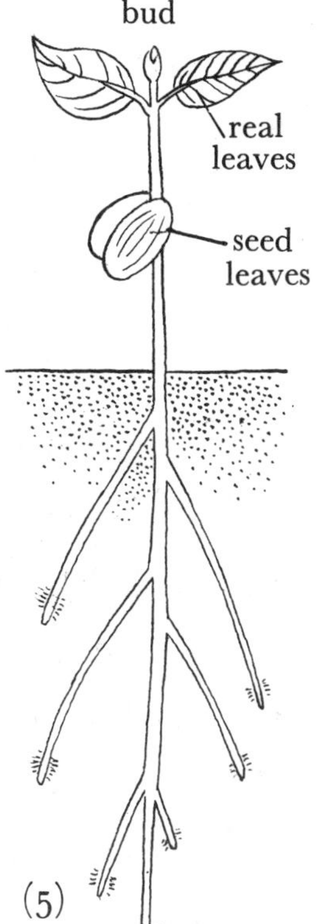

(5)

(5) Now here we see the real leaves. The seed leaves are still there on the stalk, but soon they will drop off. There is a little bud at the top. This will go on growing.

1. *Coltsfoot.* This is a queer little plant. It has a yellow flower and blooms very early in spring. It does not have flowers and leaves at the same time. The leaves come much later than the flowers.

Can you see the scales on the flower stalks? These are woolly and feel warm.

The flowers close at night and also when the weather is very cold.

Coltsfoot

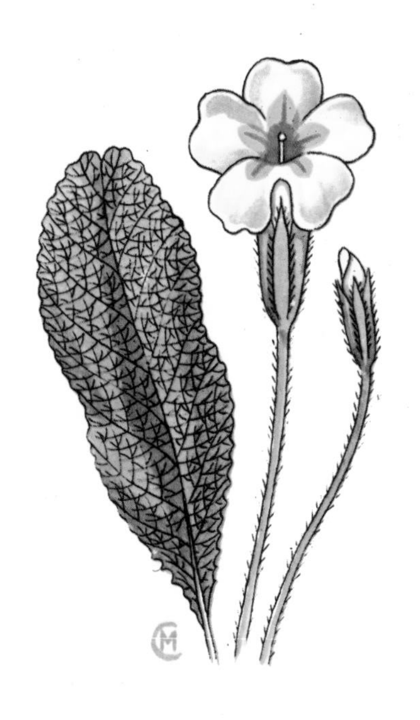

Primrose

2. *Primrose.* Many people say that primroses are their favourite spring flowers. Have you noticed what a lovely scent they have?

There are five yellow petals (count them) on each flower. Notice the bright yellow honey guides and the pin head stigma which are seen in the centre of the flower.

3. *Wood Anemone.* The flowers of this plant are sometimes called "wind-flowers". They grow on the floor of a wood and are white and delicate.

Look at the three leaves which meet together at the base of the flower-stalk. These leaves are much divided.

This plant belongs to the buttercup family.

Wood anemone

Violet

4. *Violet.* Those violets which are scented are called "sweet violets". There are other violets which have no scent. Violet leaves are heart-shaped.

There is a little spur with sweet nectar in it at the back of the flower. Can you see it? The bee takes this nectar and makes honey with it.

DIFFERENT KINDS OF ROOTS

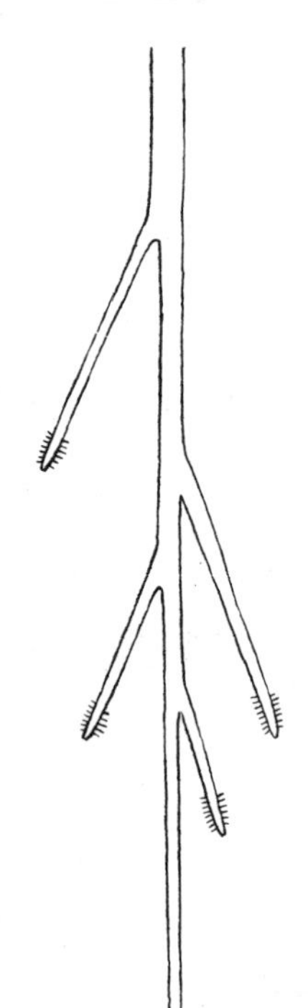

You have seen how little roots begin to grow ; now let us see what they grow into.

Some grow into long roots like this with one main root and others growing from it. Bean and pea plants have roots like this.

Some roots grow like this one below. Here all the roots appear to grow out from the base of the plant. You do not see a main root. Grass grows roots like this.

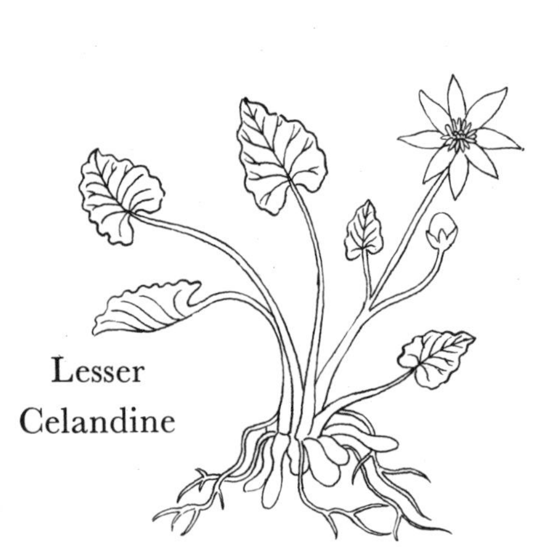

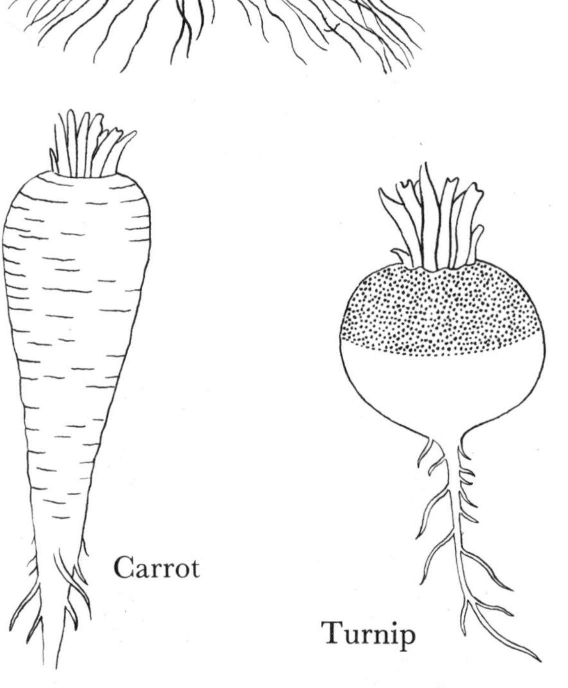

Above are the roots of the Lesser Celandine, with little swellings or tubers among them. These tubers contain stored food.

Some roots are swollen out with food. The carrot and turnip are like this.

LEAVES UNFOLDING IN SPRING

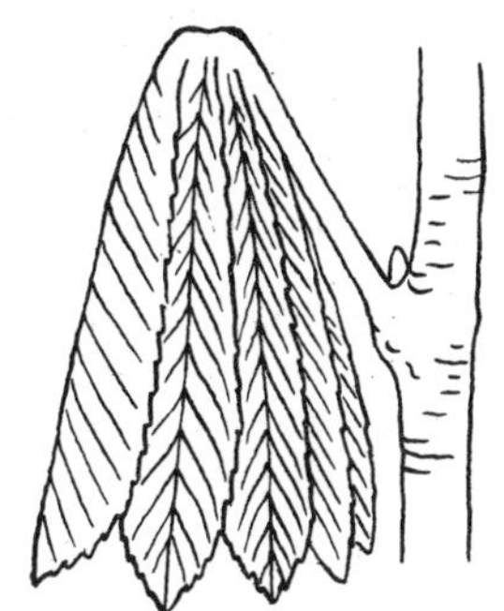

Horse-chestnut

Young leaves of the horse chestnut droop at first.

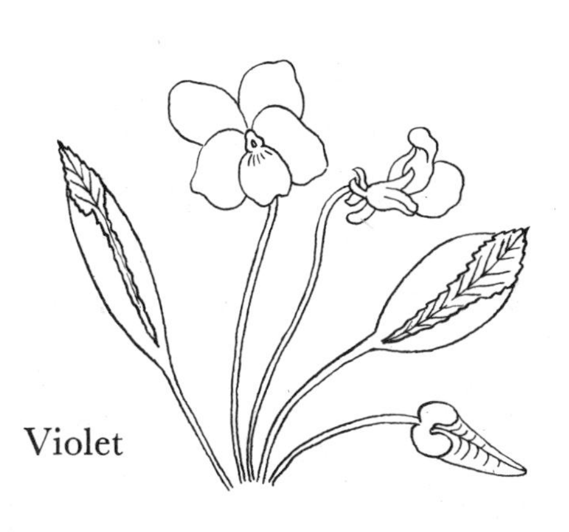

Violet

Violet leaves are rolled.

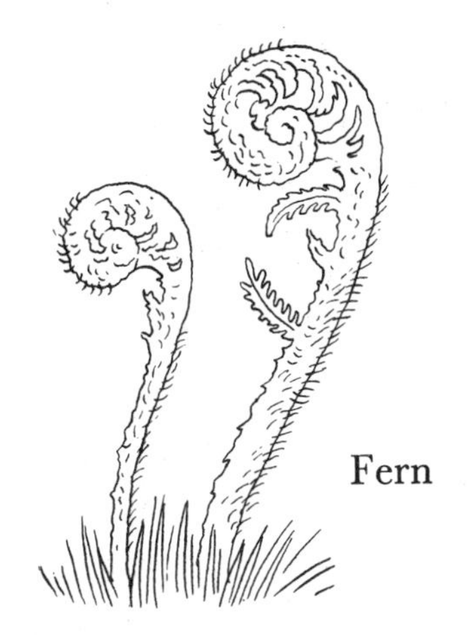

Fern

Fern leaves or "fronds" are coiled up like a question mark—?

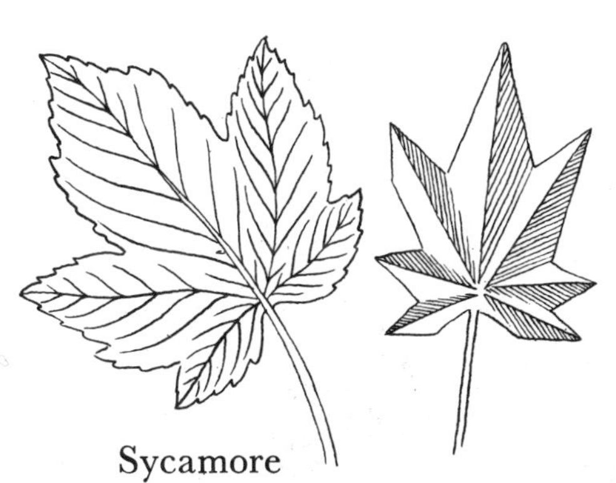

Sycamore

Young sycamore leaves are folded like a fan.

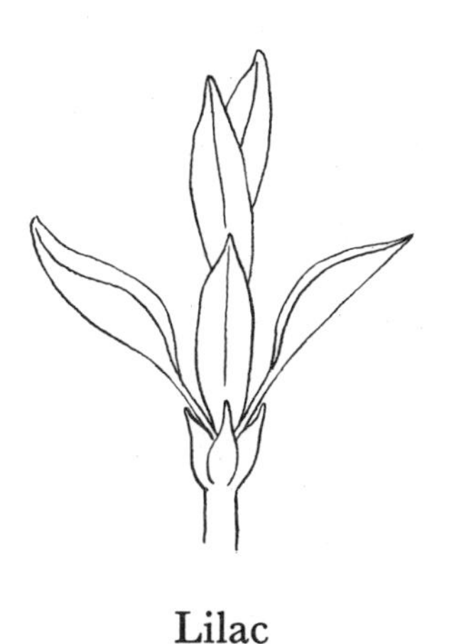

Lilac

When a lilac bud opens, we see that the leaves are folded in half.

NEARER TO SUMMER—
BLUEBELLS AND BLOSSOM

Bluebells flower later in spring than the flowers you saw on pages 32 and 33. They bloom first in April but can go on flowering until June. The correct name for the bluebell is " wild hyacinth ".

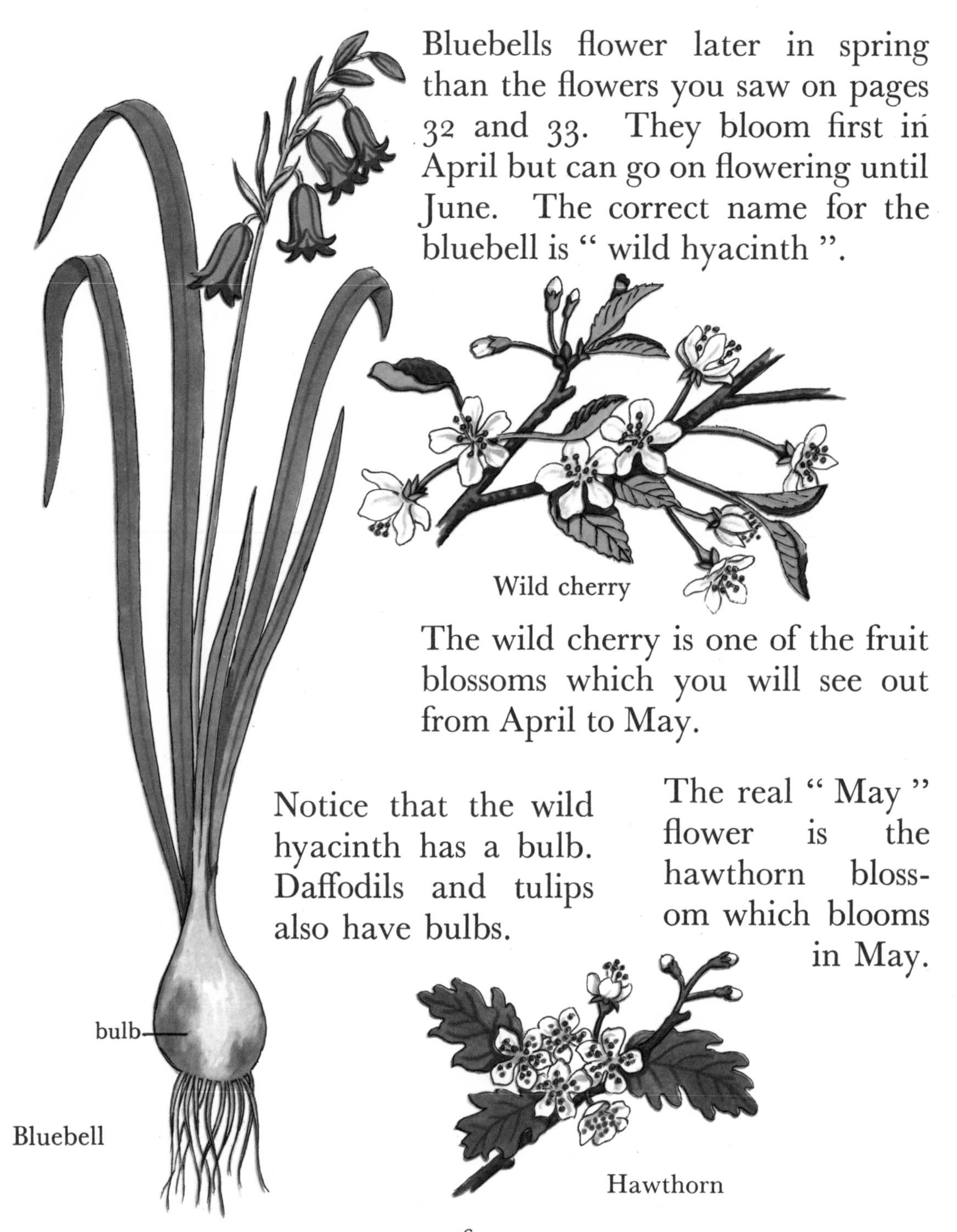

Wild cherry

The wild cherry is one of the fruit blossoms which you will see out from April to May.

Notice that the wild hyacinth has a bulb. Daffodils and tulips also have bulbs.

The real " May " flower is the hawthorn blossom which blooms in May.

Bluebell

Hawthorn

THE MERRY MONTH OF MAY

Poets have called the month of May "merry" because all Nature seems to be bursting with energy during this month.

Trees and shrubs are bursting into flower. Above are the flowers of the guelder rose.

Birds are busy rearing their families. They follow their young ones about and feed them even after they have left the nest.

There are more insects about in May. Look out for butterflies. There are many white ones about now. Here is the small white butterfly.

You will see ladybirds in your garden, often on rockery plants. Count their spots. The most common ones have seven spots (see above).

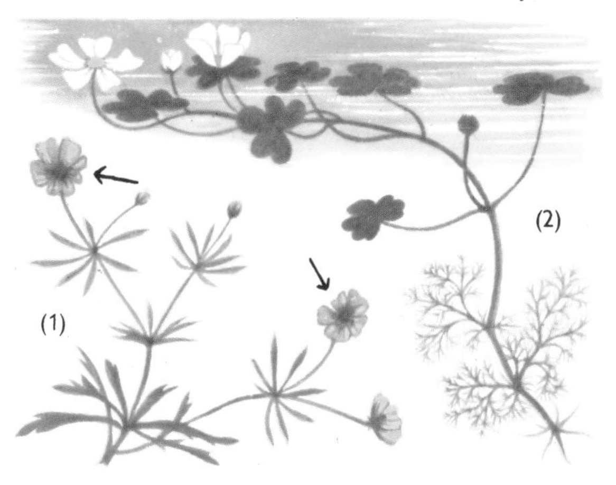

There are many interesting flowers in May. (1) is a little buttercup called goldilocks. Follow the arrows. One petal is deformed.

Ponds in May are often covered with a little water buttercup called water crowfoot (2). It has white flowers with yellow centres. Look also at its two kinds of leaves. One kind floats on top of the water and the other kind, much divided, is under the water.

June is here and it is summer. There are many things to see.

Swallows have been here since April or May.

We hear the twitterings of baby birds everywhere, but the older birds do not sing much now. They are still too busy filling these hungry mouths.

There are lots of living creatures in ponds and streams now, so get out your pond-dipping nets.

Trees are in full leaf and wild roses are on the hedgerows.

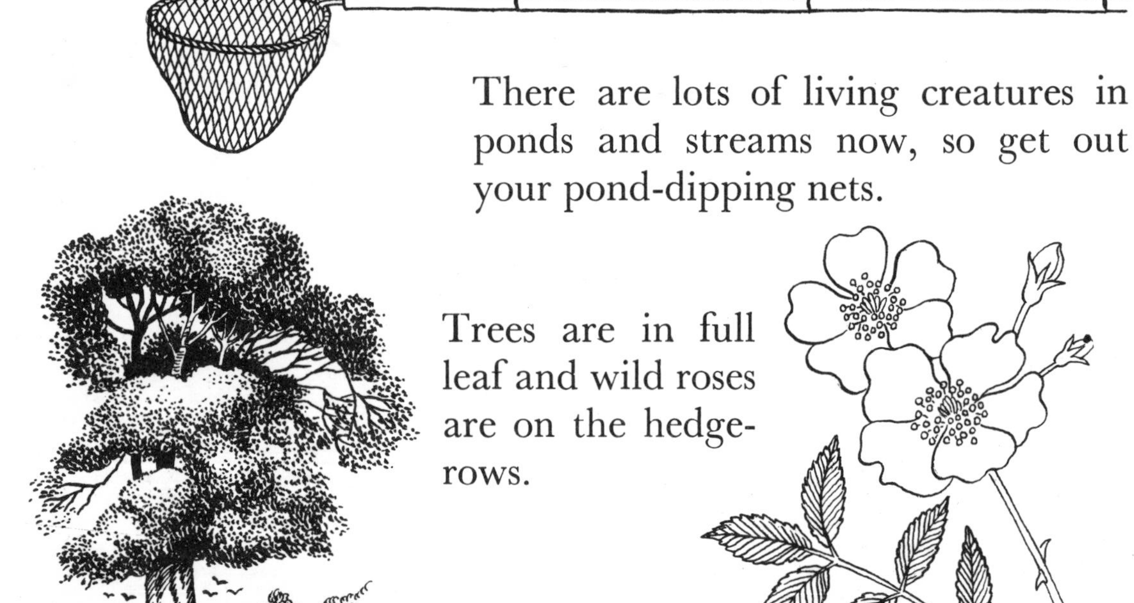

SMALL CREATURES TO LOOK FOR

Let us look for little creatures in these places :
1) on the ground ; 2) in the pond ; 3) in the air.

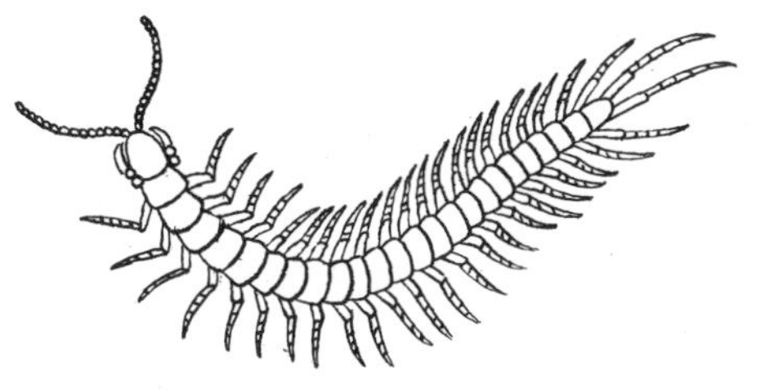

Centipede

1. ON THE GROUND

Look for centipedes—here is one. " Cent " means 100, " pedes " means feet. Centipedes have a large number of legs.

Look for ants—here are two. You will find them in light sandy soil in gardens.

Ants

Wood louse

Here is a wood louse. This creature is armour plated and its armour is jointed. Notice that legs come from each segment or part of its body.

The *silver fish* is not a fish but a primitive kind of insect, probably like the earliest kinds of insects which were on the earth millions of years ago. Do you see its three tails ?

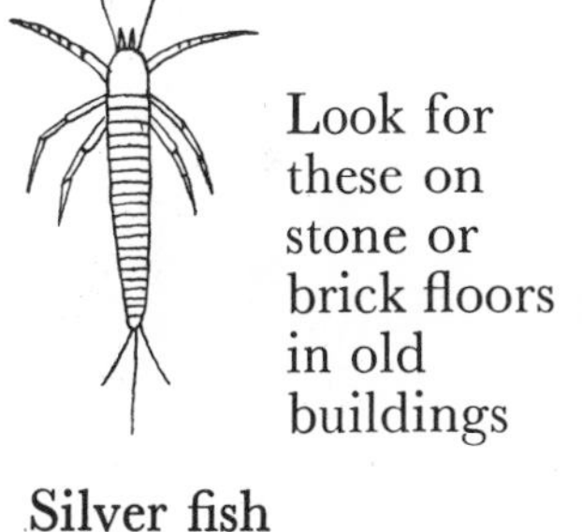

Look for these on stone or brick floors in old buildings

Silver fish

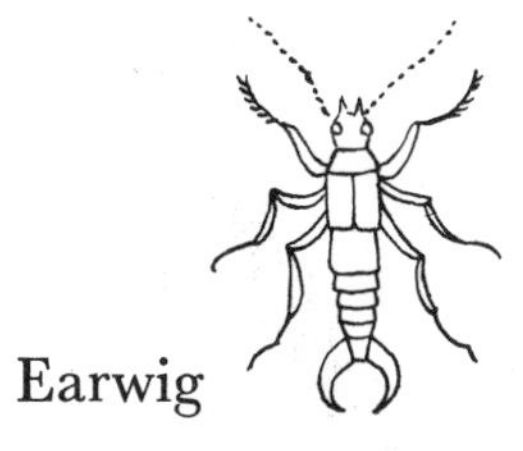

Earwig

Earwigs are found in gardens. They have pincers at the back of their bodies. Can you see them ? They can nip you with these. They can bite too !

True insects have six legs. Some of these creatures are insects and others are not. Can you spot the true insects ?

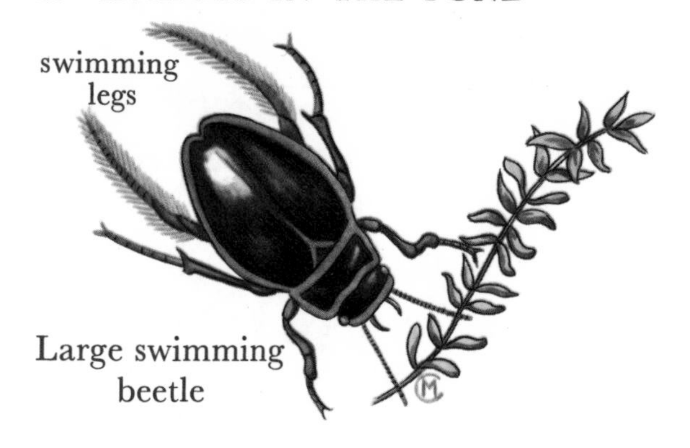

Large swimming
beetle

This is the large swimming beetle called *Dytiscus*. Can you see his back swimming-legs fringed with hairs ?

May-fly larva

This is the *larva* or " young stage " of the *may-fly*. The may-fly flies in the air but its larva lives in water. It has gills along its side to breathe by. Three tails ! Does this remind you of something on the page before ?

Here is a *water scorpion*. His pincer-legs in front make him look like a scorpion. Do you see what a long tail he has ?

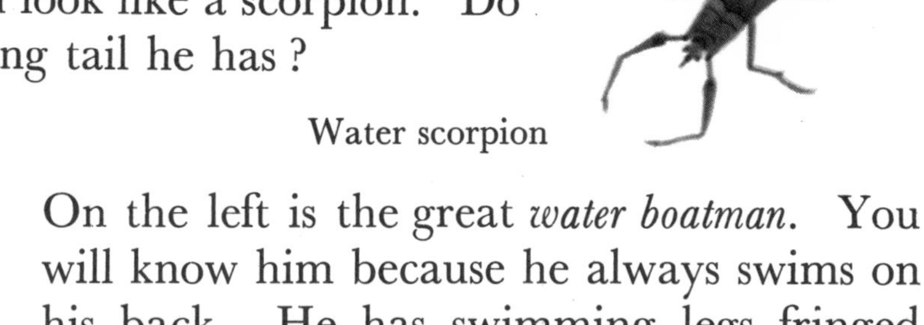

Water scorpion

Great water
boatman

On the left is the great *water boatman*. You will know him because he always swims on his back. He has swimming legs fringed with hairs like the swimming beetle (above).

This is the *dragon-fly* larva. You may think that this looks like the may-fly larva above. But look again. This insect is larger. His three tails are broader. He has no gills on his sides. Can you see his wing buds ? Later he will climb out of the water, spread his wings and fly away.

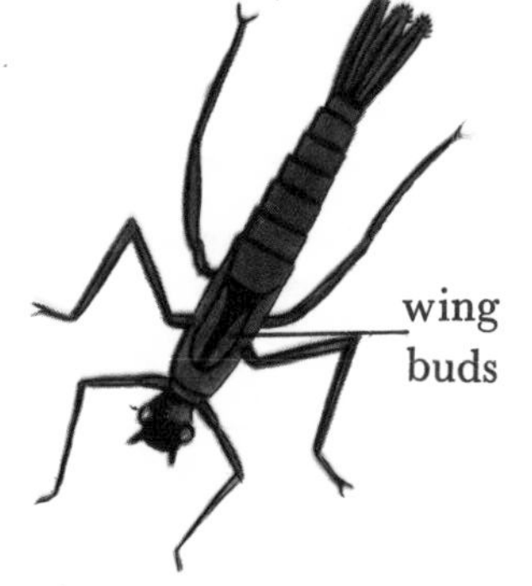

Dragon-fly larva

3. INSECTS IN THE AIR

Do you remember the very earliest butterfly, the Brimstone? (Page 26.) Now, in summer, you can see many butterflies and moths.

Meadow brown butterfly

Here is a common butterfly of the meadows. It is called the Meadow Brown Butterfly. This is an easy one to remember. It is all brown except for a lighter patch on each upper wing. Can you see the lighter patches? It also has two black eye spots with white centres. Can you see these?

Cinnabar moth

Here is a little moth to look for. It is called the Cinnabar Moth. It has a red bar and two red spots on its upper brown wings. Its lower wings are red. It likes yellow flowers.

Moths and butterflies have four wings. The pictures show differences between these insects. Moths (2) have feathery antennae (feelers) and butterflies (1) have clubbed antennae. Moths and butterflies hold their wings differently when at rest.

Can you see these differences?

FLOWERS AND INSECTS

Insects are necessary to many flowers, and some flowers are necessary to certain insects, especially insects with long tongues like bees, moths and butterflies. These all go to flowers to drink nectar, and they take the yellow pollen from one flower to another.

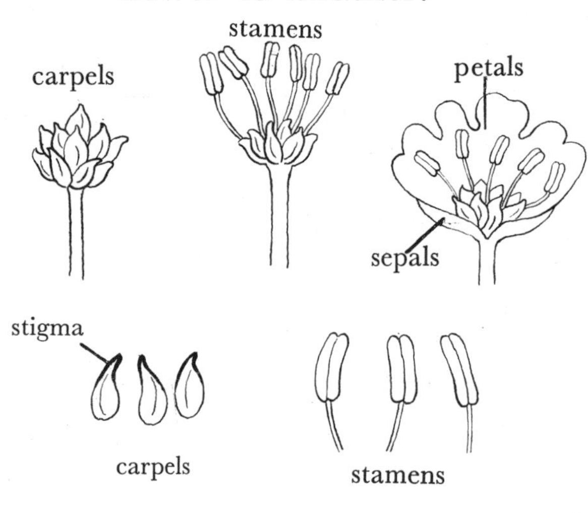

In a flower there are carpels, stamens, petals and sepals. The top part of the carpel is called the stigma. The stigmas and stamens are important because yellow pollen from a stamen has to fall on to a stigma before a seed can develop.

Some flowers have a lovely drink called nectar, which insects like to sip. The buttercup stores its nectar in a little pocket in a petal. The violet keeps its nectar in a little spur. The primrose has a little nectary at the bottom of its flower tube. The snapdragon has a nectary well down in its flower.

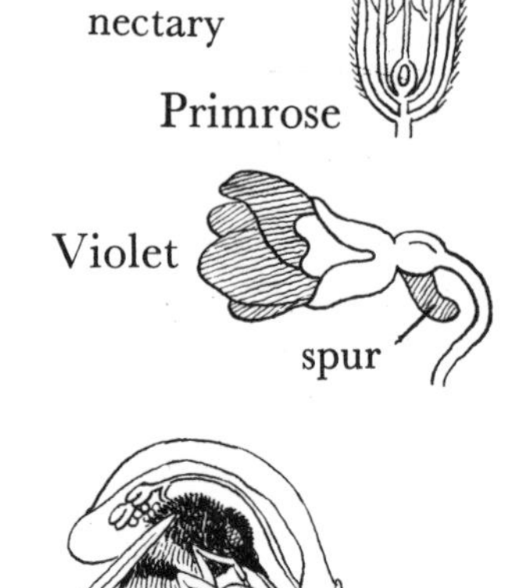

Here is a bee with its long tongue down the tube of a snapdragon flower. It is sipping nectar. As it does so, the stamens with their yellow pollen are touching its back. Can you see them? Bees make honey with this nectar.

HOW A SEED GROWS

Look again at the little worker bee sipping the nectar from the snapdragon flower. Where does it go next? It flies straight to another flower.

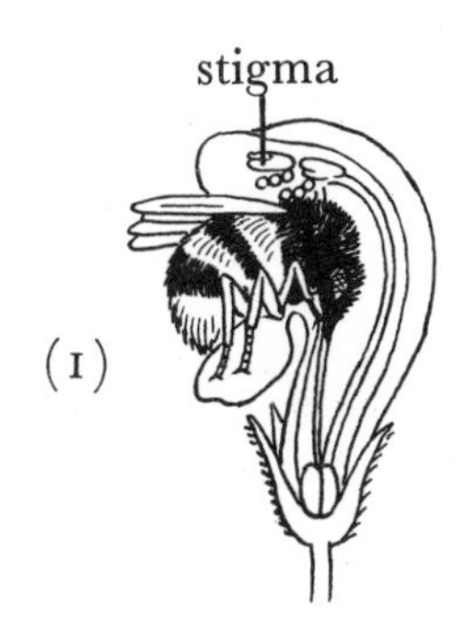

It lands on the flower and puts its tongue down to get some more nectar. While the bee is sipping the nectar, the stigma is touching its back and some yellow pollen which the bee has brought from the other flower brushes on to the stigma (1).

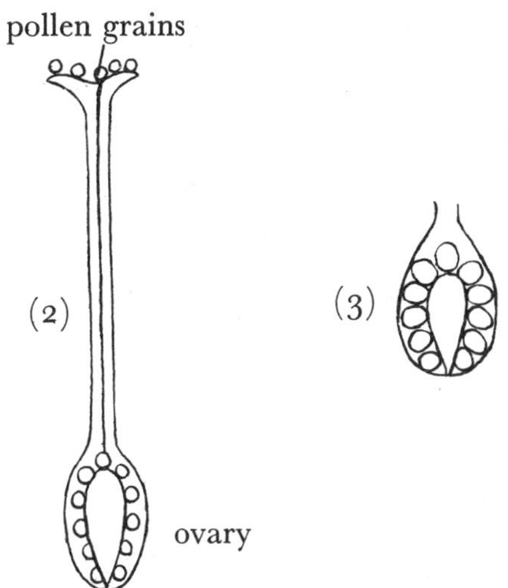

Here is another stigma with yellow pollen grains on it (2).

The pollen grains send tubes down to the ovary and when this happens the seeds grow and the ovary becomes a fruit (3).

In some plants (such as grasses) the pollen is blown about by the wind, but it reaches another grass flower just the same as the pollen carried by the bees.

FLOWER PATTERNS

Most of the flowers that you see have four parts : petals, sepals, stamens and carpels. You know how important carpels and stamens are, but petals and sepals are important too. Petals are the *prettiest* part of the flower and are usually *coloured*. They have interesting shapes and each flower has a certain *number* of petals and sepals. They have these beautiful colours and patterns in order to attract the *insects* which help them to make their seeds (see page 43).

Look at these flower patterns :

Arrowhead flower has 3 petals.

Snowdrop 3+3.

Wild mustard has 4 petals (not joined).

Speedwell has 4 petals (joined).

Primrose has 5 petals.

Lesser Celandine has 8 or more petals. This one has 10.

Ragged Robin has 5 ragged petals.

Pansy and Heartsease have flowers like a little face (5 petals).

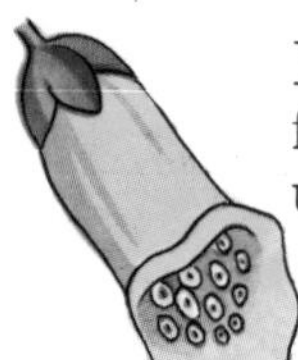

Foxglove. Bell-like flower with an uneven edge.

Harebell has flowers like bells.

DO YOU LIKE COLLECTING CATERPILLARS ?

You probably know that caterpillars change into moths or butterflies when they grow up.

Here is one of the funniest caterpillars you are likely to see. It is the Puss Moth caterpillar and it is found on poplar or willow tree leaves later in summer. It is bright green with black and purple markings and two tails.

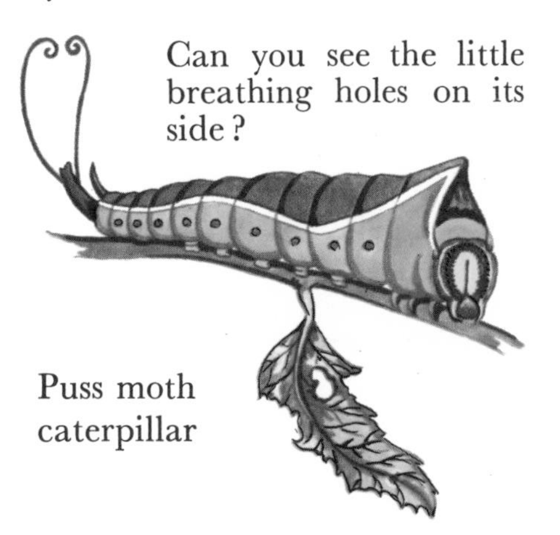

Can you see the little breathing holes on its side?

Puss moth caterpillar

Tiger moth caterpillar

Most children know this caterpillar very well. It is the " woolly bear " and is the caterpillar of the Tiger Moth. You can often find this caterpillar on hollyhock leaves in the garden. It has long brown hairs.

Puss moth

Tiger moth

The caterpillars above will develop into these moths.

Here are the flowers of two grasses :

Wild oat grass Quaking grass

You have seen grass growing in every field and on every lawn all through the year. Grass makes the countryside green. Below is a grass plant with no flowers :

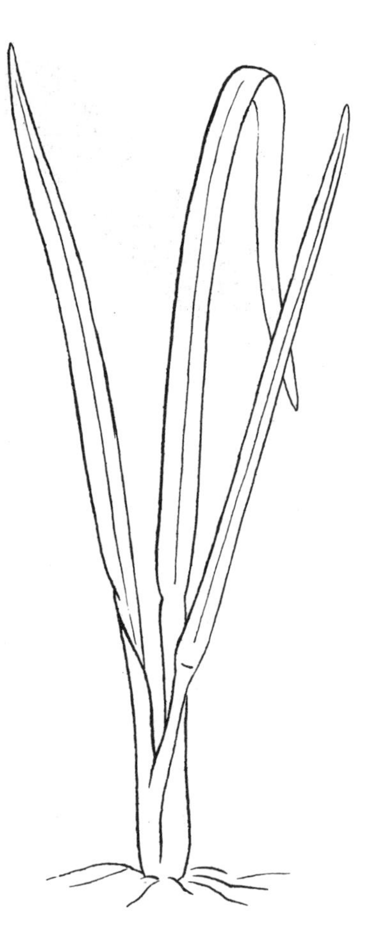

Grasses have flowers in summer time. There are three grass flowers on this page. They look different from other flowers and do not have prettily coloured petals but they have stamens and stigmas and can make little seeds.

Insects do not visit these flowers. The wind blows their yellow pollen on to the stigmas of other grass flowers and then the seeds grow just as you read on page 43.

Cocksfoot grass

Wheat, barley and oats are three important food grasses. Wheat grains (the seeds of wheat) are ground into flour and from flour we can make bread. Oat grains are ground to make oatmeal. We eat porridge made of oatmeal. From barley we get malt.

You have seen the flowers of the wild oat on the opposite page. The cultivated oat is very much like this. On this page are the " ears " or grass-flowers of wheat and barley.

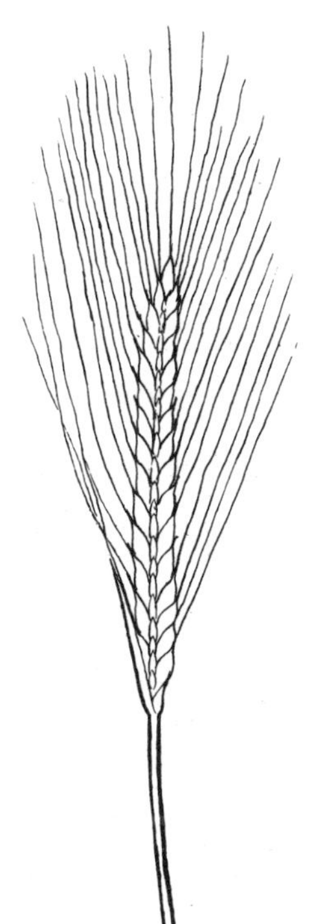

These food grasses are called cereals. Cereal comes from the name " Ceres ".

" Ceres " was said (by the Greeks of ancient times) to be the goddess of the corn.

Rye, maize and rice are also cereals.

Do you eat cereals for your breakfast ?

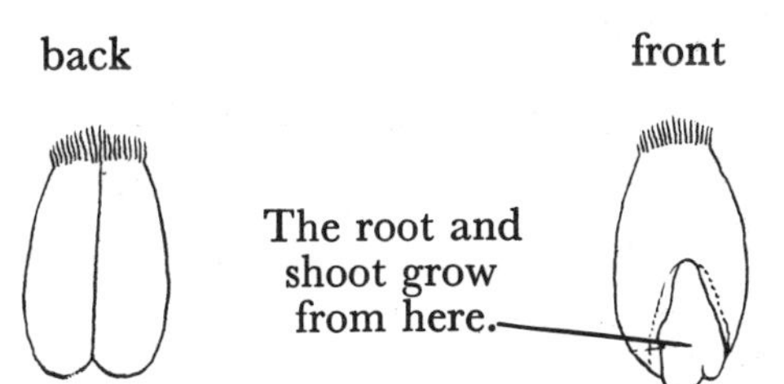

Barley

Wheat grains

Wheat

STRUGGLE IN THE CORNFIELD

Our food grasses—wheat, barley, oats and rye—are sown by the farmer in ground which he has prepared for them, and they are harvested in late summer. Sometimes, however, other grass seeds which the farmer has not sown take root in the cornfield and come up with the food grasses. These are known as "tares". Among tares is our only poisonous grass. It is called darnel or giddy rye grass.

Darnel grass

Field mouse

The farmer often has to scare off flocks of wood pigeons, rooks, jackdaws and even sea-gulls which raid his fields and steal his grain. Field mice raid the fields too. Here is a long-tailed field mouse nibbling away at a grain of corn.

Some plants are good for the farmer's fields and help to put back minerals and salts taken out of the soil by the corn crops. The little lumps or *nodules* on the roots of clover contain bacteria which help to do this. Farmers often grow a field of clover where wheat or barley has been growing the year before.

White clover

FLOWERS WHICH GROW IN CORNFIELDS

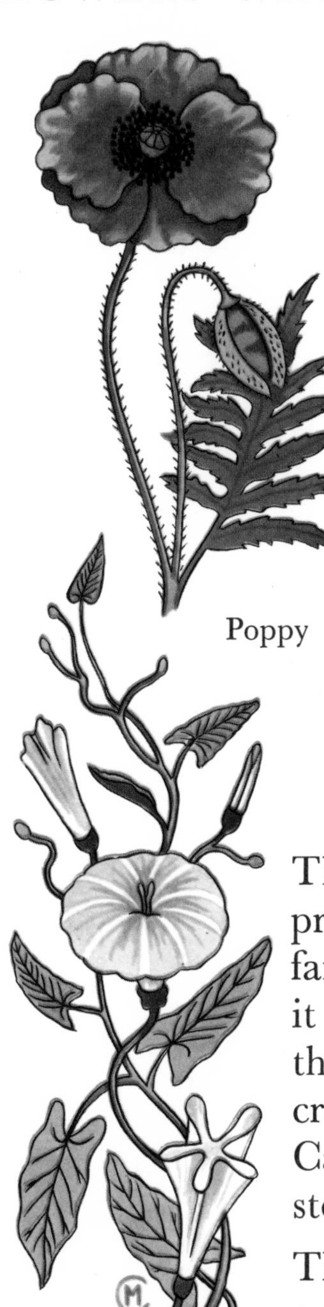

Poppy

Poppies are really weeds in the cornfield and the farmer does his best to get rid of them because they choke his crops.

This is the " poppy head " or fruit of the poppy.

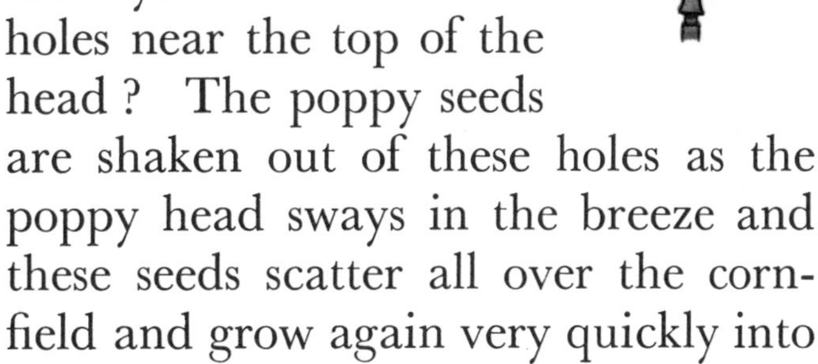

Can you see the little holes near the top of the head ? The poppy seeds are shaken out of these holes as the poppy head sways in the breeze and these seeds scatter all over the cornfield and grow again very quickly into new poppy plants.

The *lesser bindweed* has pretty pink flowers but farmers do not like it because it winds its long stem around the stalks of corn and other crops and strangles them.
Can you see its winding stem ?

The *corn-cockle* grows amongst the corn in July and August. It has purple flowers and silky hairs all over it.

Lesser bindweed

Corn-cockle

WHAT ARE WEEDS?

Do you ever weed the garden?

Weeds are common wild flowers. We call them weeds when we find them in our gardens or among our crops.

This plant on the right has fruits like little purses (can you see them?). It is called Shepherd's Purse.

Shepherd's Purse

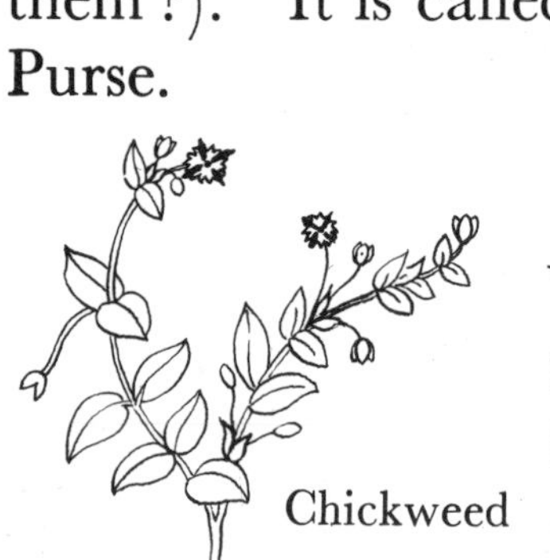

Chickweed

You must often have seen this little plant on the left. It is very fresh and green and has tiny white flowers. It is called chickweed (birds like to eat it).

Dandelion "clock"

Hop trefoil

You will have seen this plant, hop trefoil, creeping over grass. The flowers are like tiny yellow clover flowers.

Dandelions are "weeds" in the garden. You must not let them turn into "clocks" like the one above. All the little fruits have parachutes like this. The parachute carries the fruit away and the seed inside it will then come up somewhere else in your garden.

SUMMER IN THE PARK

Here are some trees you might see in your town park or in the country.

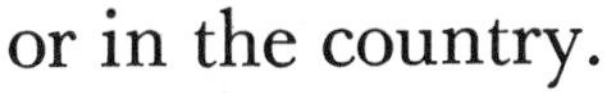

Cedar.
This is a fir tree and has cones like the spruce.

Beech. This is a large spreading tree with a smooth grey trunk. It gives much shade and does not let much light get through to the plants on the ground so that very few plants grow beneath beech trees.

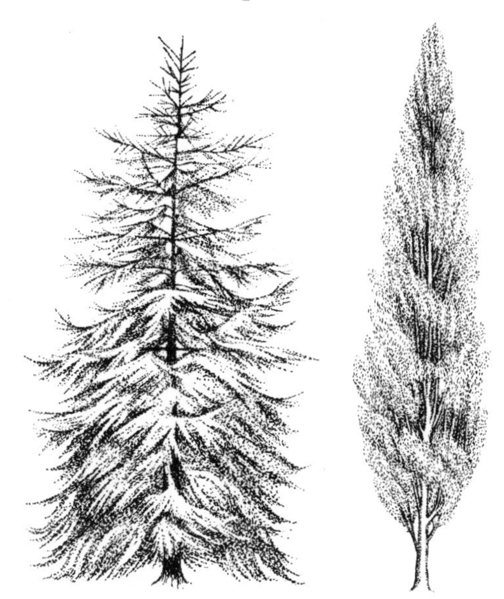

Lombardy poplar. A tall, thin tree with branches held closely to its trunk.

Spruce fir or Christmas tree.
This tree has a long, straight trunk. You have seen it before on page 8. It is often seen in parks and you will remember that it keeps green all the year round.

Horse-chestnut. This is dome-shaped. You saw it in winter on page 19.

DUCKS ON THE PARK LAKE

Mallard

The commonest duck you see on the park lake is the mallard duck. This duck has a green head and a white ring round its neck. It has a chestnut-coloured breast and two little curled tail feathers. These are black.

You will know this little duck on the right by the tuft on his head. He is black and white.

Both these are male or "father" ducks. We call these "drakes". The females or "mother" ducks are not so brightly marked or coloured.

Tufted duck

Shoveler duck

The shoveler duck has a long beak which is rather like a spoon at the end. He uses it to "shovel" up the little creatures in the mud, upon which he feeds.

On the right is the foot of a duck. Can you see the webs between the toes? The duck swims with its webbed feet.

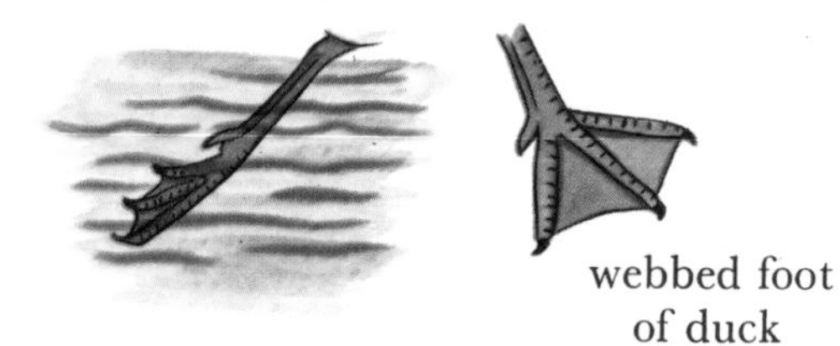
webbed foot of duck

SUMMER IN THE GARDEN

Do you remember the children on page 27? It was spring then and they were planting seeds. They planted three rows of flower seeds. Now that it is summer, let us go back to that garden and find out what flowers have grown from the seeds.

This little plant with round leaves and orange-coloured flowers was growing at the edge of the flower border.

Behind the orange-coloured flowers was a row of blue flowers with thread-like leaves.

Here are the three rows. Now guess the flowers.

Answers: (1) Nasturtiums (2) Love-in-a-mist (3) Sunflowers

In the back row were some large yellow flowers with tall stems.

THE STREAM IN SUMMER

Let us stroll along the banks of a little stream in summer. There are so many things to see. Two weeping willows lean over the water, (1) and (2). (2) has had its top cut off or pollarded and has a lot of branches (like a head of hair) springing out of the top. Two water-birds, a black one (3) and a white one (4) are swimming down stream. The black one is a moorhen, the white one is a swan. As we walk farther down we see beds of tall plants along the edge of the stream. These are rushes (5) and sedges (6). You can see larger pictures of these if you look below at (10) and (11). The other water-plants are yellow flags or yellow iris (7), water lilies (8) and water arrowhead (9).

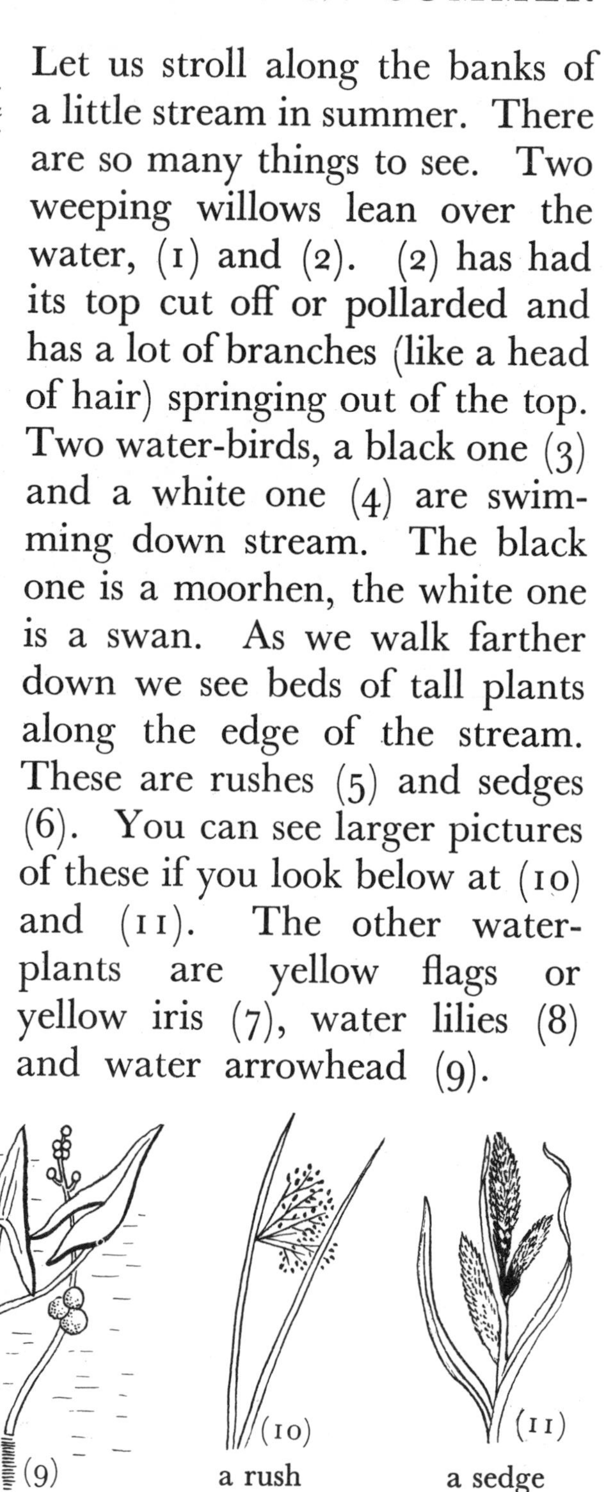

a rush

a sedge

WE GO FISHING—WITH A POND NET

Here is your pond net. The handle is made of bamboo. The net is a fine one on a wire frame.

string handle

string holder

Here is your jam jar.

Now we are off to the pond.

If we put on our wellington boots, we can step into the pond and then perhaps we can get some fishes in our net.

Here is one little fish you might find. It is the three-spined *stickleback*.

Here is the *minnow*. Unlike the stickleback, it has no spines.

Minnow

You might catch a *newt*. A newt has a long tail. It is a relation of the frog.

You will also find pond snails.

Ramshorn snail

Remember! When you have caught these, do not leave them in the jam jar all night or they will die : put them in an aquarium if you can, or in a big bucket or large enamel bowl full of water and don't forget to put some green water-weed in with them. This helps to aerate the water.

AT THE SEASIDE

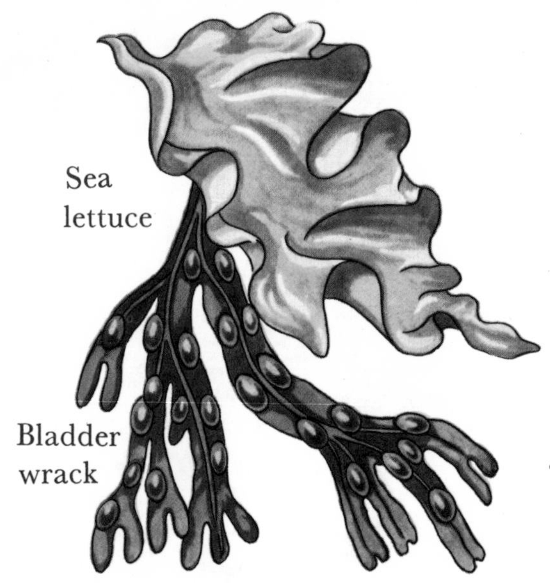

Summer holidays are here. As you walk over the slippery rocks, you will see this green flat seaweed spreading over them. It is called sea lettuce.

On the rocks you will also see this brown seaweed known as bladder wrack. Its bladders are full of air and make it float on the water.

On the sands you may see this brown object called a " Mermaid's Purse ". It is really the egg case of a fish (the skate). A baby fish was once inside this case.

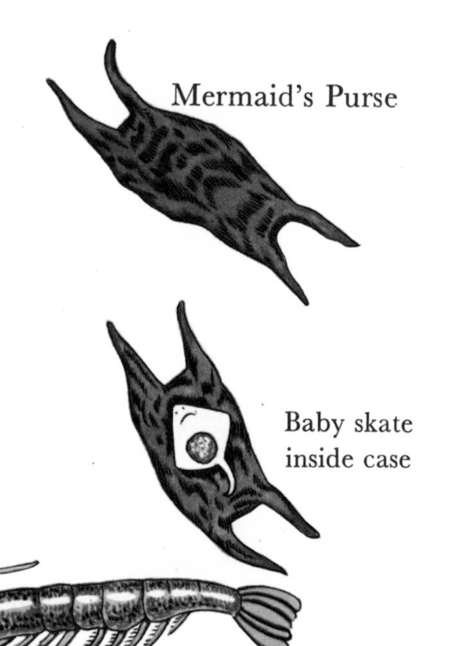

Below is a baby fish shown inside its case. You can see it is a little flat fish with a tail. Its mouth is underneath its head and it has a round sac full of food attached to its " tummy ". This is its " yolk sac ".

Shrimp

Have you got a shrimping net ? It is shaped like this with one side flat so that you can scrape it along the sand at the bottom of a pool. Above is your shrimp—waiting for you to catch him. His body is jointed.

SEA-GULLS

What do you know about sea-gulls ? Did you know that there are many different kinds ? Here are two kinds :

This one, with the slate grey back, is the Lesser Black Backed Gull. Do you notice a little red spot on his lower beak ? What do you notice about his feet ? (They are yellow and webbed.)

Lesser black backed gull

Black headed gull

Here is a gull with a black head. We call him the Black Headed Gull. He has a grey back and red legs and red beak.

Young gulls do not always have the same coloured plumage (feathers) as older gulls, so that sometimes it is difficult to recognise them.

Here is a gull flying. Look what a tremendous wing-span it has ! Compare the gliding flight of gulls with the flapping flight of such birds as the lapwing.

STILL AT THE SEASIDE

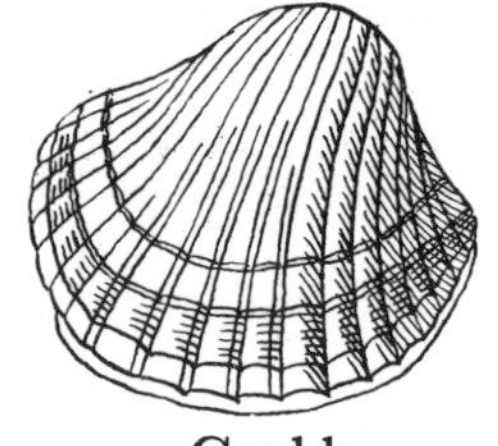
Cockle

Look for shells on the shore. This is the cockle shell. The cockle has another shell under the top one.

Here is the mussel shell. It is a deep, dark blue. The mussel attaches itself to little stones by threads.

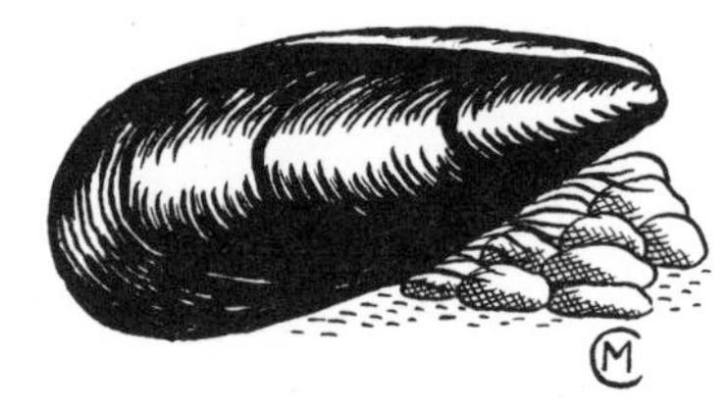
Mussel

Limpet

This is the limpet shell. It sticks closely to the rock and never gets washed away by the waves. If you try to pull it off, you will find it very difficult to do so.

What else will you find on the shore or among the rocks and rock pools ?

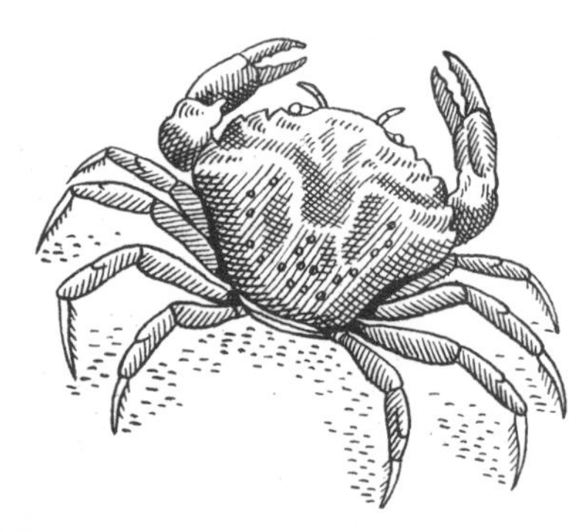
Crab

You will find crabs on the shore as the tide goes out. Baby crabs are like the little drawing on the right. They are found in the sea and are no bigger than a pin head.

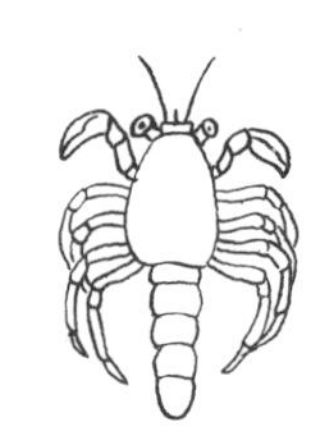
Baby crab

Here is the whelk shell and by its side are some whelk's eggs. You may find both on the seashore.

COLLECTIONS

Things you might collect during your summer holiday. Bring them back to school and label them.

Flint tool

You might pick up a flint tool which was once used by prehistoric men.

You might find a fossil like this one. It looks rather like the ramshorn shell on page 55. It is called an ammonite.

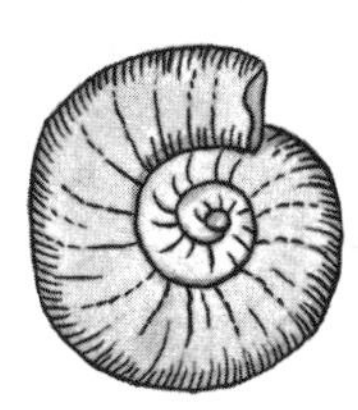

Ammonite

Fluor-spar

You might find some rock crystal. This is fluor-spar. The crystals are in cubes.

Perhaps you collect birds' feathers. Do you know the parts of a feather? Here they are.

Or perhaps you collect shells? (There are several shells shown in this book.)

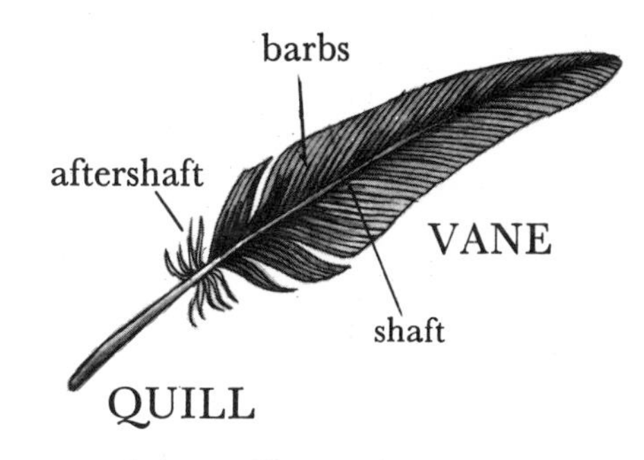

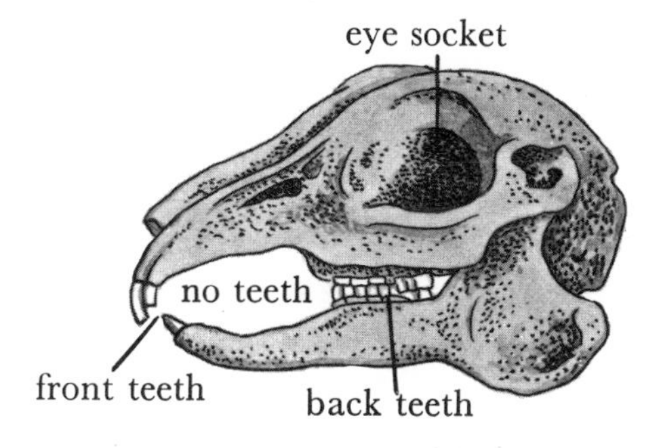

You could collect bones. Here is a rabbit's skull. Can you see its gnawing teeth in the front? It has back teeth but there is a space in between with no teeth. This is where its cheek pouches are.

(It is better to collect dry, rather hard, things which will keep. Soft things may decay.)

NEWS-FLASHES OF THE YEAR

AUTUMN

The leaves are falling.
There are wild fruits in the hedgerows.
Swallows leave Britain for the south.
Wild geese fly in.
Hedgehogs, dormice and bats, etc., prepare for winter.

WINTER

We look at the sleeping animals. We track animals in the snow.
We feed the birds in our gardens. We hear the song of the robin.
We try to recognise trees in winter. We look at stars.

SPRING

The sleeping animals wake up.
We plant seeds. We watch how roots and leaves develop.
Birds sing, lay eggs and bring up their families.
Bluebells and fruit blossom come out.

SUMMER

Summer brings the insects.
We look for insects—on the ground, in water and in the air.
We learn about insects and flowers.
The seeds which we planted are now flowering plants.
We learn about flowers, grasses, weeds.
We go to the park, the stream and the seaside.

Do you remember these words ? They are all in this book.

The word	*What it means*
A leaflet	A little leaf—part of a bigger leaf.
Albumen	The white of an egg.
Fronds	The leaves of ferns.
Larva	The young stage of an insect.
Adult	Grown up.
Stamens	The parts of a flower which contain the yellow pollen.
Pollen	The yellow powder which must reach a stigma before a seed can form.
Stigma	The top of a seed box. The yellow pollen falls on a stigma or is put there by an insect.
Ovary	Another name for " seed box ".
Carpel	The name given to the complete seed box with the stigma at the top and the ovary at the bottom.
Tentacles	The " horns " of snails, used for seeing and feeling.

The word	*What it means*
Cereals	Food grasses.
Ceres	Goddess of the corn.
Parachute	Flying apparatus like an umbrella. (Dandelion fruits have parachutes.)
Duck	Name given to a female or " mother " duck.
Drake	Name given to a male or " father " duck.
Webbed feet	Feet with skin between the toes (as in ducks).
Plumage	Birds' feathers.
Domed or dome-shaped	Rounded at the top ∩—like this.
Gnarled	Twisted, knobbly or rugged like the branches of an oak tree.
Nectar	Sweet substance in flowers—the bees make honey from it.
Nectary	The pocket in a flower where the nectar is kept.
Brittle	Easily broken.
Radiating	Coming out from the centre.

DO YOU REMEMBER?

(Answers are below—but look for them first in the book if you do not know them.)

1. What flower in this book has three petals?
2. Name a black swimming-bird (in this book).
3. Name a small furry animal which can fly.
4. What does the hedgehog do in winter?
5. What is inside an egg-shell?
6. What kind of feet has a duck?
7. What is the flat part of a bird's feather called?
8. What does a bee get from flowers?
9. What long-tongued insects do you know?
10. Name two birds which we call "winter visitors".
11. What is the difference between a pond net and a shrimp net?
12. Name a tree which has needle-like leaves.
13. Name a bird which sings while it is flying.
14. Name a large beetle which can swim.
15. Name a water plant which has two kinds of leaves.

ANSWERS

1. Arrowhead. 2. Moorhen. 3. Bat. 4. It goes to sleep. 5. Two thin skins; an air space; white of egg; yolk of egg; and possibly a fertile spot from which a chicken can grow. 6. It has webbed feet. 7. The vane. 8. Nectar. 9. Bees, butterflies, moths. 10. Fieldfare and redwing. 11. The pond net has a round frame; the shrimp net has a straight edge. 12. Spruce fir. 13. Skylark. 14. Dytiscus beetle. 15. Water crowfoot or water buttercup.

PART 2

NATURE AT HOME

CONTENTS OF PART 2

Page

Autumn and Winter

Spring and Summer

ABOUT PART TWO

If you really wish to know about Nature, you must go out to look for plants and animals at the right seasons and in the right places.

You must be like trackers in the wild, and learn to know the places where wild plants grow and where animals live (their " habitats ").

Plants and animals have their own special homes, just as people do. This book tells you where to find living things in their own homes.

When you go out on fieldwork you will need to take collecting apparatus with you. Here are some of the things you may need.

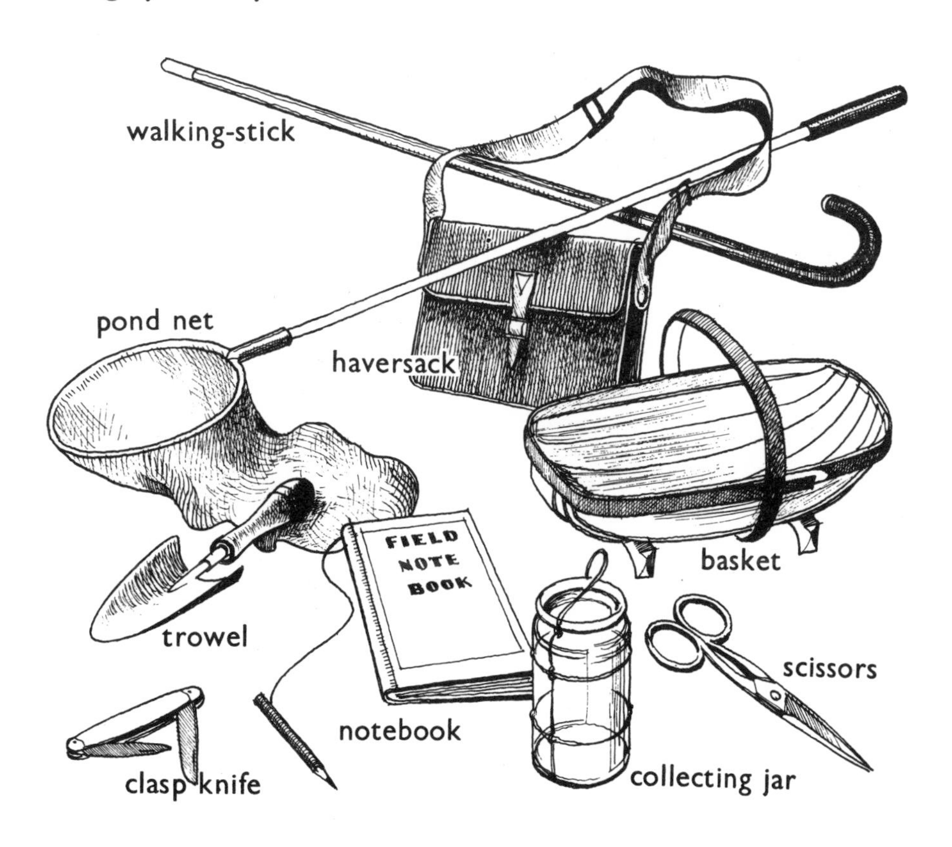

OUR FIRST EXPEDITION—IN AUTUMN—IS TO COLLECT WILD FRUITS

You need a basket for this. You also need your field notebook to record what you find, and you may need a knife or some scissors to cut twigs. We are going to look at hedgerows.

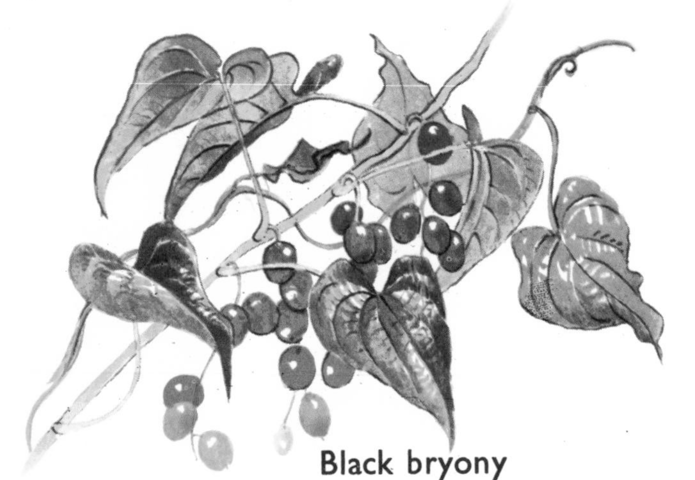

Black bryony

Twining around the hedgerows you may find these wreaths of the black bryony. The fruits of this plant are poisonous, so be careful.

Here is another plant which twines and makes wreaths around the hedgerows. Strangely enough this is called the white bryony. Can you see the difference between the leaves of the black and the white bryony? You will see, too, that the white bryony has curly little tendrils to help it to climb.

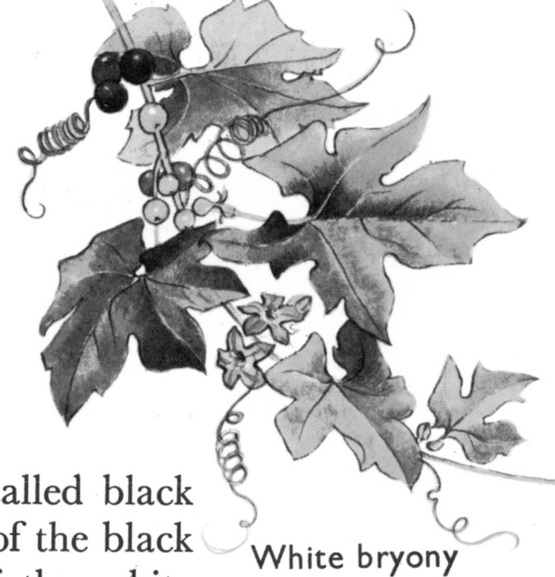

White bryony

The reason why one plant is called black and one white, is that the root of the black bryony is black while that of the white bryony is white.

A BASKET OF FRUIT
(from the hedgerows)

Bring your wild fruits home in a basket and make a list of them in your field notebook.

Here is a list of the wild fruits shown on this page.

(1) Snowberry
(2) Woody Nightshade
(3) Guelder Rose
(4) Sloe or Blackthorn
(5) Blackberry
(6) Hazel
(7) Hop
(8) Hawthorn

MORE FRUITS FROM THE HEDGEROW

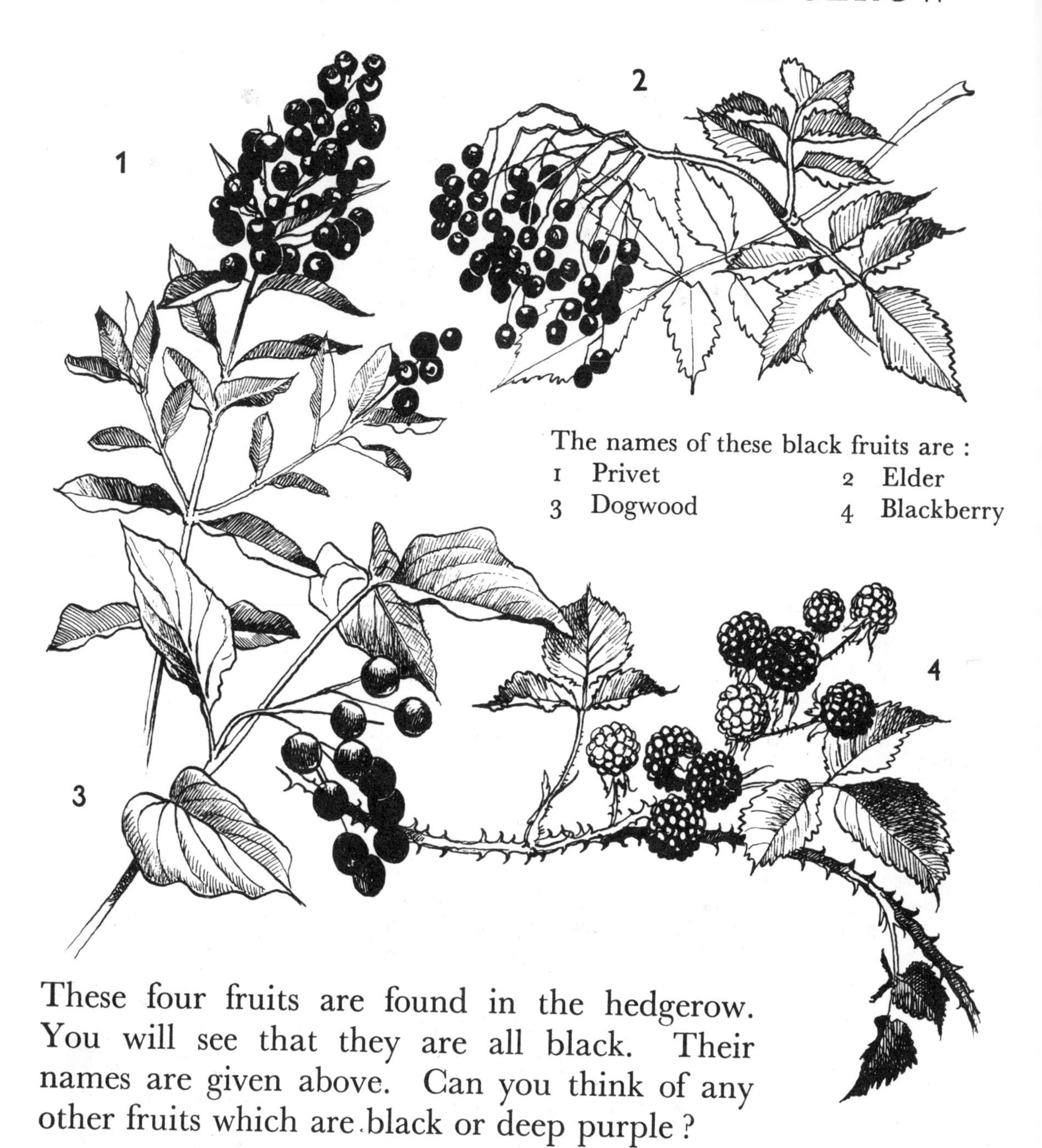

The names of these black fruits are :

1 Privet
2 Elder
3 Dogwood
4 Blackberry

These four fruits are found in the hedgerow. You will see that they are all black. Their names are given above. Can you think of any other fruits which are black or deep purple ?

Blackberries and elderberries are useful wild fruits. We collect blackberries for making jam and elderberries for making wine.

Here is a young naturalist making notes in her book. She is walking across some heathland which is covered with bracken. Can you see the fronds of bracken in the photograph? The soil is dry and sandy here.

A frond of bracken

What is bracken?

Bracken is a fern. We call the leaves *fronds*. The fronds are much divided as you can see in the picture on the left. These fronds grow from an underground stem. When the fronds first grow from their underground stem, their tips are curled under as in the picture below.

As it is now autumn, our young naturalist above will see the bracken in its autumn colours. These are tawny brown and sometimes pale orange.

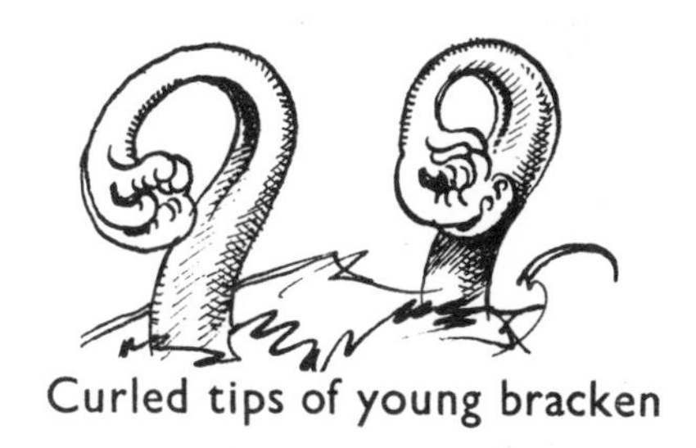

Curled tips of young bracken

The brown leaves of the bracken in autumn are very dry and crisp, and they easily catch fire and burn. This is dangerous and may cause big fires to start on bracken heathland.

COLLECTING FUNGI

Autumn is a good time to go looking in the woods for toadstools. At this season there are many different kinds of fungi to be found. You can see pictures of some of them below.

Collect specimens or make notes and coloured sketches in your book. You can find out the names of your discoveries by looking at books on fungi. Coloured pictures will help you.

Parasol mushrooms

Scarlet flycaps

If you look underneath the head of a parasol mushroom, you will find delicate skin membranes stretched out like the spokes of a wheel. These are called *gills*. The spores, which are a toadstool's " seeds ", are kept in these gills.

As you go deeper into the woods you may notice some fungi with an unpleasant smell.

Stinkhorn

Look at the picture on the left. This shows a *stinkhorn*. The stinkhorn begins life as a soft white egg [you can see one in the picture (a)]. The stinkhorn grows and bursts out of the " egg " and as it becomes tall, the top is covered with green slime which gives it an unpleasant smell.

WOODLAND AND HEATHLAND

Screw moss

Looking for Mosses Mosses are fresh and green all the year round. You will find them in many places such as in the woods and on the heath and also on banks and walls and on tree stumps. This moss, known as screw moss, grows on walls and on woodland banks.

Can you see some things which look like pixie caps on stalks? These are the spore capsules. Spores are rather like seeds. They grow into new plants.

On the right are the capsule (1) and the leaf (2) of screw moss.

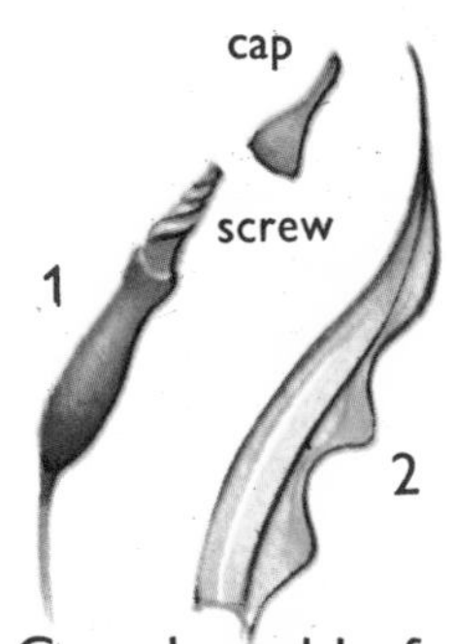

Capsule and leaf of screw moss

Can you see the screw at the top of the capsule of screw moss? This gives the moss its name. The screw helps to push off the little cap when the spores are ripe.

Do you remember what we said about heath fires on page 9?

These two mosses, hair moss and cord moss, often grow on heathland which has been burnt. Cord moss is sometimes called "fire moss" for that reason.

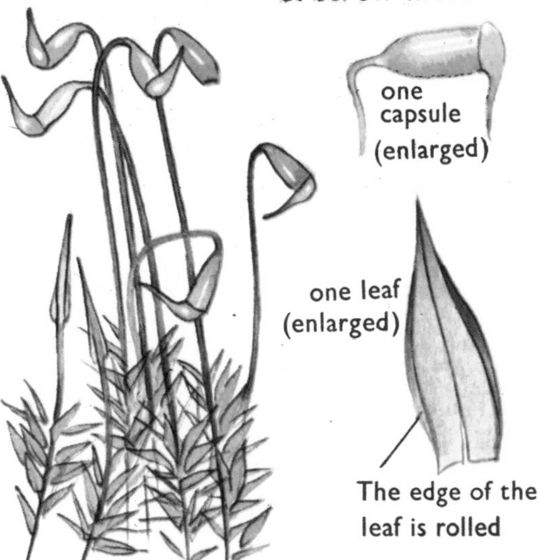

Cord moss

Hair moss is one of our largest mosses, and is sometimes called "juniper-leaved hair moss" because its leaves are said to be like those on a sprig of juniper. Look at the juniper sprig on page 13. Do you think so?

Hair moss

FIR TREES

Leaving the bracken, we climb up into the pine woods. Can you see them at the top of the picture? Fir trees grow in dry places—and often in cold and windy places. When we are in the pine woods we shall be able to collect some fir cones. Here is the cone of the Scots pine.

Scots pine

Fir trees are called conifers because they bear cones in which their seeds develop. The picture above shows a fir tree called a Scots pine. It is a tall straight tree and the trunk is often used for pit props and masts. It bears cones like the one in the picture on the right.

Cone of Scots pine

The narrow leaves of Scots pine are always in pairs

This cone is two years old. The cones take three years to ripen. When they are ripe, the scales of the cone open and the seeds fall out.

LOOKING AT GARDEN HEDGES

You may not be able to go as far afield as our young naturalist, who went to the woods and to the heath. Perhaps you know a nice garden or a park where hedges are planted to divide the garden up, or to provide shelter from a strong wind.

Here is a beautifully long, thick hedge in a garden. This is a yew hedge. The yew is an evergreen. Gardeners often use evergreen plants for hedges in gardens, because they look fresh and green all through winter as well as in summer.

Hedges may also be of privet, holly, box, laurel and sometimes of cypress or juniper. All these are evergreens.

A SUNSET EXPEDITION—to see rooks and starlings

A rook

Watching rooks in late autumn

This time we are going on a *sunset* expedition to watch the rooks coming home to roost. When rooks have finished rearing their families, they like to be together in large numbers. Because of this we say they are *gregarious*. In late autumn and winter, they go to sleep all together in a rook roost.

Here are the rooks flying to their sleeping quarters at sunset. They usually go to a wood or to clumps of tall, old trees in some big estate. In some rook roosts there are thousands of birds collected together every evening. The rooks come from miles around.

The rooks fly home to roost

A flock of starlings

Starlings are gregarious (" like to be together ") birds, too. They have their own " starling roosts " in late autumn and winter in the same way as the rooks. Have you ever watched a flock of starlings in the air? They " wheel " and " tower " and " fall " in the air, all moving together as one bird. They often give a flying exhibition like this just as they are setting off to their roost.

LOOKING AT DUCKS

Here are some ducks which you might see on a pond or lake, perhaps in a town park. They are:

1. Male mallard
2. Female mallard
3. Tufted duck
4. Ruddy shelduck
5. Pochard

How can you recognise them?

Here are some clues to help you:

1. Male mallard—bottle green head.
2. Female mallard—mottled brown and white.
3. Tufted duck—white and black.
4. Ruddy shelduck—orange-brown, with pale head.
5. Pochard—grey and black with red-brown head.

This is the webbed foot of a duck

The tufted duck
(a) swimming
(b) flying

Ducks sleep a good deal during the day. It is best to go in the morning or later in the afternoon to see them. They often put their heads under their wings at midday and have a " nap ".

Here is a picture of a beautiful duck called the Garganey. It spends the winter in Africa and comes to us in spring.

Garganey duck

LOOKING AT DEER

A fawn (baby deer)

In some of our large, forested estates and park-lands, herds of deer are kept. There is a young deer, or fawn, in the picture on the left.

We call deer "cloven-hoofed" animals because their foot is a divided, or cloven, one. Their foot-print is like the one in the drawing.

You may see this footprint in the snow if you visit a deer park in winter.

Oxen, sheep and goats all have hoofs like the deer and all can run well, but the deer is the swiftest runner of all.

When they are older, the male deer (stags) grow antlers on their heads.

They often fight with these antlers as you can see in the photograph.

Two stags fighting

Late in the season the stags shed their antlers but they grow again in the spring.

This is a photograph of deer taken on the Duke of Bedford's estate at Woburn in Bedfordshire.

These deer are waiting to be fed in the evening near the stables. At other times of the day they roam about the park quite freely.

Deer are gregarious ("like to be together") animals and stay with the herd. Where else have you seen the word gregarious in this book?

You saw on page 80 that a young deer is called a fawn, and a male deer is called a stag. The male deer is also called a buck, and the female deer is called a hind or a doe.

LOOKING AT GAME BIRDS

Cock pheasant

We must go back to the cover of the woods to see where pheasants live although we may see them feeding in open fields and parklands.

The cock pheasant drags his tail on the ground. If you go out in the snow to track pheasants, you may see the track mark, like the one shown here. The long line between the foot tracks is made by the tail.

The plumage (feathers) of the red grouse is reddish-brown.

Red grouse

This bird is found on heather moors in more northerly districts of Britain. It runs to take cover when frightened.

Partridge

The male partridge has a horse-shoe shaped patch on its chest. Its plumage is grey and brown and it has a reddish-brown head.

The partridge lives near cornfields but, like other game birds, it runs under bushes to take cover.

EARLY MORNING AT THE EDGE OF A WOOD

The woodcock

Here is a very strange bird, the woodcock, rising early in the morning—at dawn on the edge of a wood. As he flies in the air, he keeps his long beak pointing downwards.

The woodcock spends most of the day under cover in the wood and only comes out in early morning or in the evening.

The woodcock is a game bird like the pheasant, the grouse and the partridge.

EVENING AT THE EDGE OF A WOOD

The barn owl

At dusk the barn owl may often be seen flying silently, like a white shadow, between the tall trunks of a woodland glade. His feathers are sandy coloured but he looks white in the evening light.

Here is a picture of a wood on a November evening. Can you see the owl?

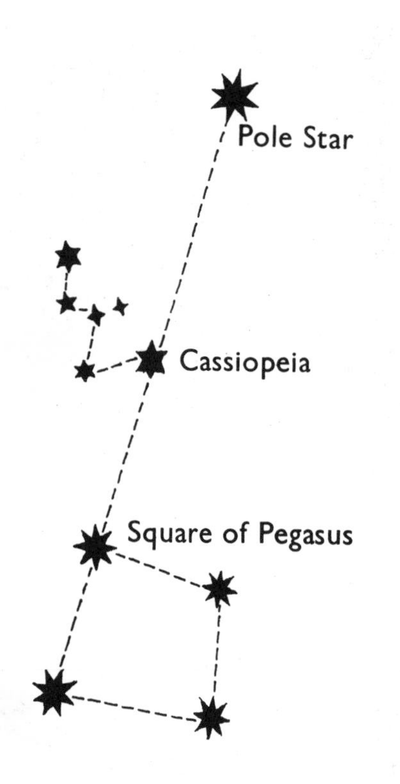

A group of stars is called a constellation. We are able to discover directions (north, south, east and west) by looking at constellations. The first thing we must do is to find the north. From this constellation in the drawing, you can see that if a line was drawn from Pegasus through Cassiopeia, it would lead straight to the North or Pole Star. In the days before the magnetic compass was invented, sailors used the Pole Star as a guide when they were navigating. To find this constellation in the sky, look for the four big stars forming a square (the square of Pegasus) and the big W of Cassiopeia.

The Greeks had stories about stars. They had one about Pegasus. Here is Pegasus. You can see the square which forms part of its body. Pegasus was a winged horse (so the Greeks said). This beautiful winged horse was supposed to have arisen from the body of a terrible monster, the Medusa, which was killed by Perseus. There is a group of stars called Perseus.

Pegasus

Here is the star group called Perseus.

Here is Perseus as the Greeks imagined him, struggling with the great sea monster.

Perseus was in love with a beautiful princess called Andromeda. He found her chained to a rock at the mercy of a sea monster sent by Neptune.

Perseus killed this monster. You can see him struggling with the sea monster in the picture above. The picture on the right of the page shows how the Greeks saw the princess Andromeda chained up.

Do you see how the stars form part of the pictures ?

Andromeda

LOOKING AT THE GROUND—PLANT ROSETTES

Shepherd's purse

Ground plants are often found in the form of a rosette in winter. Their upstanding stems, leaves and flowers have died off, leaving the plants much smaller.

Here is shepherd's purse in its winter rosette form. It will grow tall again in spring and summer.

On the right is the winter rosette of the dandelion. Can you see how closely it is pressed to the ground? It gets more moisture in this way. The dandelion will grow from the point in the middle of the rosette—in spring and summer.

Dandelion

Primrose

Here on the left is a plant whose leaves always form a rosette, but it is a tighter and flatter rosette in winter.

It is the primrose.

WHERE HAVE WE BEEN? (Autumn and Winter)

In autumn we collected wild fruits, toadstools, fir cones and mosses. Nearer home we examined the plants of our own garden hedges, or perhaps looked at the ducks on the park lake. We have watched rooks or studied game birds or owls, or have gazed at the stars on winter evenings. Now the days are beginning to lengthen; the door of winter is closing and we come to the door of spring.

The door of spring might be your own front door. Does your door lead into a garden?

These crocuses on the right might be in your own garden.

Crocuses

Aconite Grape hyacinths

On the left is a very early flower, the aconite, which comes out in January.

These grape hyacinths came out on April 1st.

Make a note of new flowers with the dates on which they appear.

SPRING IN A COUNTRY LANE

Blackthorn

" Oh, to be in England
Now that April's there . . . "

(from *Home Thoughts from Abroad*
R. Browning)

What can you find in a country lane in spring ?

Blackthorn is an early blossom ; it is often out in March. Another name for blackthorn is sloe. You will see the purple fruits of sloe if you look at page 7.

Listen as well as look.

The thrush may be singing.

Thrushes soon begin building nests and rearing families.

You may see birds going in and out of hedgerows to their nests.

A little later in spring, blackbirds will sing. You can see a female blackbird on her nest on page 34. Blackbirds and thrushes are relatives.

Look under the hedge too. You may find violets like these in the picture.

Violets

Hawthorn is a common hedgerow plant. You can see its leaf here. It is said that there are no two leaves alike on a hawthorn bush ; look at some to see if this is true.

Hawthorn

The flower of the hawthorn is called " may ". It will bloom in the month of May.

Ash is another tree which often helps to make a hedge.
Here is an ash twig.
You can see its black sooty buds and its feathery flowers.

Often elm trees have been planted along the road and have become hedge trees.
You may see green things like these on elm trees in spring. They look like small round leaves. They are not leaves, however, but green fruits. They are flat and thin.

Elm fruit

Ash Elm

Find out what other trees and bushes make up our hedgerows.

Baby thrushes

Many young birds are reared in spring. As soon as they can, they leave their nests. They never go back to the nest but their parents follow them about and feed them until they can feed themselves. Here are two baby thrushes looking for their parents. They are rather like baby blackbirds as you will see on page 98.

There are many young animals about in spring and they are full of energy and like to play.
Here is a young fox cub. His home or " earth " is in the wood sheltered by trees. He enters it through a hole in the ground. Now he comes out to explore the world in spring.

Fox cub

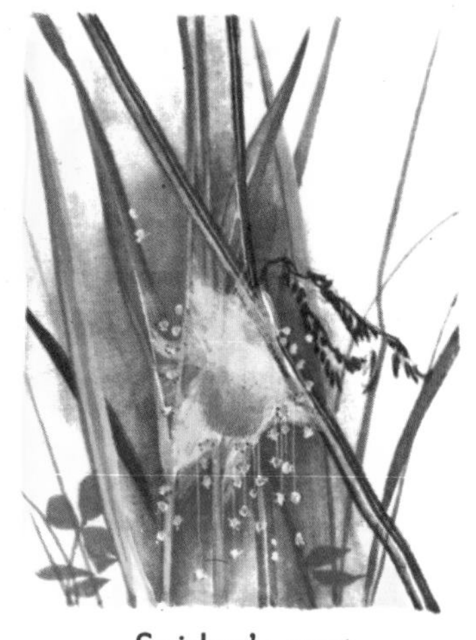

Spider's nest

Even little creatures like spiders have their spring nurseries. Here is the spider's nursery, a tight web between two stalks. The centre is white and more tightly woven than the rest. You can see the baby spiders escaping on silken threads. They also are going to explore the world.

Fresh young leaves are unfurling. Some of the most beautiful of all young green leaves are those of the lime tree.

The flowers of lime will not be out until midsummer.

You may have a cherry tree in your garden. This is the white blossom of the wild cherry or *gean* from which all other British cherry trees have been developed.

The lilac is one of the most popular flowers in spring. (Is it *your* favourite?) Our garden lilacs all came from the wild lilac of south-eastern Europe.

Do you see that every little flower has four petals and the leaves are heart-shaped?

Make a note in your field notebook of all the leaves and flowers which you see opening as spring grows into summer.

SPRING-TIME BY POND OR STREAM

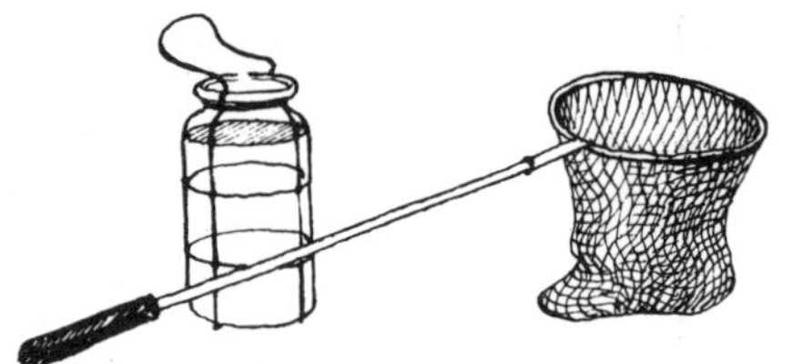

It is time to get your nets and collecting jars ready

Now it is " tadpole time ". There may be three kinds of tadpole in your pond :

(1) frog tadpoles ;
(2) toad tadpoles ;
(3) newt tadpoles.

Frogs, toads and newts are related. These animals are all amphibious: that is, they live partly in water and partly on land.

Frogs lay eggs in a mass of jelly like this.

Frog spawn

Toads lay eggs in a string like this.

Toad spawn

Newts lay little white eggs. They put each egg in the leaf of a water plant and then fold the leaf over like this.

Newt's eggs

FROGS, TOADS AND NEWTS (differences)

Notice that a toad has warts

A toad sitting in the sun

Newts have a long tail

A newt climbing out of the water

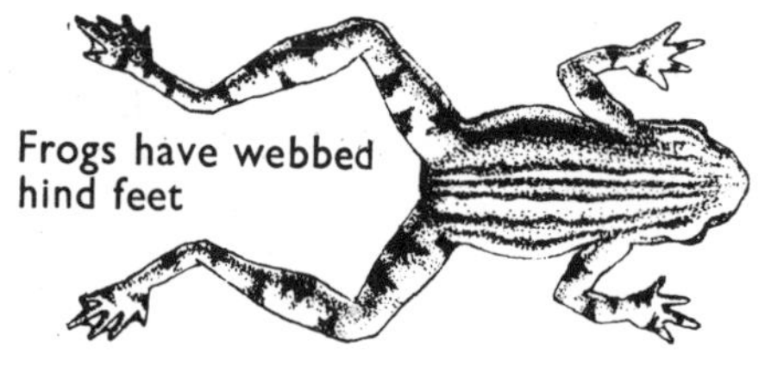

Frogs have webbed hind feet

A frog swimming

Frogs and newts like being in the water more than toads do, but all these lay their eggs in water.

WATCHING YOUR FROG TADPOLES GROW UP

Start reading from the bottom of this page

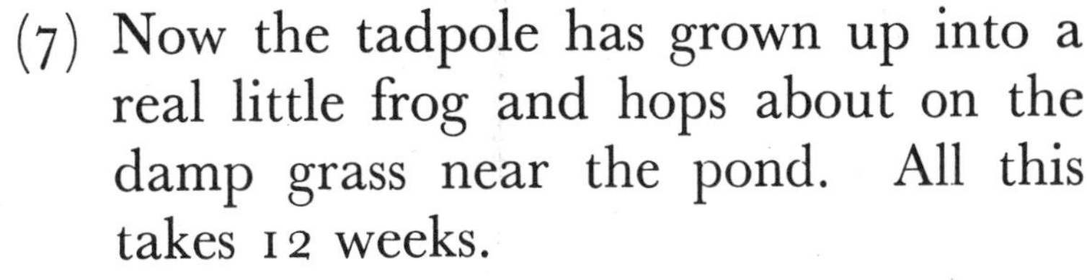

(7) Now the tadpole has grown up into a real little frog and hops about on the damp grass near the pond. All this takes 12 weeks.

(6) The tadpole is much more like a frog now and often it is out of the pond, but it still has a little tail.

(5) The tadpole has now got all its legs. Do you notice that its tail is getting shorter ?

(4) The tadpole has lost its feathery gills and has now got its back legs. The front legs will soon grow.

(3) The tadpole can swim about now. It has no legs yet. It has some feathery gills.

(2) The tadpoles are out of their jelly. They hang on to water weed with their little sucker mouth.

(1) The tadpoles are still in their jelly but they have a little shape—like a comma.

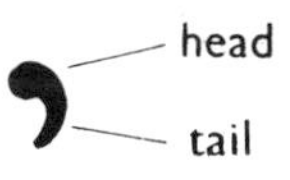

7 UP

6 UP

5 UP

4 UP

3 UP

2 UP

1

GROWING UP

SPRING IN THE WATER MEADOWS

Some flowers which live in marshy places

If you go to the pond for tadpoles, look around the marshy land near the pond and you may find some golden kingcups. Their correct name is marsh marigold. You may find marsh marigolds out as early as March.

This is the yellow flag. It flowers a little later than the marsh marigold but you will often find it in the same place. Look out for the sword-like leaves of yellow flag.

With the marsh marigold you may find the cuckoo flower, which has four lavender coloured petals.

The ragged robin grows near these other marsh plants. It has pink flowers, each with five ragged looking petals, and blooms in spring.

BIRDS WHICH NEST IN THE MARSHES IN SPRING

Redshank

Here is a bird which nests in the marshes in spring. The redshank is a wading bird. You can see his long legs. He is brown above and lighter below. He has orange-coloured legs and a long beak for digging into mud and sand.

The redshank is in the marshes in spring. He makes his nest in a little hollow in the ground and the eggs are laid in April or in May.

Here is a baby redshank

Lapwing

The lapwing is sometimes called the peewit because he makes a call which sounds like " Peewit ". He is greenish-black with a black bib and white underparts. He has a crest on his head.

The mother lapwing lays her four eggs in grassy hollows on the ground in April. The eggs are olive green with brown blotches.

Here is a baby lapwing

CATCHING THINGS FOR YOUR AQUARIUM

Dip your net among the rushes and sedges at the edge of the pool. Move very gently—or you will frighten the water things away.

Scoop the net along in the shallow water and then bring it out. Look in it to see what you have caught.

These girls have been pond-dipping. They are now emptying the water creatures into white pie-dishes. Then they can see their little animals more easily.

Here are some of the things they have caught :

Ramshorn pond snail (see page 55)

Common pond snail (see page 55)

Caddis worms (see page 128)

Leeches (see page 128)

Sticklebacks (see page 55)

and all the things shown on these two pages.

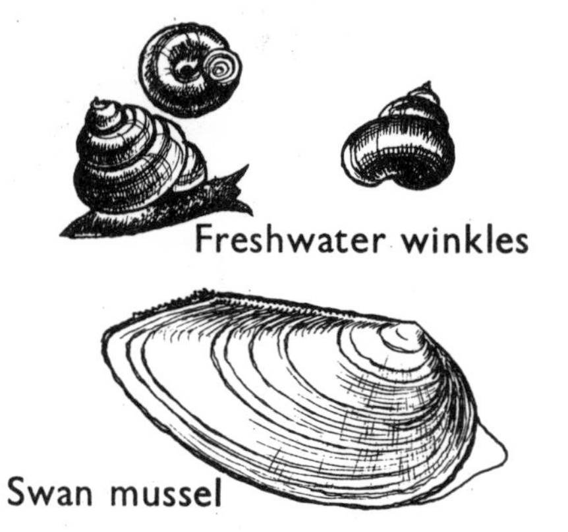

FIERCE WATER CREATURES

You can keep water snails in your aquarium, but here are some creatures you had better keep separate. They may eat all your other water animals.

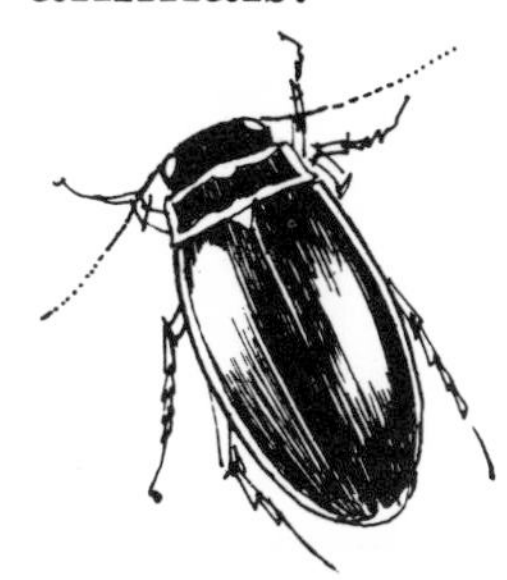
Dytiscus beetle

Here are the Dytiscus beetle and the Dytiscus larva. The larva grows up and becomes the beetle.

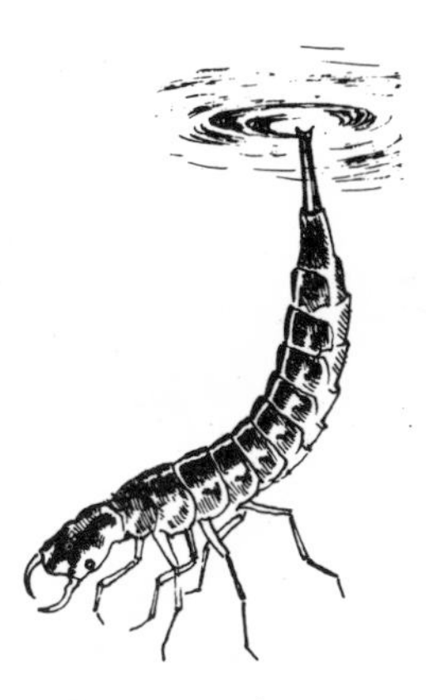
Dytiscus larva

Both of these creatures will feed on soft things such as tadpoles. If you keep them away from their natural food you could feed them on very small pieces of raw meat such as liver.

Demoiselle

This demoiselle larva will grow into a small type of dragon fly. The larva can be dangerous to other water creatures. So can the water boatman which swims on its back, using its hind legs as oars.

Water boatman

CREATURES WHICH ARE GOOD FOR THE AQUARIUM

Water slater

These two water creatures are good to have in your aquarium because they eat up waste food materials and so act as scavengers.

Freshwater shrimp

WATCHING BIRDS BUILDING NESTS AND REARING YOUNG

Willow warbler

When you see birds going about with their mouths full of straw or feathers or little twigs, you know that they are building nests, and when they are carrying little grubs or other food about, you know that they are feeding their young.

Here is a willow warbler with its mouth full of caterpillars.

Birds build their nests of many different materials. Thrushes' and blackbirds' nests are rather alike, made of grass and long fibrous roots or twigs. The blackbird uses more moss, often with an inner layer of dried grass. Mud is used to bind the soft grass and moss. You may see some of it in this picture of a female blackbird on her nest.

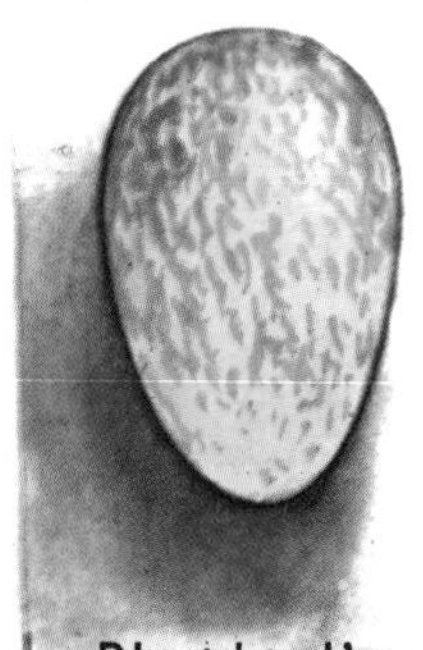
Blackbird's egg

On the left is a blackbird's egg. Its shell is a greenish-blue with light brown spots all over it.

The baby blackbird is rather like a little thrush and has a spotted breast.

Baby blackbird

WATCHING THE BIRDS RETURN FROM ABROAD

Swallows

The swallow and the cuckoo both return to us about the middle of April. They spend the winter in Africa.

Cuckoo

Soon after arrival the swallows build nests of mud and straw. They often build on ledges of buildings, or under roofs.

In Holland the stork comes back from Africa and builds a nest on a chimney stack.

The female cuckoo puts its egg in another bird's nest.

When the young cuckoo hatches out it pushes the other eggs out of the nest.

The nightingale comes back from abroad too. Listen for the cuckoo and the nightingale and write down the date when you first hear them.

Many young creatures are born in spring.
Here are lambs, piglets and chickens.

Chickens are hatched out of eggs. Lambs and piglets are born like puppies and kittens.

Lambs are cloven-hoofed animals. Where else have you read about such animals in this book ?

Animals which live on a farm are not wild animals any longer. They are looked after by human beings and are called domestic animals. Most farm animals are useful to us because they provide us with some of our food.

Chickens grow up to be hens or cockerels. We eat hens' eggs and use these birds as food.

Piglets grow up to be pigs (boars or sows) and we eat their meat, which we call pork.

Lambs grow up to be sheep (rams or ewes) and we eat their meat, which we call mutton. Sheep also give us wool.

There may be horses on a farm.

These are young carthorses. They are stronger and heavier than other horses.

A group of young horses or colts is called a " rag " of horses.

The smaller kinds of horses are called ponies. There are several breeds of ponies such as Shetland ponies, Dartmoor ponies and New Forest ponies.

Here is a pony from the New Forest in Hampshire. The ponies are friendly but they roam about, quite wild, all over the forest.

A horse does not have a cloven hoof like a sheep or a pig or a deer. It walks on one toe. This toe has a strong, hard " nail " on it which we call the horse's hoof.

Horses are shod with an iron shoe.

This is the print in mud of a horse's shoe.

THE FIRST BUTTERFLIES

Brimstone butterfly

We often see butterflies early in the year and perhaps we wonder why. Where have they come from ?

Many butterflies which we see late in summer, or in autumn, go to sleep for the winter and then wake to life again on the first spring day.

Comma butterfly

The three butterflies you can see on this page are often seen in spring.

(1) The *brimstone* butterfly may be seen as early as February. This butterfly has a small orange spot on each yellow wing. Can you see it ?

(2) This is the underside of the *comma* butterfly. Can you see the comma mark on its lower wings ?

Small white butterfly

(3) This is the *small white* butterfly. During the winter it was a sleeping grub or pupa. In April it became an adult or " grown-up " butterfly.

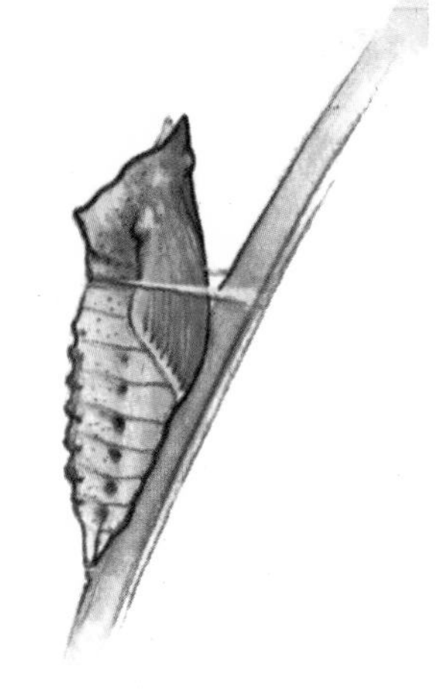

Butterfly chrysalis

Pupa means "little doll". Do you think that the pupa in this picture looks like a doll? The pupa of a butterfly is often called a chrysalis.

This is a butterfly chrysalis. Can you see that it has attached itself to a twig by a silken thread?

When the butterfly comes out of the chrysalis, it lays eggs.

This small white butterfly has laid eggs on a cabbage leaf. What happens to these eggs? Look at the picture at the bottom of the page.

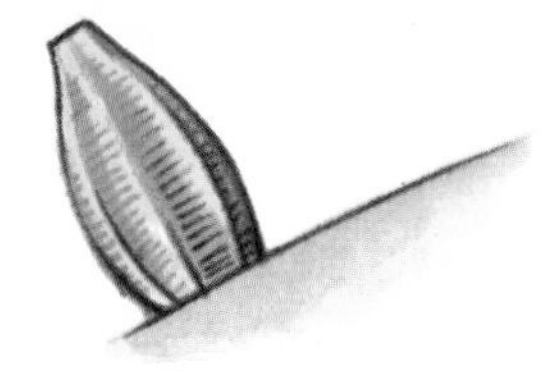

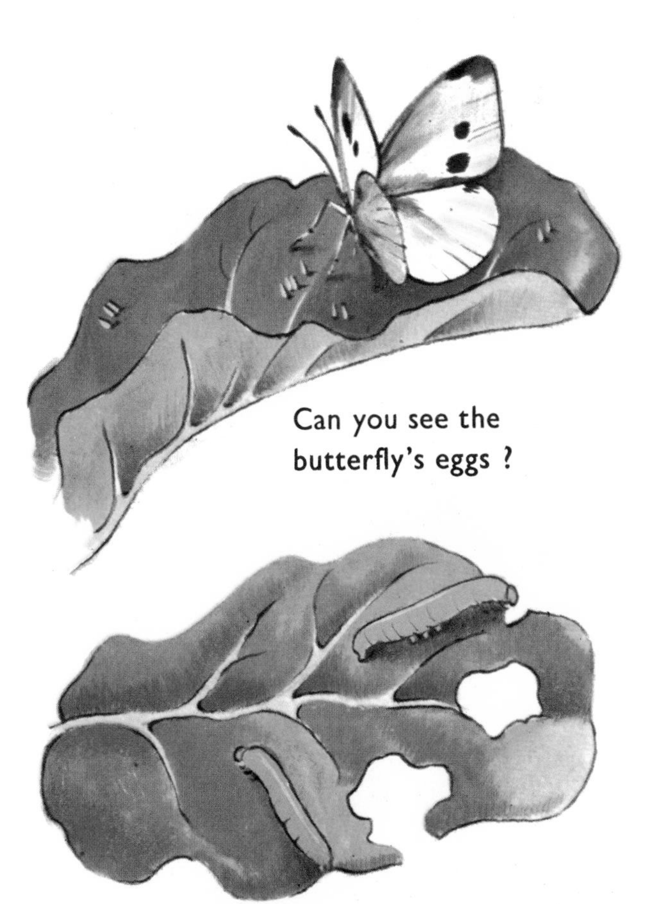

Can you see the butterfly's eggs ?

Look at the holes which the caterpillar has eaten in these leaves

Here is one egg (enlarged). The eggs of white butterflies are usually yellowish in colour. Out of each egg comes a caterpillar. The caterpillar feeds on the leaf. Caterpillars of the small white butterfly are green with a yellow stripe along their back.

IN THE MEADOWS

Summer-time in the grass jungle

In early summer-time there is more sunshine and we see that the grass is growing long in the fields. Let us lie in the grass of a field and keep very quiet. What happens in this grass jungle ?

We shall see many different kinds of grasses and many common weeds. Insects and other little animals often make their homes here.

Do you see these patches of bubbly white stuff which stick on to grass stalks ? We often call them " cuckoo spit " but they are not made by the cuckoo.

Froghopper

In the bubbles of the " cuckoo spit " are the green larvæ or grubs of an insect called the frog hopper (Picture 1). The young frog hopper pierces the plant to get sap for its food. With some of the sap it blows a froth of bubbles around itself. Later this green bug develops into a fully grown insect with wings (Picture 2).

If we remain quietly in our grass jungle we may hear something. It is a little song, but it is not the song of a bird. It is the song of a grasshopper.

The grasshopper does not sing through his mouth. He makes a clicking noise by scraping his wings over his back legs. His back legs have saw-edges like this one in the drawing.

The picture below shows how this saw can quite easily rub against the long straight wings.

Leg of grasshopper

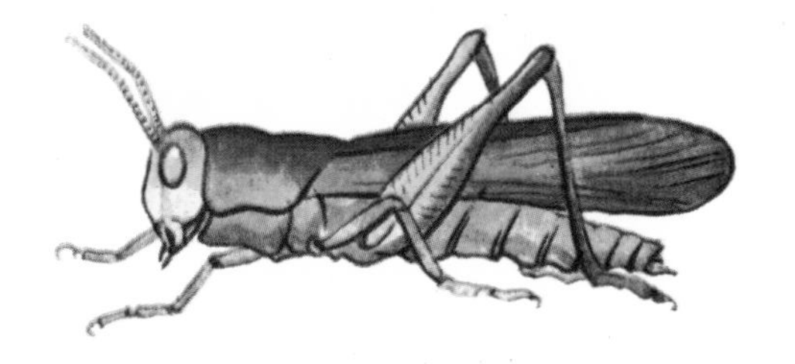

Grasshopper

The grasshopper is a jumping insect. Its long back legs which you can see here are specially long for jumping.

Long back legs and straight bodies are found in other creatures which are relatives of the grasshopper, such as crickets and earwigs.

Field cricket

Here is a relation of the grasshopper —the field cricket. You may find him in the grass jungle too, for he feeds on the roots of grasses.

The cricket makes a little noise too— rather like chirping.

The field cricket has a relative which lives in old houses—the house cricket.

Now let us look at the grasses themselves.

Here are two you may see.

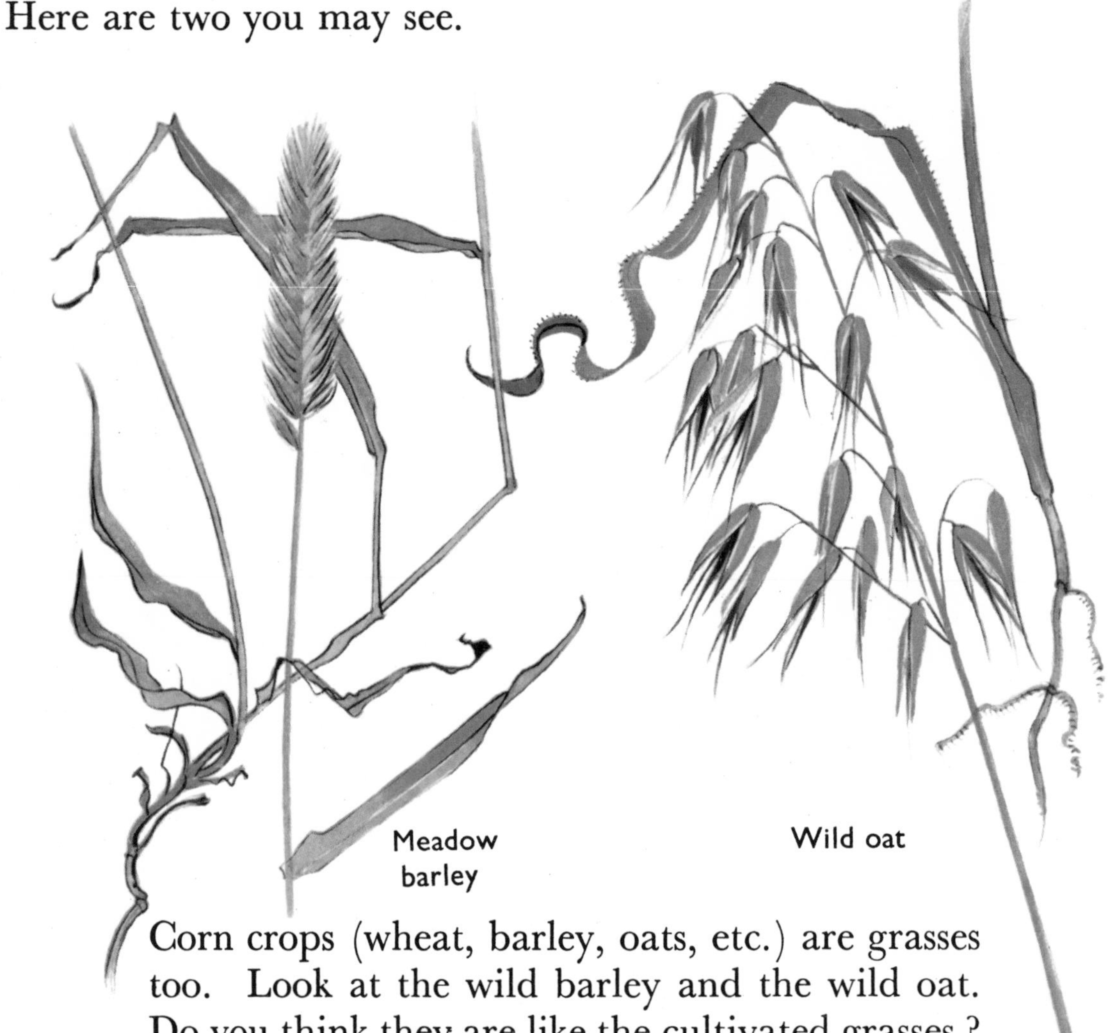

Corn crops (wheat, barley, oats, etc.) are grasses too. Look at the wild barley and the wild oat. Do you think they are like the cultivated grasses ?

Barley has a " beard " of little hairs which protect its spike of fruits. These hairs are called *awns*.

You can see some awns sticking out from the green flowers of wild oat.

Would you recognise these grasses again ?

IN THE CORNFIELD

Many flowers come up among the corn. Here are some in a wheatfield. Do you know them?

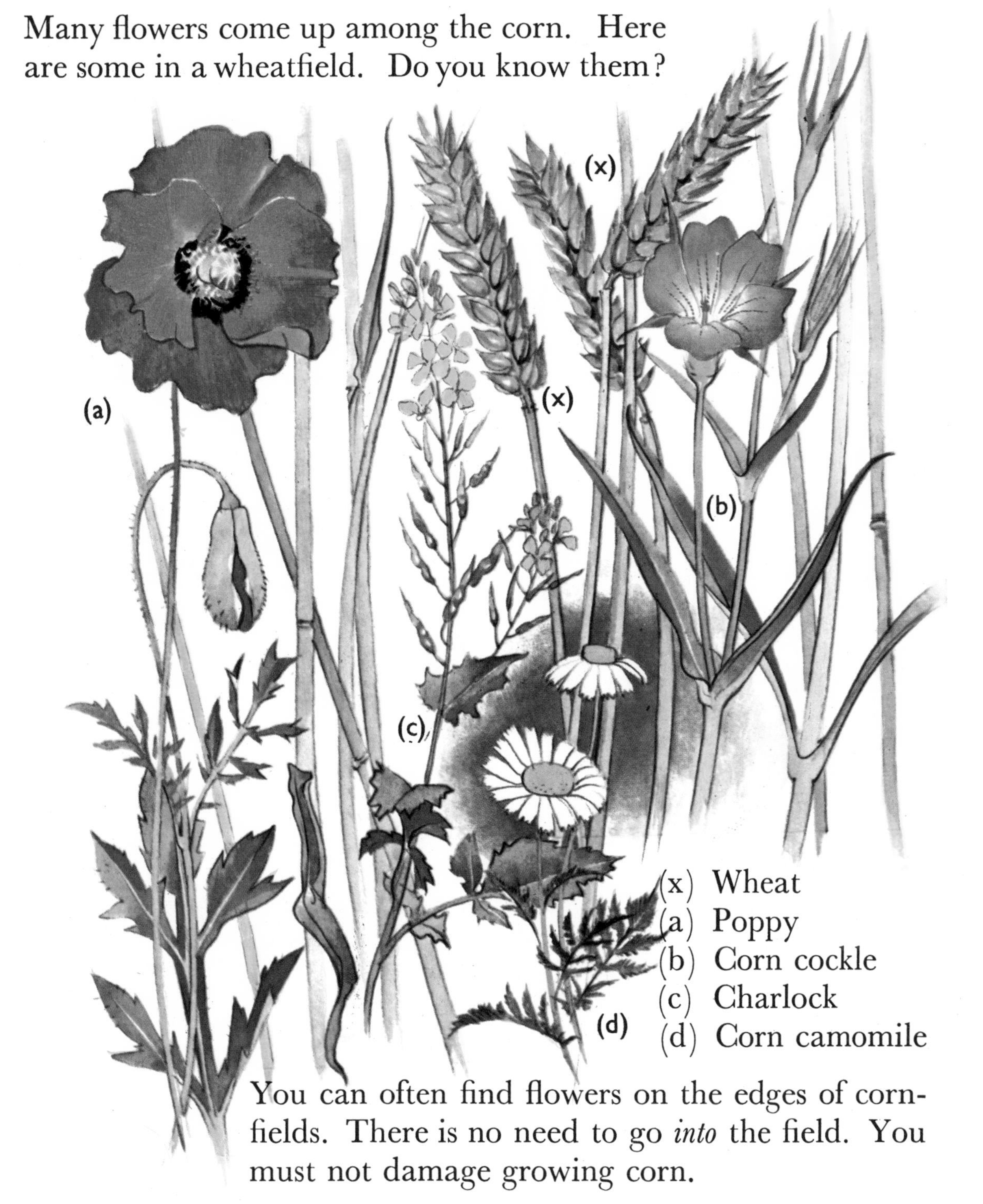

(x) Wheat
(a) Poppy
(b) Corn cockle
(c) Charlock
(d) Corn camomile

You can often find flowers on the edges of cornfields. There is no need to go *into* the field. You must not damage growing corn.

The bog is a *wet* place : the moor is a *dry* place, but the soils in both places are very much alike and sometimes a bog may become a moor or a moor may become a bog, depending on the amount of water which the land receives.

If there is much water we have a bog. Here are some bog plants.

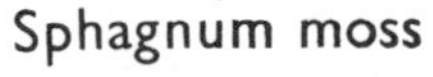
Sphagnum moss

Sphagnum moss often grows in boggy places. It has leaves which swell out with water.

Butterwort

Above is a very strange little plant which grows in a bog. It is called the butterwort. Its leaves are light green and sticky. Insects sometimes stick fast on these leaves.

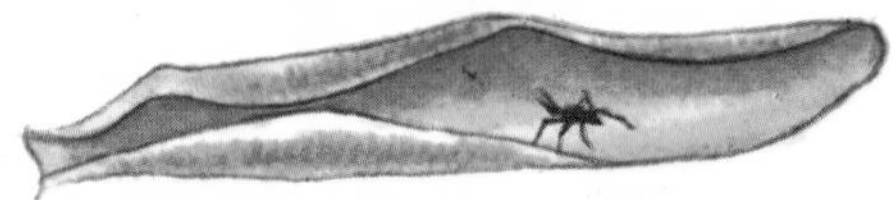

You can see a tiny insect on this leaf. When this happens the edge of the leaf rolls over. The sticky stuff on the leaf dissolves the insect's body and this is absorbed into the leaf.

Cotton grass

Cotton grass grows on water-logged soil. It has heads of little fruits which look like cotton wool.

ON THE MOORLAND

Heather

Have you heard of the Yorkshire Moors? Heather grows on moors.

This is the true Scottish heather or *ling*. It has a very small flower which is light purple in colour. There are many of these little flowers on one spray of heather.

There is not much water on the windy moors and heather is a dry plant with tiny leaves which are rolled to prevent the escape of moisture.

One flower of heather (enlarged)

Here is another little plant you will find on the moors—the bilberry, sometimes called the whortleberry.

It grows low near the ground and has pink flowers which become purple fruits.

(Bilberry pies are very good to eat!)

Bilberry

Cross-leaved heath

If you compare this heather plant on the left with the one above, you will see that the two are different. The leaves of the cross-leaved heath are in little groups or " whorls " all the way down the stem and the purple flowers are in a big group at the top.

LET US LOOK AT A CHALK HILL

You will see that it has one long side and one short side.

Chalk hills are usually rather bare. They are covered with short grass. Sheep like to feed on this grass.

Sometimes trees grow on the short side or escarpment (scarp). You can see where the trees are in the picture. These woods often seem to be clinging to the side of the hill and are called " hangers ". Beech trees like to grow on a chalk hillside and a beech wood on a scarp is called a " beech hanger ".

Beech leaves

Below the hill, as you can see, there are some farm fields and some haystacks. Chalk mixed with clay soil is good for corn crops and other crops too. This kind of chalky soil is called " marl ".

Be careful where you sit on chalk downs.
You may sit on this stemless thistle

FLOWERS YOU MAY FIND ON THE CHALK DOWNS

The rock rose above is one of the commonest and most beautiful of the flowers which grow where the soil is chalky. It is sometimes called a "chalk-indicator", which means that it shows that there is chalk in the soil where it is growing.

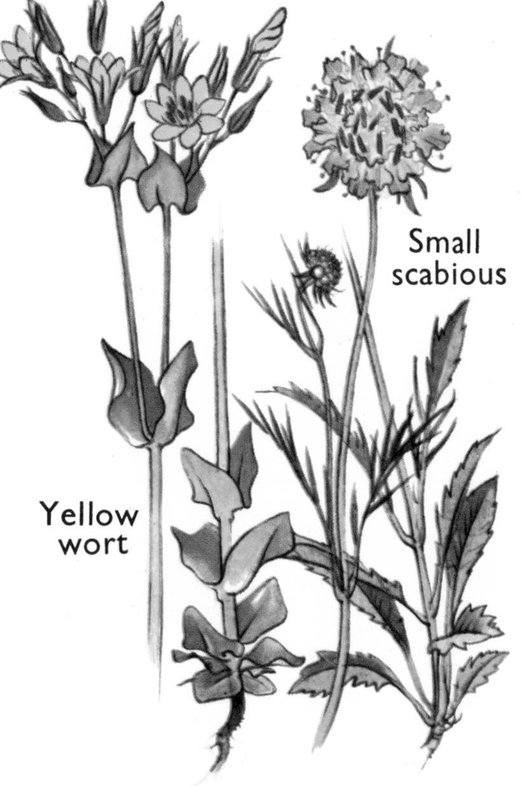

Another true chalk plant is the yellow wort. Notice the yellow flowers and also the way the leaves are in pairs and close around the stem.

The third flower is the small scabious. It is tall with a head of purple florets (little flowers).

REMEMBER Very many beautiful flowers, including some rare orchids, grow on chalk. It is better to see these flowers growing naturally. Do not collect them or you may spoil this pleasure for others.

Gannet

Here is a gannet flying. Look at his wing span. It is nearly two metres wide. You can see that he has dark wing-tips. The gannet is a fishing bird and dives into the water after its prey.

The cormorant has a long neck and a long beak. In this picture he has bent his neck in a curve like the letter S.

He can swim and catches fish with his long, hooked beak.

Cormorant

Puffin

Puffins are sometimes called " sea parrots ". They have brightly coloured bills. The end of the bill is red and the part nearest to the face is blue-grey. The legs are orange-coloured.

The puffin dangles these orange legs when it flies. It likes to catch shrimps to eat.

Another bird which can be seen on rocks above the sea is the guillemot. This bird spends most of its time at sea, diving for fish. It lays its eggs on cliff ledges.

Guillemot

The egg of the guillemot is a large, mottled one. It cannot roll off the cliff because it is so well balanced. Instead, it rolls around in a circular fashion like this, when disturbed.

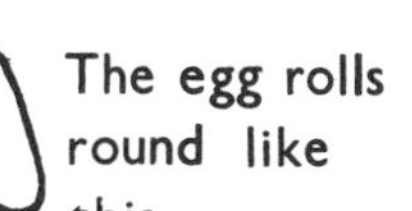

This beautiful sea-bird, who is sitting on her nest on top of the cliffs, is called the kittiwake. The kittiwake is a type of gull. It is white with a soft grey " mantle " on its back. Its wing-tips are black. Can you see these in the picture ?

Kittiwake

You may not be able to climb up high cliffs to see all these sea-birds, but you may see them in flight or hear their call. Look out for them if you go out in a boat on your summer holidays.

Sea thrift grows on cliffs in rocky places.

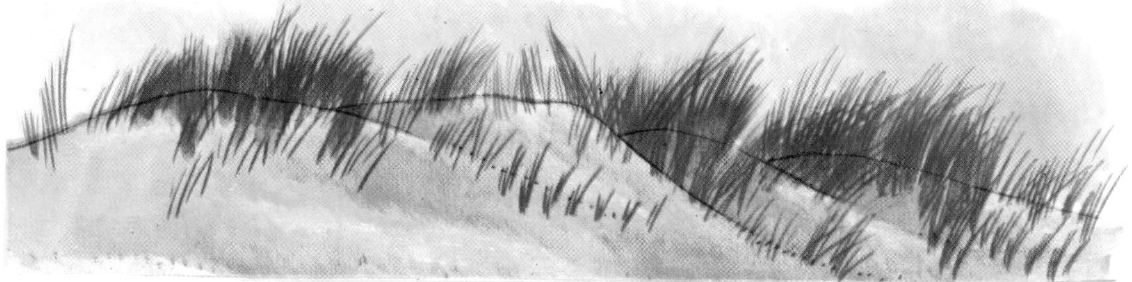

On sand dunes (sand hills) we find coarse grasses like marram grass and spartina grass. These grasses have long roots which help to bind the sand and to keep the dunes firm. Otherwise the sand of the dunes blows about like this :

ON A SALT MARSH

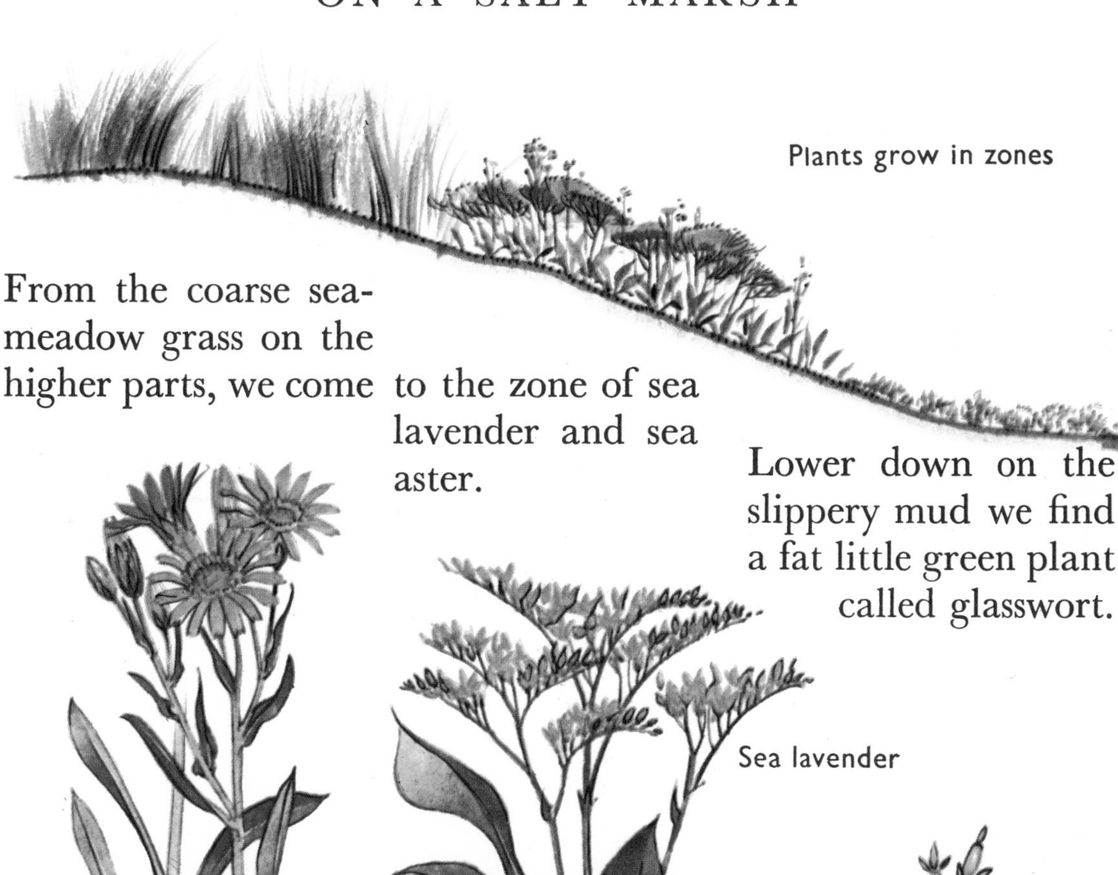

From the coarse sea-meadow grass on the higher parts, we come to the zone of sea lavender and sea aster.

Lower down on the slippery mud we find a fat little green plant called glasswort.

A salt marsh is found in low coastal regions where the salt water of the sea creeps over the mud of the low-lying coastal fields or the estuary of a river. You can find very special plants here. These are quite different from those of a fresh-water marsh.

Here is the juicy green plant which grows on the mud flats of the salt marsh, the glasswort.

ON THE SEA SHORE

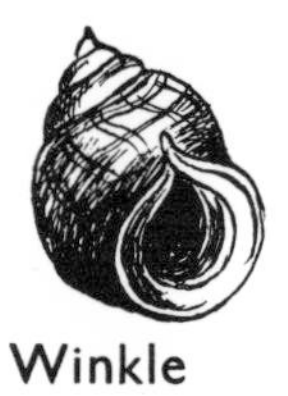

Winkle

Mussel

Cockle

Top shell

Look for these four shells and for other shelled creatures such as the limpet which sticks closely to a rock. Even when it moves away to feed it comes back to its own spot.

Limpet

Look also in rock pools and on the shore for seaweeds.

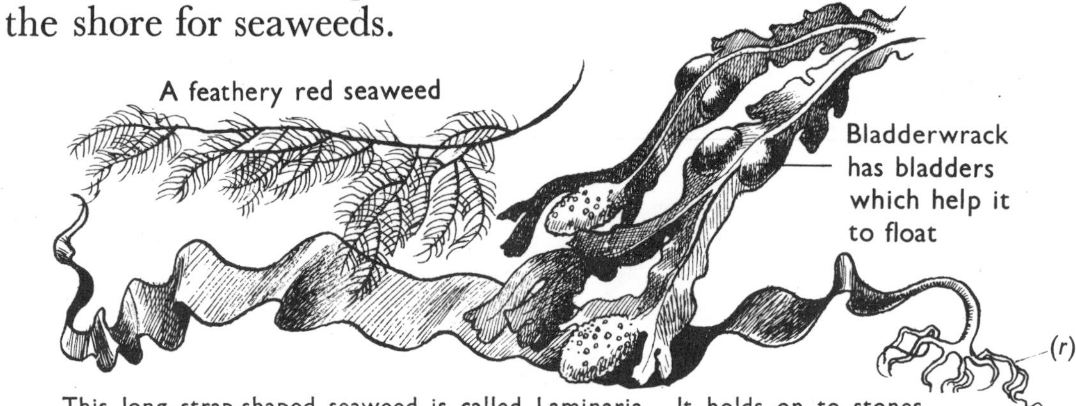

This long strap-shaped seaweed is called Laminaria. It holds on to stones with its root-like end (*r*)

You should find three kinds of seaweed: red, green and brown. Above are coral weed (the feathery red seaweed), bladderwrack (brown) and laminaria (green changing to brown).

Shrimp

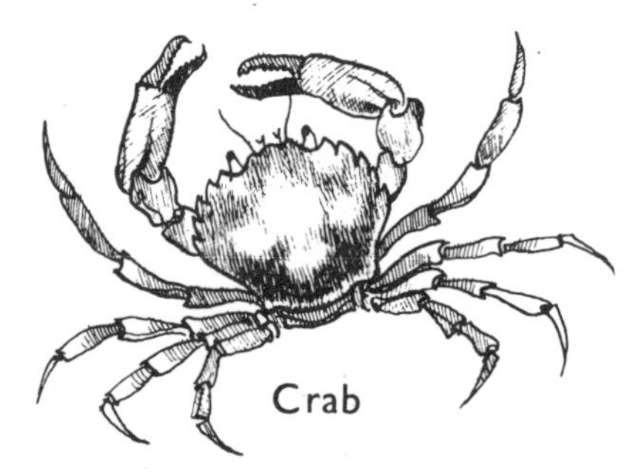

Crab

Get out your shrimping net and look for shrimps and crabs in the rock pools.

WHAT CAN WE SEE ON ROCKS UNDER THE SEA?

Have you ever looked down into the sea on a clear, still day from the side of a boat—as you cross over the rocks below? Imagine yourself looking down into this water in the picture.

Here are some of the things you might see:

(1) and (2) jellyfishes. Jellyfish (1) is swimming.
(3) and (4) anemones. Anemone (3) is opened right out.
(5) Starfish and (6) sea-urchin crawling on the sea bed.
(7) Big edible crab, claws open, waiting to capture prey.
(8) Bull head fish swimming about.

LOOK FOR THESE FLOWERING PLANTS NEAR THE SEA

Sea holly is very prickly, just like real holly. It grows on sand dunes and sometimes on the shore and it has a lovely blue head of flowers.

The sea pea is a strong little plant and grows on shingle banks where no other plants can grow. It is called a " pioneer plant ".

Sea pea

Here are three more plants which grow on sandy shorelands.

They are the sea heath, the sea rocket and the samphire. You will see that the sea rocket resembles the cuckoo flower on page 30. The sea heath is a creeping plant, rather like heather, and the samphire has greenish flowers and fat leaves.

TWO MORE SEASIDE PLANTS

Scurvy grass

This plant, scurvy grass, has a good reputation. It was well known to sailors in early days when long voyages without modern methods of storing food meant that the men suffered from lack of the vitamins found in fresh vegetables. This plant was often eaten by ships' crews as a remedy for scurvy, a disease which resulted from this lack of fresh food. Scurvy grass is a fleshy plant about 30 centimetres high. It has little clumps, or panicles, of white flowers and it is common on muddy sea-shores and on damp sea cliffs.

The leaves of this plant, the buck's-horn plantain, are supposed to be like the antlers of a deer, and this explains its name. It is a fine plant and these tall, straight stalks you can see bear heads of tiny flowers—which are rather hairy when young.

This plant often grows on the top of cliffs where the ground is stony or sandy.

Buckshorn plantain

SOME FRUITS WHICH FORM IN SUMMER

Let us see how many little fruits are getting ripe as the flowers begin to fade.

Dandelion fruits have little parachutes to carry them away.

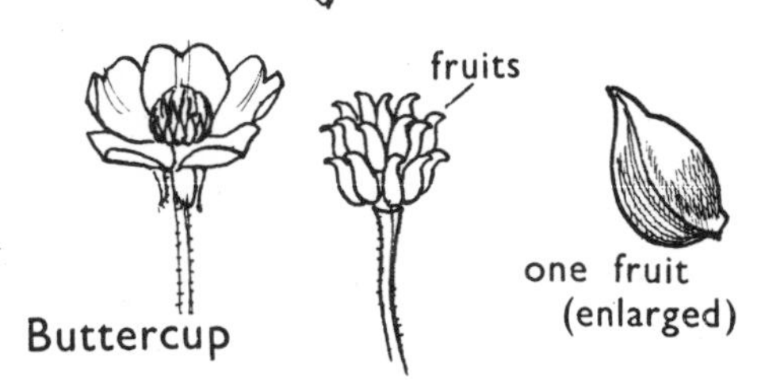

When the buttercup flower dies and the petals fall, a head of little fruits like this is left. Each fruit has a seed inside.

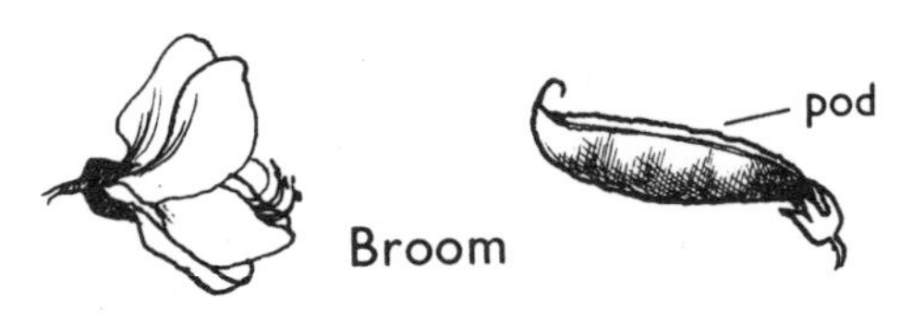

The yellow flower of broom has a fruit called a pod. The pod opens as you see below and the seeds fly out.

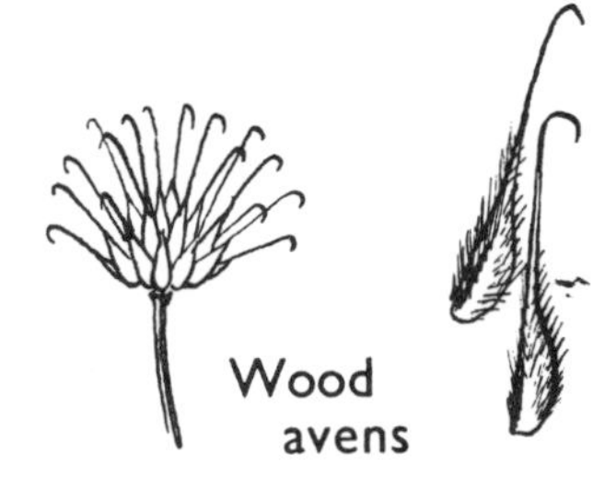

Wood avens has a head of little fruits like the buttercup but each fruit has a hook on it. What do you think this hook is for ?

This is the white flower of greater stitchwort. When it dies it leaves a small round fruit like this

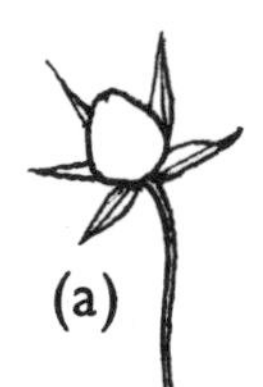

Inside the fruit the seeds are arranged on a central rod (b).

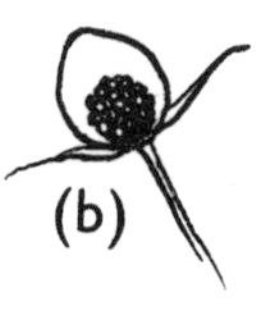

Look at some other fruits and see how they are made and how the seeds are scattered.

HOW ARE SEEDS SCATTERED?

Some are blown by the wind, like these dandelion fruits.

Some fruits, such as cherries, are good to eat. The cherry stone, containing the seed, falls to the ground. The seed may grow.

Some plants, like the poppy, scatter their seeds from little holes as the fruit stem sways in the wind.

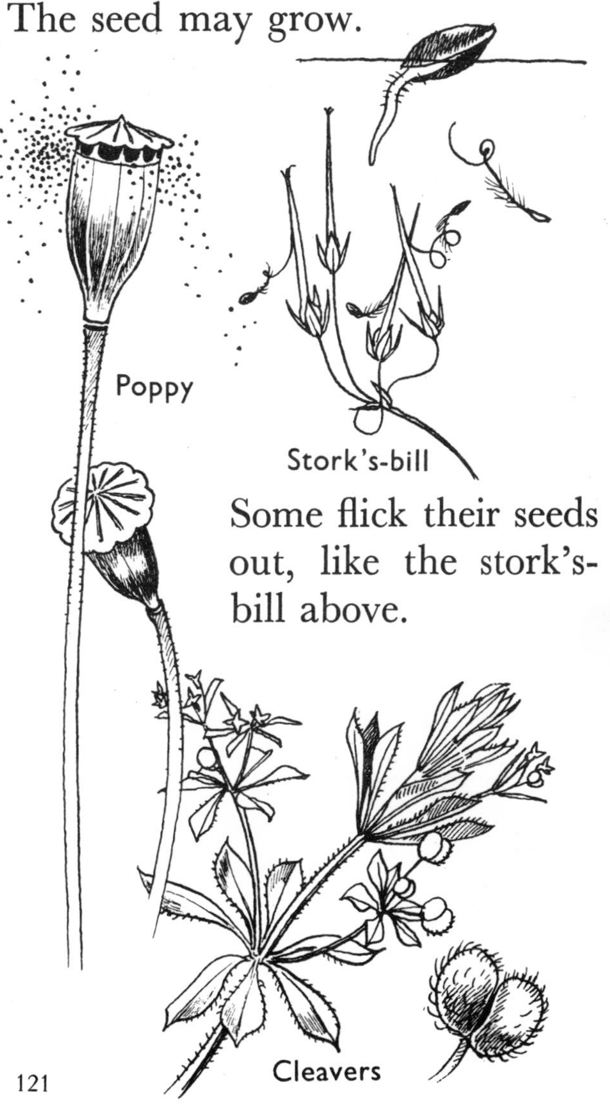

Some flick their seeds out, like the stork's-bill above.

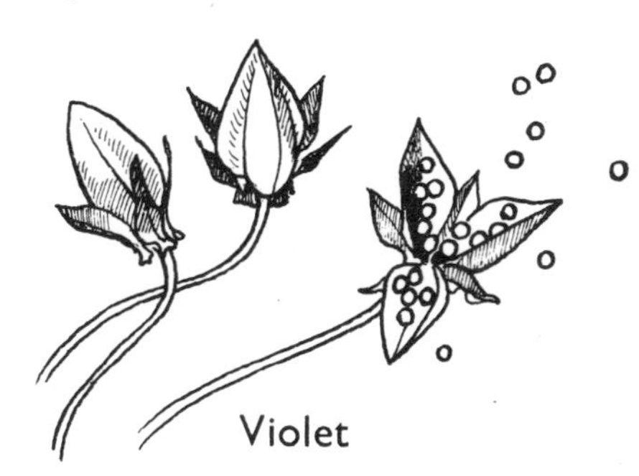

Some split open and their seeds fly out. The violet fruit splits into three parts.

Some fruits or seeds cling on to the coats of animals or of people. Cleavers has little green fruit balls which cling.

At the greengrocer's we find fruit and vegetables. There are apples, pears, oranges, bananas, plums, cherries, strawberries and other fruits—in their right seasons. Let us examine some of these fruits.

(1)

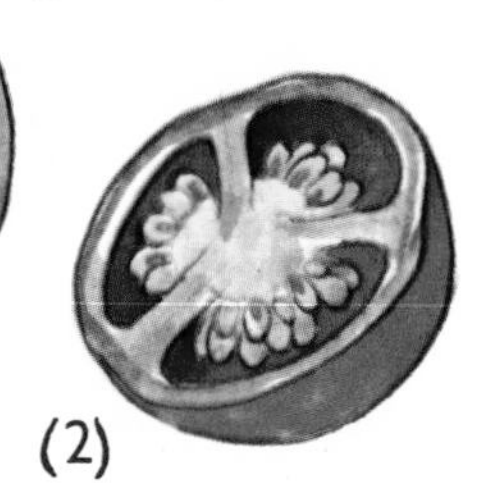

Tomato

We choose first the tomato, which is a fruit though you may have thought of it as a vegetable. You can see the seeds in their own compartments inside the fruit in drawing (2).

You will easily recognise the apple as a fruit and yet it is really a false fruit. The part of the apple that you eat has grown round the real fruit. You can see the real fruit, marked (F), which is the core of the apple. Inside the middle part of the fruit, you can see the seeds.

(1) F

You can see that the core looks like a little star when you cut the apple the other way as in drawing (2).

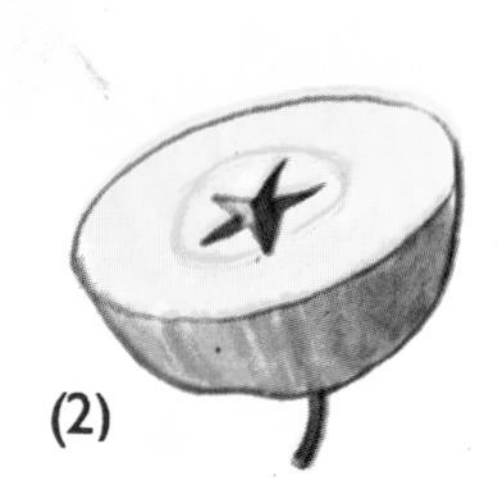

Apple

Strawberries

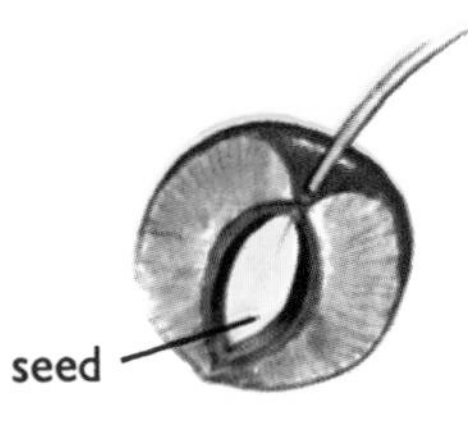

Cherry

Strawberries are " false fruits " too. You can see the small yellow fruits on the outside.

Cherries are " stoned fruits ". Inside the " stone " is the seed of the cherry.

NOW LET US LOOK AT SOME VEGETABLES

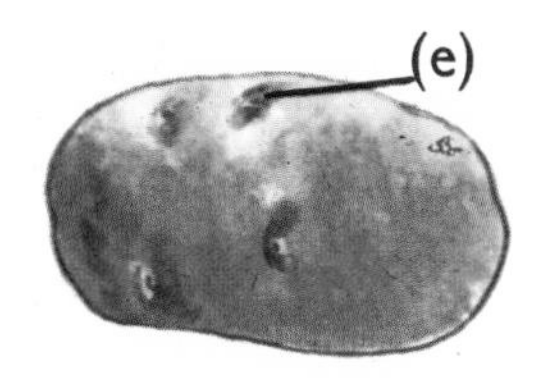

Potato

What are potatoes? They are swollen stems. Like all stems they have little buds, which you can see in the marks called the "eyes" of the potato (e).

Another vegetable, the cabbage, is a very large bud. The lower picture shows you a section through the cabbage. Inside the cabbage bud are other smaller buds on a central stem. Each bud is protected by one of the large cabbage leaves.

Cabbage

Carrots, turnips and parsnips have swollen roots which are full of food. You can see in the picture that leaves are growing from the upper part of the carrot.

Carrots

The onion has a store of good food too. It stores its food in thick fleshy leaves. Inside the bulb is a bud. If you put the base of the onion in water, roots will spring out and grow, and the bud which is inside the bulb will send out green leaves.

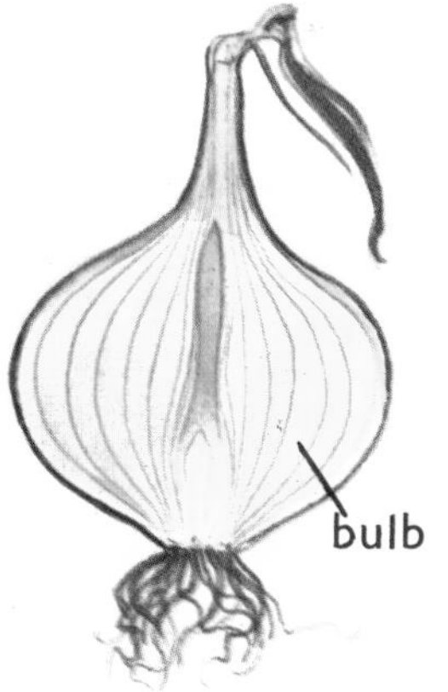

Onion

Later in the season we can buy nuts. Nuts are fruits with a hard shell and one seed inside.

Hazel nuts

LET'S GO TO THE FISHMONGER'S AND POULTERER'S

Here are four fishes which you might find at a fishmonger's shop.

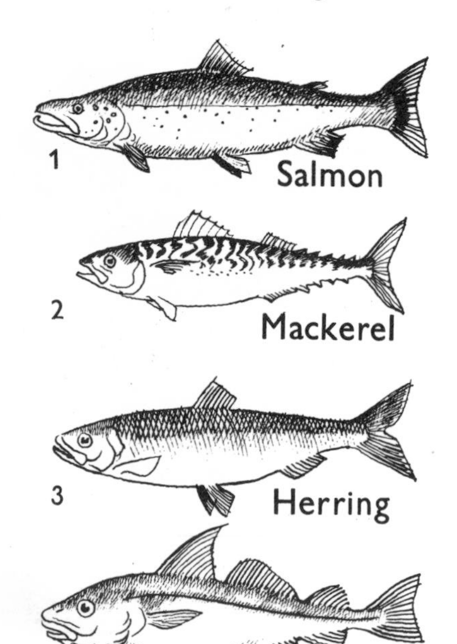

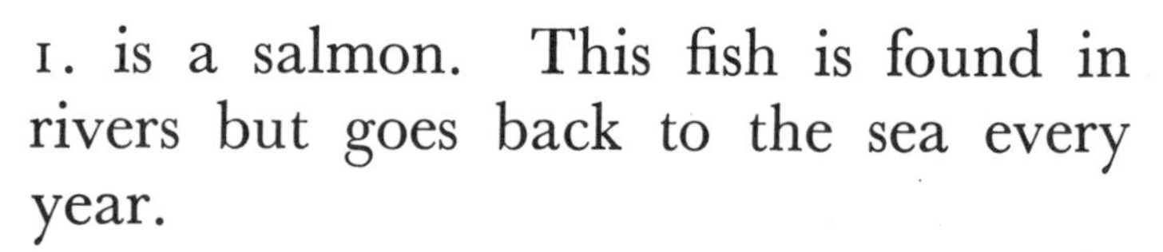
1. is a salmon. This fish is found in rivers but goes back to the sea every year.

2. is a mackerel. Can you see the "mackerel markings" on its back? The mackerel is a sea fish.

3. is a herring. One of the commonest fishes which lives in the Atlantic Ocean and is caught around our coasts.

4. is a haddock. Haddocks are sometimes smoked. You can buy fresh and smoked haddock at the fishmonger's. The haddock is a relation of the cod. It is caught by trawl nets in northern seas.

Fishmongers often sell poultry and game. Here are two game birds, (5) partridge and (6) grouse, which you saw before on page 18. You can buy these as food in a poulterer's shop. You can also buy ducks, geese, chickens and turkeys. You may have seen all these hanging up for sale in a fishmonger-poulterer's shop especially at Christmas time.

WHERE DO DUCKS, GEESE AND TURKEYS COME FROM?

All domestic ducks are descended from the common mallard (see page 15—where you will see mallards together with many other ducks). The pure white Aylesbury duck appeared as a table bird in the early years of the eighteenth century. These ducks were bred in great numbers in the Vale of Aylesbury and are still the most important "table ducks" in Britain today.

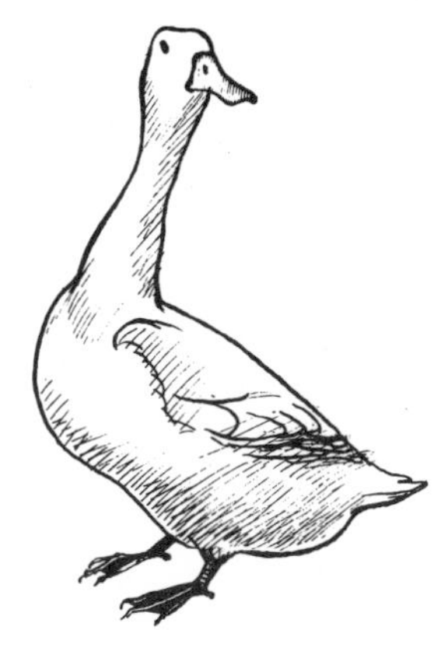

Aylesbury duck

Grey lag goose

All the geese we see in poulterers' shops are descended from the grey lag goose. You can see one in this picture. It has pink feet.

The goose was a sacred bird in Egypt 4000 years ago, and in Rome long before the birth of Christ.

In the Middle Ages people used to write with quill pens made from goose feathers.

Turkeys came originally from America (and not from Turkey). Wild turkeys were eaten by Red Indians about A.D. 1000. We found out about turkeys after Columbus discovered America.

Look at the poultry and game birds in your fishmonger's shop. You will see that they are all heavy and fattened. They are not good fliers.

Turkey

LOOKING AT PEBBLES—We go to shingle beaches

Have you ever collected pebbles from the beach? A beach which has many pebbles is called a *shingle* beach. Pebbles are made of many different kinds of rock and are of many colours, though you would not find all the pebbles on this page on one beach. Pebbles become round by being rolled over and over in the water.

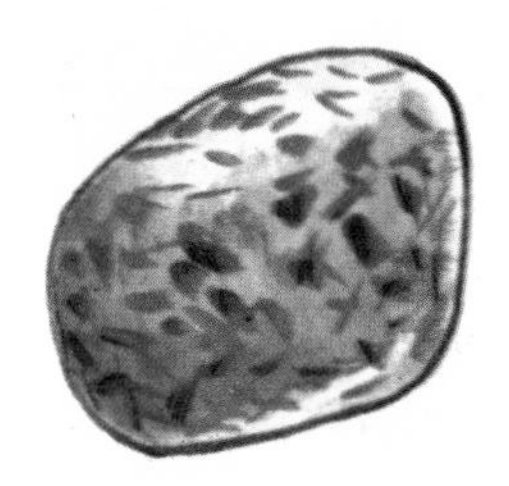

A granite pebble

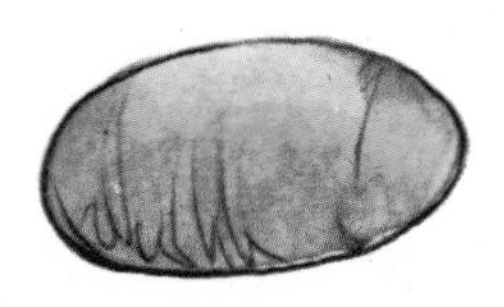

If you find a green pebble, it is probably made of a rock called serpentine.

Reddish-brown pebbles are usually made of sandstone.

This pebble has many different colours in it.

On a Dorset or Yorkshire limestone coast, you may find grey limestone pebbles like this one.

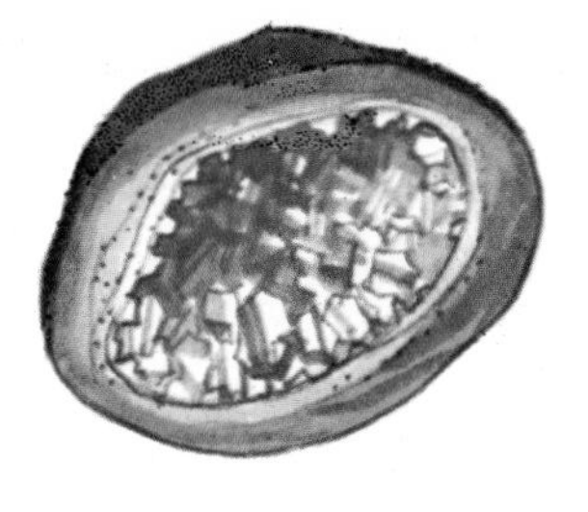

Sometimes you may find a very large brown pebble. When this is cracked open, you can see the quartz in the middle of it. Quartz is hard and crystalline.

These bluish pebbles are made of slate. One has a stripe of quartz in it. This stripe is called a " striation ".

SOME SPECIAL ROCKS AND FOSSILS

Pudding stone

Sometimes pebbles become cemented into the sand or silt in which they lie, and form a solid rock with pebbles inside it. This is called pudding stone. Here is a pudding stone, with the pebbles embedded in it.

Coal fossil

Some old rocks show imprints of plants or animals which lived long ago. You might find a print of a fern on a piece of coal in your own coal shed.

Long ago big rivers and seas carried down many little grains of sand and clay which hardened into rock as other layers pressed down on them. These layers of rock are called strata. Here is a cliff which shows the strata or layers of rock.

Sometimes these strata have become tilted as in the lower drawing.

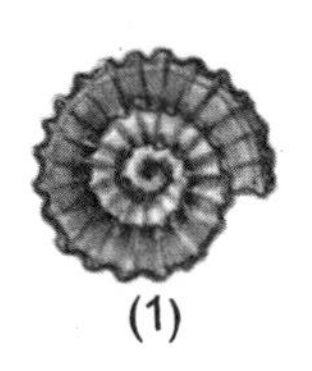
(1)

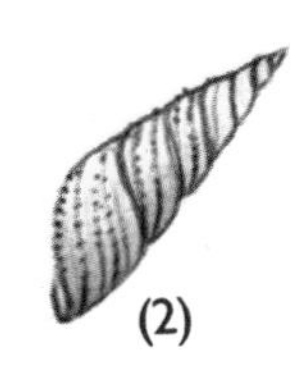
(2)

(3)

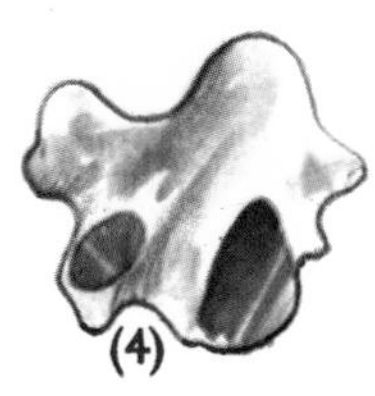
(4)

Interesting fossils are found embedded in rocks, especially in chalk rocks. Here are some things you may discover in chalk. (1) is an ammonite—the fossil of a creature with a coiled shell. (2) is the fossil of a snail. (3) is a fossil sea-urchin. (4) Flint is also found in chalk rock. From these flint stones primitive men made spear-heads.

Have you ever waded in a shallow stream in your Wellington boots?

Turn up some of the stones on the stream bed. You may find some interesting creatures on or under them.

(1)

Turn over a big stone. You may find caddis worms. These are not really worms. They are larvae (young stages) of the caddis fly. They build little homes for themselves out of tiny stones or little sticks or hollow stems.

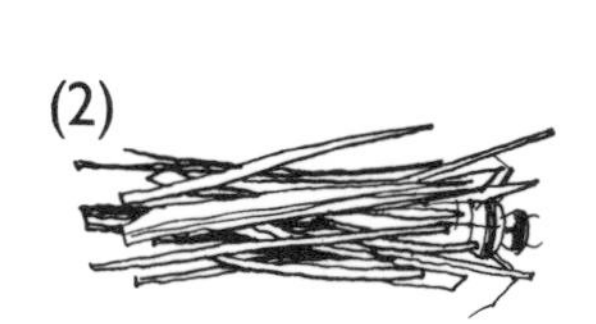

(2)

You can see that each one has a different kind of case. (1) is made of very small shells, (2) is made of grass and (3) of many little sticks. You can see the head of each insect sticking out from the tube.

(3)

Three caddis larvae

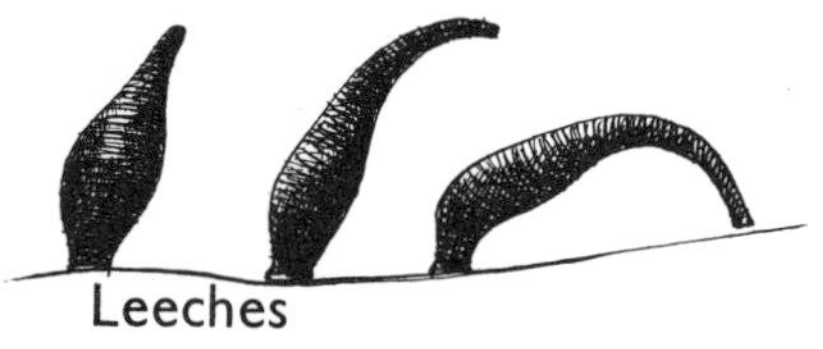

Leeches

You may find these water creatures moving slowly about on your stone. They are leeches which look like short worms. These have suckers at both ends. Leeches move about in a " looping " fashion, using their suckers to help them.

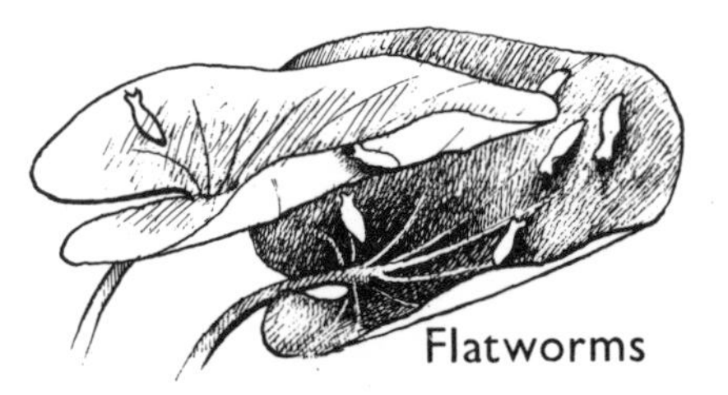

Flatworms

These little creatures called flatworms may be found adhering (sticking) to pond leaves and to stones. They can move about on these with a gliding movement, using a sucker to help them. They can swim.

LOOKING UNDER STONES ON THE GROUND

If you turn over a stone on the soil, you may find many little creatures underneath it.

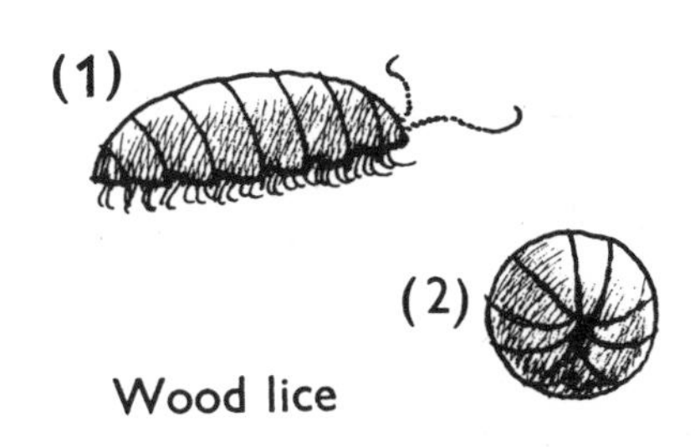

Wood lice

Here is a wood louse. Some wood lice can roll themselves up as in (2) until they look like shiny pills.

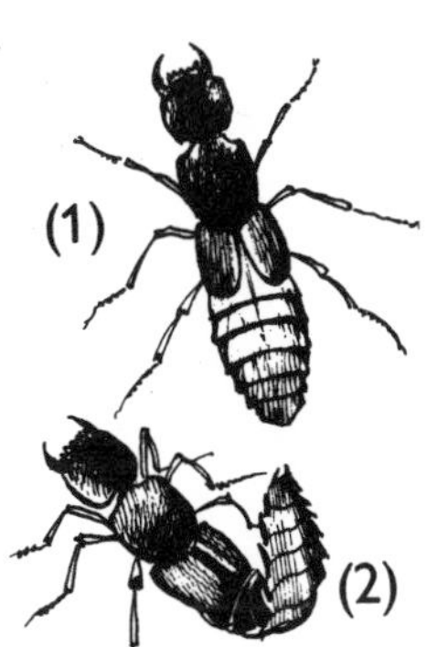

This insect on the left is called the devil's coach-horse. It tilts its tail as in drawing (2) when angry or disturbed.

Devil's coach-horse

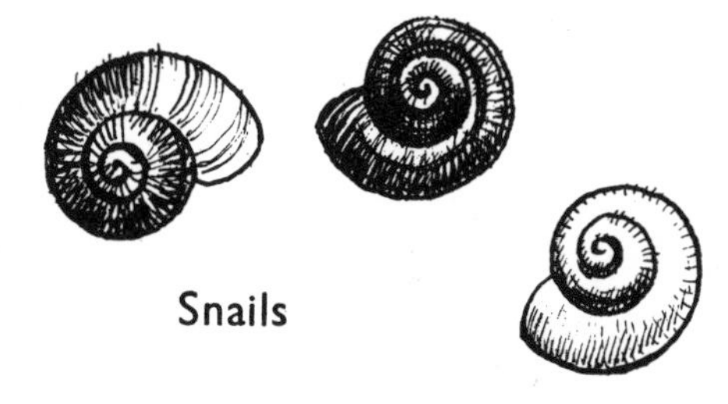

Snails

You may also find some little land snails. These snails often have pretty shells.

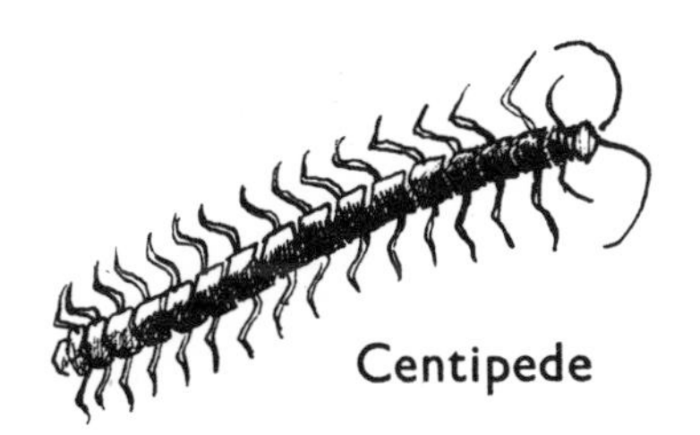

Centipede

This is a centipede. It has legs from every segment of its body and two antennae at the front of its head.

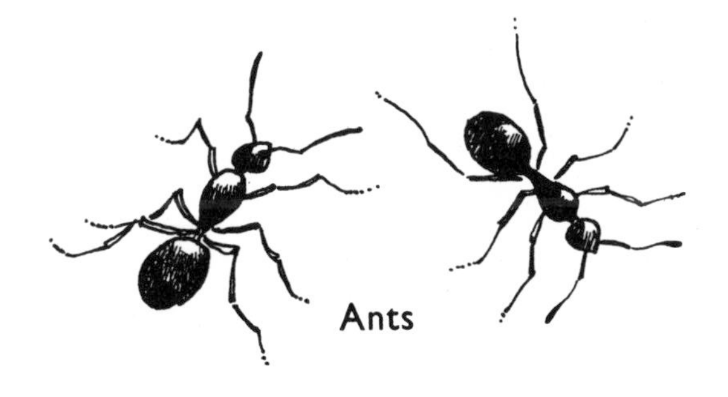

Ants

Here are some ants which you may find under your stone.

LOOKING FOR ORCHIDS

People often think of orchids as very beautiful and rare flowers which grow only in tropical countries, but we know that we can find beautiful little orchids in our own country. These two orchids grow best in chalky soil.

The *early purple orchis* flowers in May. This orchis grows on chalk downs or in woods which are not very dense. It does not grow well on acid soil. The stem, which bears purple flowers, is almost 30 centimetres high. Notice the dark blotches on the leaves.

The flowers of the *butterfly orchid* look like small white butterflies.

Early purple orchis

Can you see the long spur at the back of each flower? This spur contains nectar. Moths can reach this nectar with their long tongues.

Butterfly orchid

Orchids often look like insects. As well as the butterfly orchid, there is also a bee orchid, a spider orchid, and a fly orchid. Here are two views of the *bee orchid* flower. You can see that it is very like a bee's body perched on a flower. Perhaps bees mistake this for a rival bee and fly away, for we know that this flower pollinates itself.

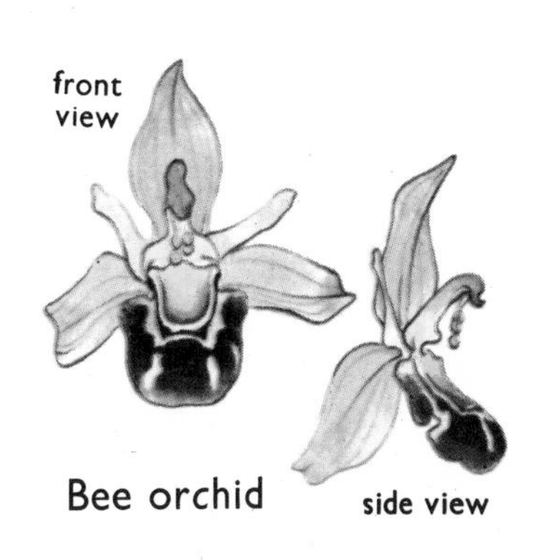

Bee orchid

Bees do, however, go to many orchid flowers. In fact, they sometimes come away from these flowers with the pollen sacs or *pollinia* (p) stuck on top of their heads as you see in this picture of another orchid flower with a bee emerging from it.

Here is a rare and beautiful orchid, the *lady's slipper*. This is so rare that it is almost extinct and may be found only in some parts of north-eastern England. We must take care that our rare plants are not all lost to us, so do not pick them. Instead,

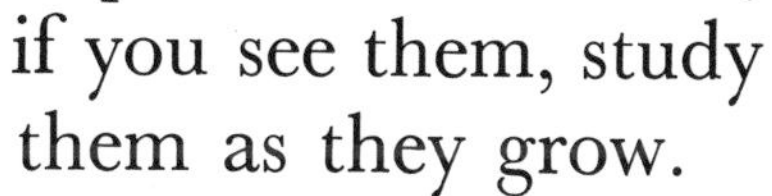

if you see them, study them as they grow.

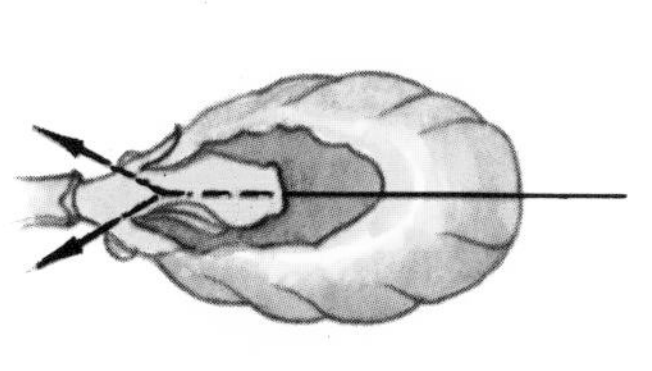

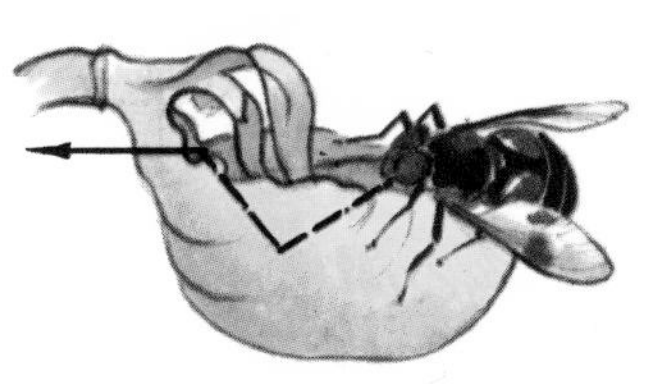

This flower has an enlarged lip. It is rather like a slipper. The pictures show how a pollinating insect passes through.

Lady's slipper

FLOWER FAMILIES

What flower families do you know?
On the last two pages you met several members of the orchid family. Now let us look for other flower families.

One of the commonest and most successful flower families is the daisy family. The daisy has many tiny little flowers in a head, so that it looks almost like one single flower. There are two kinds of little flowers in this head (pull some off to see). There are white, strap-shaped flowers on the outside of the ring and yellow tube-like ones inside.

Flowers belonging to the daisy family are: ox-eye daisy, dandelion, thistle, marigold, coltsfoot, ragwort and cornflower. (Find others for yourself.)

Now let us look at some buttercups. This is the meadow buttercup or *Ranunculus acris*. The last name refers to the bitter taste of its stem. There are several different types of buttercup, but all have a large number of stamens and carpels (seed boxes) in their flower as you can see in these diagrams.

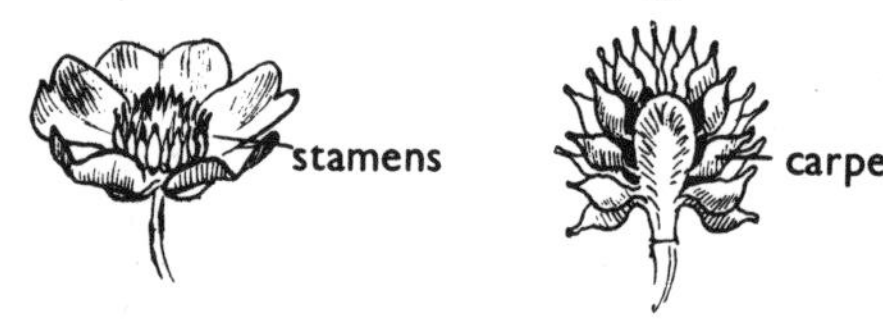

Some members of the buttercup family are: marsh marigold, hairy buttercup, bulbous buttercup, lesser celandine and water crowfoot.

Below are two vetches, the bush vetch and the tufted vetch. These plants belong to the sweet pea family.

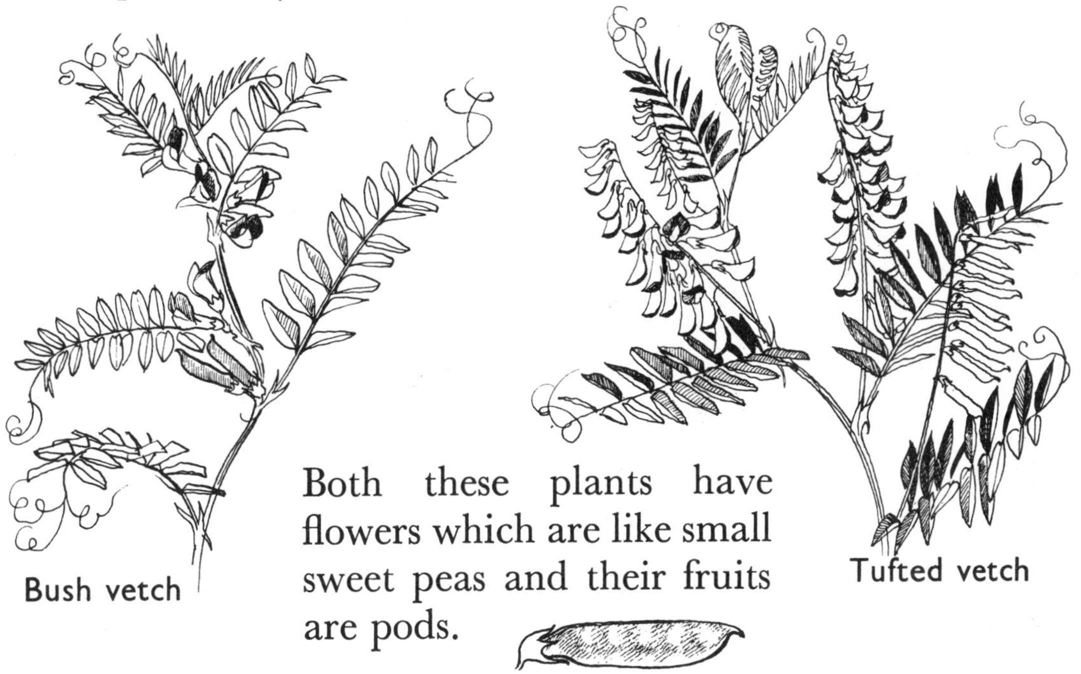

Both these plants have flowers which are like small sweet peas and their fruits are pods.

Below is a vetch flower (enlarged).

Here is the pod or fruit of a vetch. You can see that both flower and pod are similar to those of the pea plant.

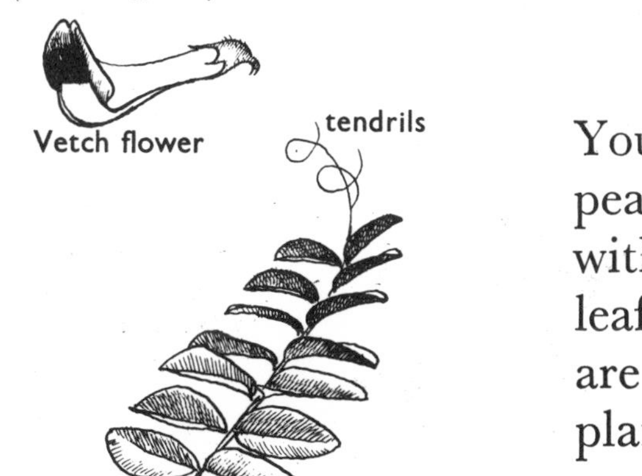

You will see, too, that members of the pea family have leaves like this one with leaflets in pairs. Instead of leaflets at the end of the stem, there are tendrils by means of which the plant climbs.

Other members of the pea family are gorse, broom, yellow vetchling, bird's foot trefoil, clover and rest harrow. There are many others. Try to find some.

Beating the boughs

You will need a stick and something such as a piece of cloth in which to catch the caterpillars. You use the stick to beat the boughs of the trees or bushes in the wood or hedgerow where you go to seek caterpillars. The caterpillars should then fall off their leaves and they will fall on to your cloth if you have placed it under the boughs.

You have seen the first butterflies which come early in the summer (p. 102) and you have seen the chrysalis or pupa from which they come, the eggs which they lay, and the caterpillars which come out of those eggs (p. 103).

Caterpillars are coming out of moth and butterfly eggs all the summer. Let us go and collect some.

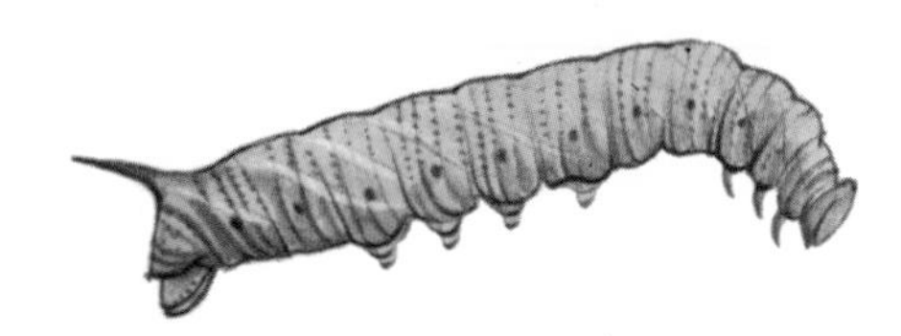

Here is the big caterpillar of the poplar hawk moth. You can see the moth in the drawing below.

The caterpillar of this moth feeds on the leaves of the poplar tree.

THREE INTERESTING CATERPILLARS

This beautiful green caterpillar with the rings round its body is the caterpillar of the emperor moth. You will find it feeding on blackberry leaves. Look at the picture of the emperor moth below the caterpillar.

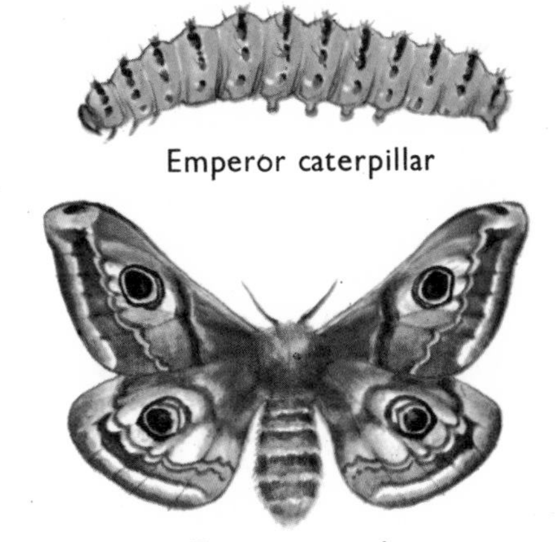

Emperor caterpillar

Emperor moth

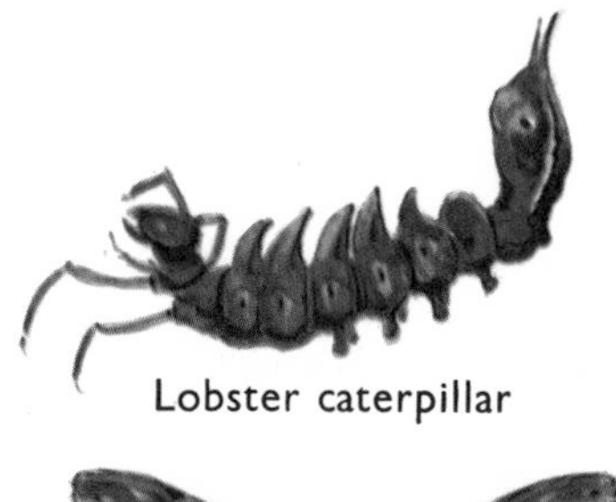

Lobster caterpillar

Lobster moth

The interesting caterpillar on the left is called the lobster caterpillar because it looks like a little lobster. It will not nip if you pick it up. You may find the caterpillar on oak or beech trees in August. The lobster moth is shown below its caterpillar.

If you search among nettle leaves, you may find the caterpillar of the peacock butterfly. It is black with little white dots all over it and is covered with bristles. Below is the peacock butterfly.
Can you see from the pictures the difference in the antennae of moths and butterflies ? .

Peacock caterpillar

The feathery antennae of a moth

The clubbed antennae of a butterfly

Peacock butterfly

LOOKING AT CLOUDS

A Roman poet, Lucretius, wrote this poem about clouds about 60 B.C.

" Now clouds combine and spread o'er all the sky
When little rugged parts ascend on high
Which may be twined though by a feeble tie.
These make small clouds which, driven on by wind,
To other like and little clouds are joined
And these increase by more ; at last they form
Thick, heavy clouds and thence proceeds a storm."

It is over 2000 years since Lucretius lived. Although he did not know as much about clouds as we do today, he was a good observer.

The clouds in this first picture are called *alto-cumulus*. *Alto* means high. *Cumulus* means that they are heaped together as you will see in the picture.

Little rugged clouds are collecting in this lower picture, just as Lucretius said in the second line of his poem.

These little wind-swept clouds are called *cirrus* clouds. These are probably the ones Lucretius meant in the fourth line of his poem.

These clouds above look as though they might develop into storm clouds.

Here is a picture of a storm cloud. Storm clouds are large, heavy-looking ones and cast a dark shadow on the land below. The top of the cloud is shaped like an *anvil*, the metal platform used by a smithy when he hammers horse's shoes.

During a storm, the atmosphere is charged with electricity and we see electric flashes and streaks in the sky between cloud and cloud, and sometimes coming down to earth. Sometimes the lightning comes down to earth by means of trees, so do not stand under a tree in a thunderstorm.

HOW DOES A RAINBOW FORM?

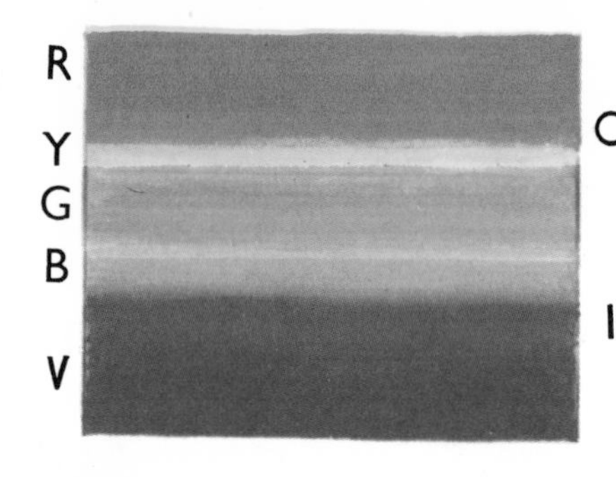

The main colours in a rainbow are red, yellow, green, blue and violet. You may see more than these five colours because where the red overlaps into the yellow we see orange (O), and where blue overlaps on to the violet we see indigo (I).

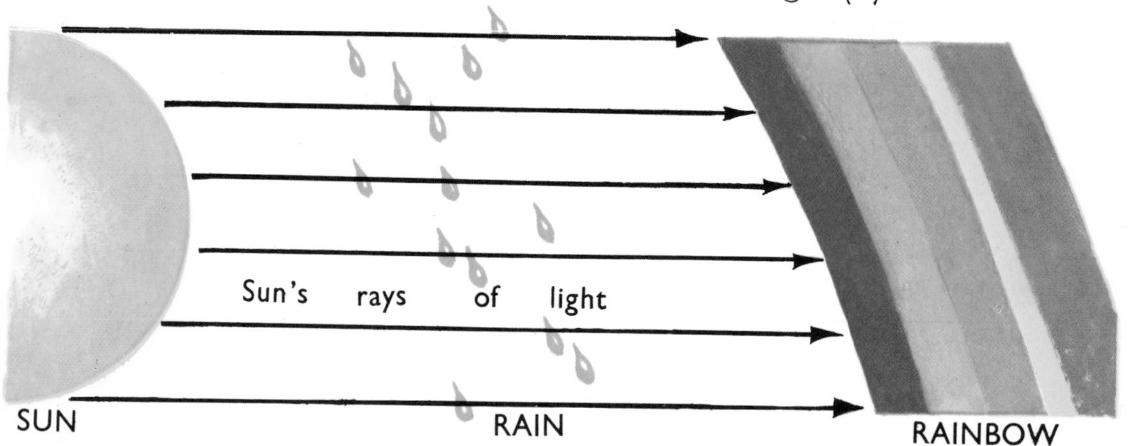

Whenever you see sunshine and raindrops at the same time you are likely to see a rainbow. A rainbow forms on the clouds when the light rays from the sun are split up into many different colours by the crystal raindrops.

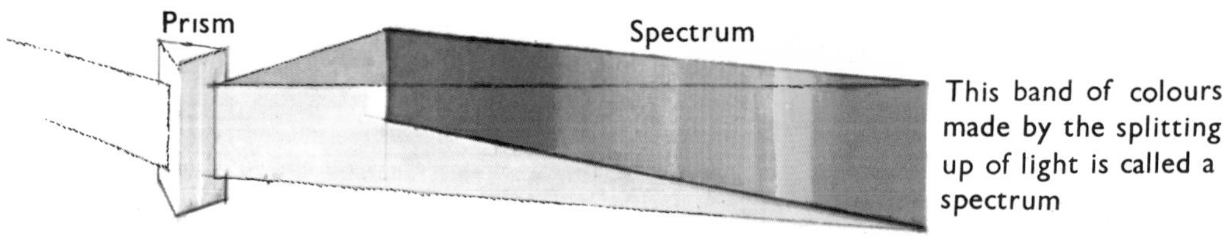

This band of colours made by the splitting up of light is called a spectrum

You can make your own rainbow if you allow rays of light to shine through a glass prism. (Find out what " prism " means in the word list on page 144).

Sometimes we see the sun or moon through a mist. This mist gives the sun a " solar halo ". The mist is made up of fine water-drops.
(Look, too, for a halo round the moon.)

Sometimes the moon passes between the earth and the sun. This only happens occasionally, but when it does, the sun is hidden, or partly hidden, from our view. We call this event an eclipse or a partial eclipse.

Here is the sun partly eclipsed by the moon. The moon does not look bright because its other side is facing the sun.

Where does the sun go when it sets ? It seems to go below the horizon. At sunset, the earth is turning away from the sun. As it does so, the sun seems to be dropping and moving away from us. Here is a picture of the sun, setting in the west.

HOW IMPORTANT IS THE SUN TO US?

The sun is important to all living things. It gives them light, warmth and energy.

Energy from the sun is called " solar energy ". Plants can make use of this energy. It helps them to make their food.

Plants store some of their food. You can see stored plant food in roots and stems. You see it in carrots and turnips and also in seeds like peas and beans. When we eat these foods we are eating stored energy from the sun.

Have you heard the words—" All flesh is as grass " ? These words were said by St. Peter. You can find them in the Bible.

Grass is important to grazing animals, and all kinds of green plant foods are important to vegetable feeding animals.

All animals are not vegetarians however. Some eat the flesh of other animals. Usually grass-eating or vegetable-feeding animals are the prey of flesh eaters. The grass or green plant of some kind must be there first, otherwise no animal would get any food.

Look at the pictures on the next page. The rabbit eats dandelion leaves or nibbles in the grass. The fox waits and watches for an opportunity to spring on the rabbit and make a meal of it.

Grazing animals, like deer, eat grass, but there is often a lion or panther or tiger ready to spring on them.

Enemies of these quiet, grazing animals are not found now in this country, but in the forests and jungles of some tropical countries these large, cat-like flesh eaters are always waiting for their prey.

Animals which prey on the flesh of other animals are called " carnivores ".

Animals which live on green foods like grass or on vegetable matter which comes from green plants are called " herbivores ".

Animals which can live on either flesh or vegetable foods are called " omnivores ". We are omnivores.

WHERE HAVE WE BEEN?

AUTUMN

We have collected wild fruits from the hedgerow and fungi from the woods.

We have found bracken on heathland and mosses on woodland and heathland.

We have been to the pine woods and to look at hedges in the garden.

WINTER

We have looked at rook roosts and starling roosts ; at ducks on the pond ; at deer in parklands and zoos ; at game birds and owls near woodland cover ; at stars in a winter sky ; and at plant rosettes on the ground.

SPRING

We have looked for signs of spring in a country lane, in woods and sheltered places, in ponds and streams, and in the water meadows.

We have seen water creatures and collected them for the aquarium ; birds building nests and coming back from abroad ; farm animals and new babies on the farm ; horses and ponies—the wild ponies of the New Forest.

SUMMER

We have looked in meadows and cornfields and have seen early summer butterflies, insects of the " grass jungle ", wild and cultivated grasses, and flowers of the cornfield.

We have found plants of moorland, bog and chalk hill country.

We have seen birds and flowers of rocky coasts, and plants of a salt marsh as well as seashore plants and animals.

We have seen how plants scatter their seeds.

We have been to the greengrocer's for fruit and vegetables and to the fishmonger's for fish and poultry.

We have studied pebbles on the seashore ; we have looked under stones in water and on land and found many creatures.

We have looked for rare plants—orchids—and found that there are many different plant families.

We have studied clouds, the rainbow, and learned how important the sun is to all living things.

SOME NEW WORDS

and the pages where you will first find them

Habitat—69	The place where a plant or animal lives.
Tendrils—70	Climbing threads which sometimes grow on plants in place of leaves.
Naturalist—73	A person who studies plants and animals.
Frond—73	The " leaf " of a fern.
Fungi—74	Mushrooms, toadstools and moulds.
Spores—74	The " seeds " of toadstools, ferns and mosses.
Capsule—75	A " seed box " or " spore case " which splits to free the seeds or spores.
Conifer—76	A tree which bears cones.
Evergreen—77	A tree or shrub which has green leaves all winter.
Roost—78	A sleeping place of birds.
Cloven—80	Divided into two parts—like the foot of a deer.
To take cover—82	To run under the shelter of trees or bushes.
Game birds—82	Birds shot by sportsmen, such as grouse and pheasants.
Constellation—84	A group of stars.
Rosette of leaves—86	Plant leaves near the ground radiating out in circular fashion.
Gills—93	The thin membranes by which some water creatures breathe. The membranes on the underside of toadstools and mushrooms are also called gills.
Larva—97	The young stage of an insect.
" Rag "—101	A group of horses.
Chrysalis (Pupa)—102	The sleeping grub of a butterfly before it comes out as a winged insect.
Whorl—109	A ring of leaves or of flowers.
Beech hanger—110	A beech wood growing on a hill-slope.

Dune—114	A sand hill.
Estuary—115	The wide part of a river nearest to the sea.
Pioneer plant—118	One of the first plants to grow in any place.
Bulb—123	The fleshy underground part of some plants, which is mainly made up of swollen leaves full of food. Inside the bulb is a bud.
Pudding stone—127	A kind of stone which is made by the cementing together of rock and pebbles.
Pollinia—131	Bags full of pollen—found in orchids.
Stamens—131	The more usual pollen holders found in most flowers.
To pollinate—131	To take pollen from the stamens of one flower to the stigma of another. Insects may do this or pollen may be carried by the wind. Pollen causes ovules to grow into seeds. (Ovules are undeveloped seeds.)
Stigma—131	The top part of the ovary (seed box) upon which pollen must fall if seeds are to develop.
Acrid (*acris*)—132	Bitter. The stalk of a meadow buttercup (***Ranunculus acris***) tastes bitter.
Spectrum—138	Rainbow colours made by the splitting up of white light as it passes through raindrops or a glass prism.
Prism—138	A solid wedge-shaped piece of glass, which splits up rays of light into colours.
Prey—140	The living food of flesh-eating animals.
Carnivores—141	Animals which eat flesh.
Herbivores—141	Animals which eat vegetable matter.
Omnivores—141	Animals which can eat flesh or vegetable food.

PART 3

WORKING WITH NATURE

CONTENTS OF PART 3

ABOUT PART THREE

Natural Science is a scientific study and now that you are older you should try out scientific methods of working.

You are learning to look at, or to *observe*, Nature and, to work scientifically, you must also learn to *record* what you see, and to *experiment* and investigate in order to find out more about Nature's mysteries. There are many interesting ways of recording your discoveries as you will find out.

The basic materials of Nature are earth, air and water. Plants cannot live without these. We cannot live without them either, nor can we live without plants, so we should try to find out all we can about them. You will learn much about them in this part of the book and you will find many experiments described for you to try.

Weather studies form part of our work. In autumn and in spring we observe noticeable changes in weather conditions and we use interesting methods of recording them.

When plants and animals mature (grow up) they multiply and give rise to new individuals. Plants have spores or seeds, and animals have eggs. From these develop new plants and animals. You can learn more about how this happens.

You may decide that you want to observe living things at close quarters. Keeping pets helps you to do this, and you will learn how to look after them and how to keep different animal types under observation.

If you wish to become a scientist in the field of Nature, there is much work to be done.

Setting out on an expedition

APPARATUS

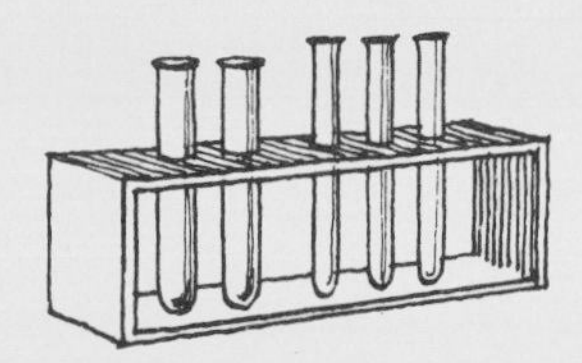

You will need some test tubes and a rack to stand them in.

Collect medicine bottles, tumblers, old milk bottles and other similar containers.

Try to get some glass funnels and beakers (or glass jars) for your experiments.

Some lengths of rubber tubing will also be useful.

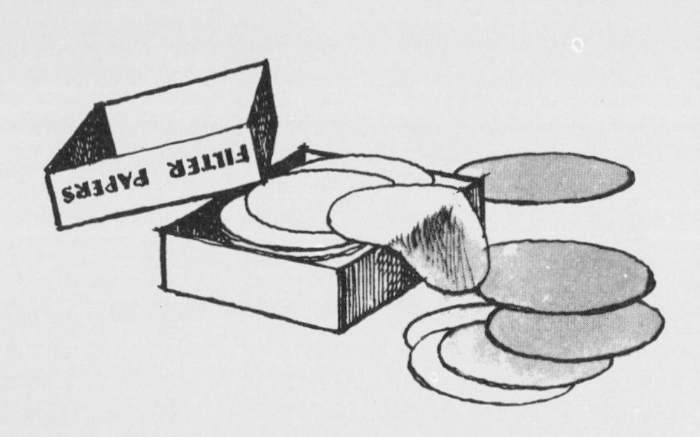

You will need filter papers for placing inside the glass funnels.

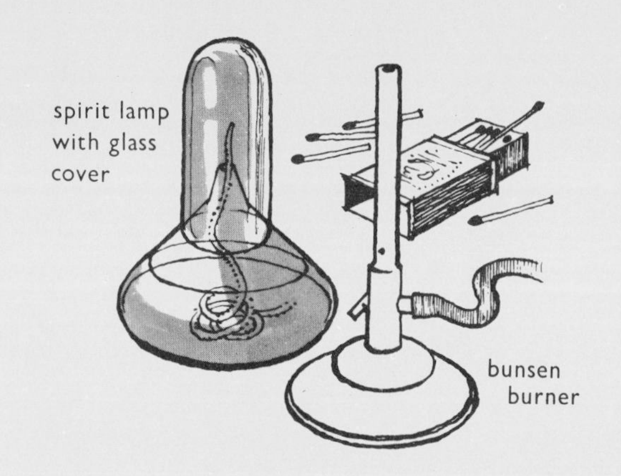

You may also need a spirit lamp or bunsen burner and some matches.

Always have your record book and pencil ready.

Observations and experiments must *always* be recorded.

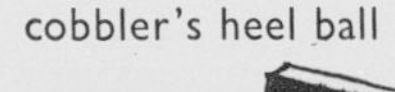

For *bark rubbings* you will need cobbler's heel-ball, or some big oil crayons.

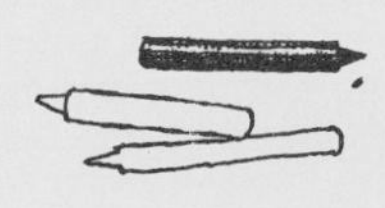

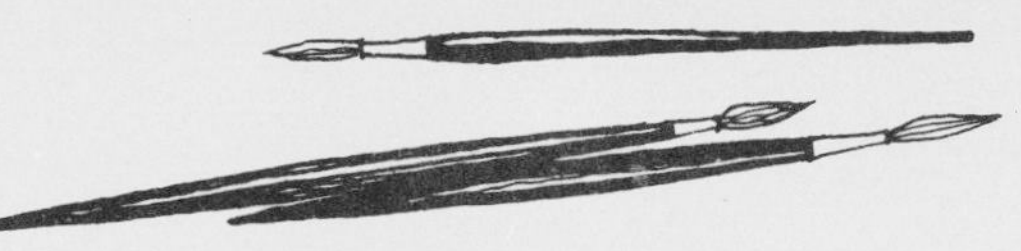

You will need saucers,

and paint brushes,
and tins or tubes of shoe polish for making *leaf prints*.

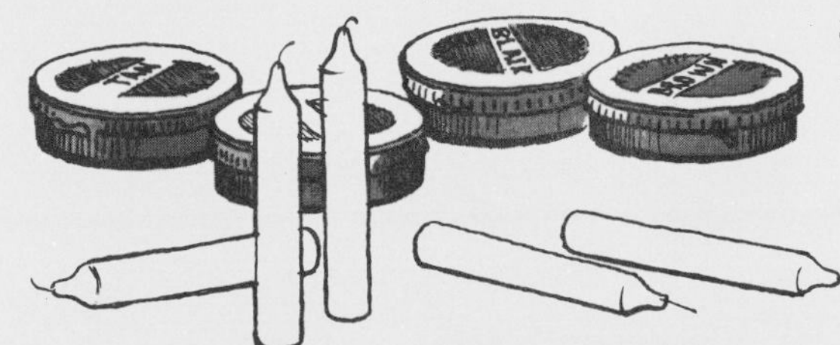

You will also need wax candles and some enamel plates.

You will need plaster of Paris and clay for making plaster casts.

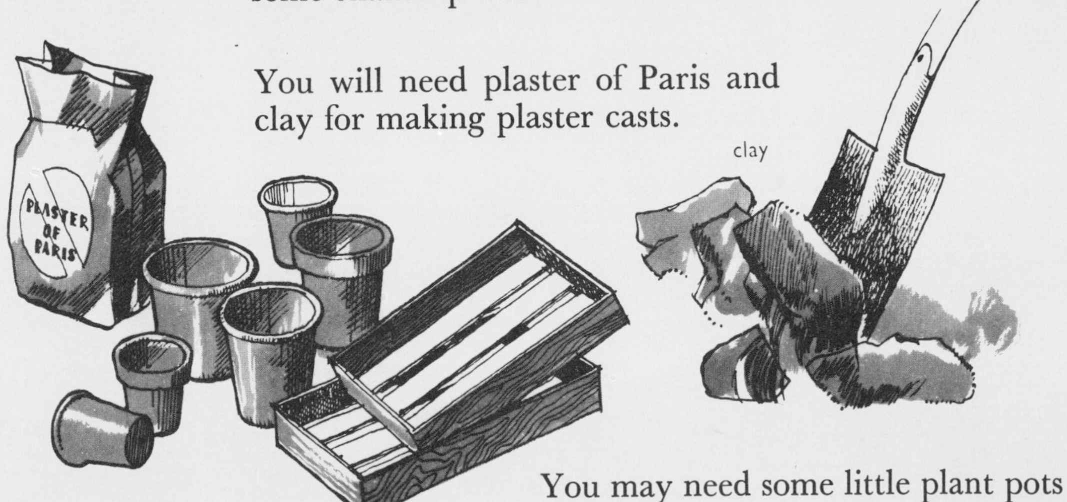

You may need some little plant pots and some wooden seed-boxes for germination experiments.

Hand lens

You will need a good hand lens so that you can see the tiny parts of plants and animals.

You will need other things too, but you will know what these are as you read the pages which follow.

WAYS OF RECORDING—LEAF PRINTS

Whatever you study in Nature, there is much to remember and you can only remember things well if you make records.

In autumn, there are many fallen leaves and these make good material for study. You can remember the name of a tree by remembering the shape of its leaves. Let us make some records of these shapes. On the opposite page you will discover how this can be done with a black paint which you can make for yourself.

Below are some leaf prints which were made in this way.

Can you recognise them ?
They are :

1. Hawthorn
2. Oak
3. Birch
4. Lime

It is not a good thing to make leaf prints with fresh leaves because the tree needs these. The leaves help to make food for the tree.

Dry autumn leaves are dying and the trees which shed them do not need them again.

Apart from this, because they are dry and hard, you can make good prints with them, for their veins stand out well, as you can see in these pictures.

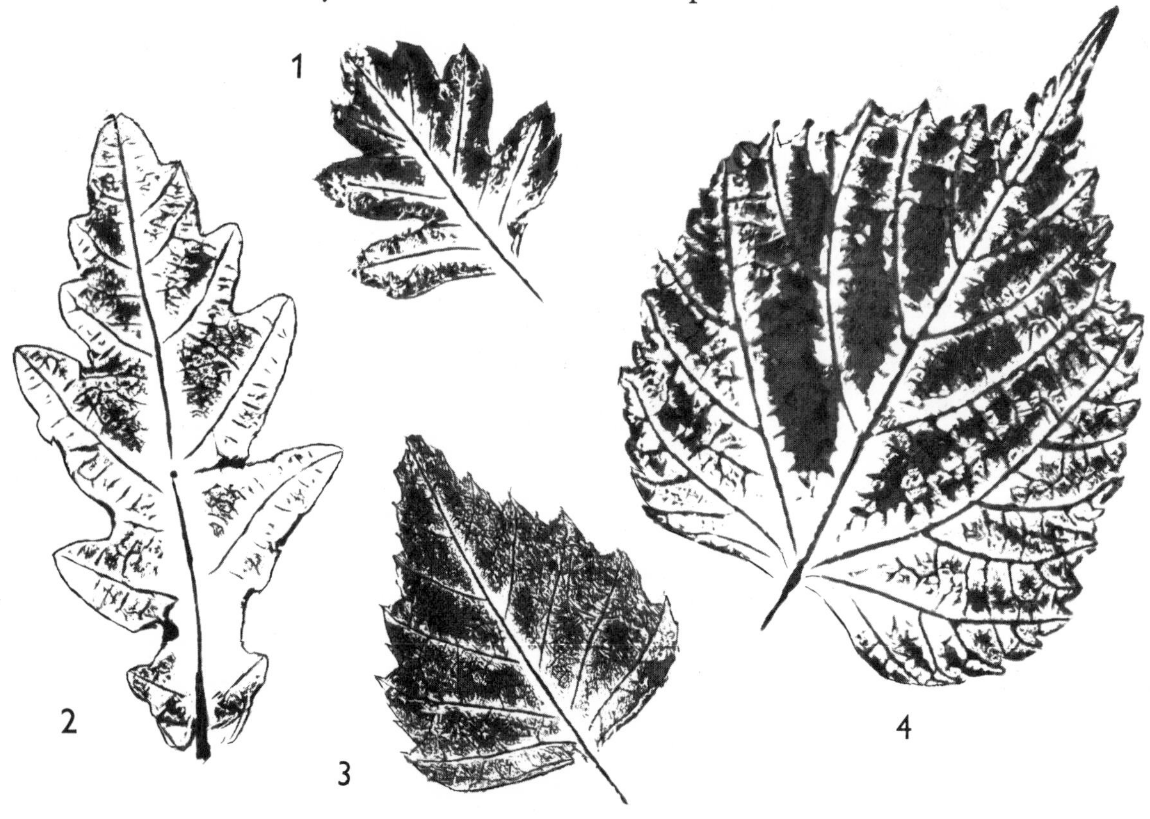

This is how the prints were made:

1. Light a candle.

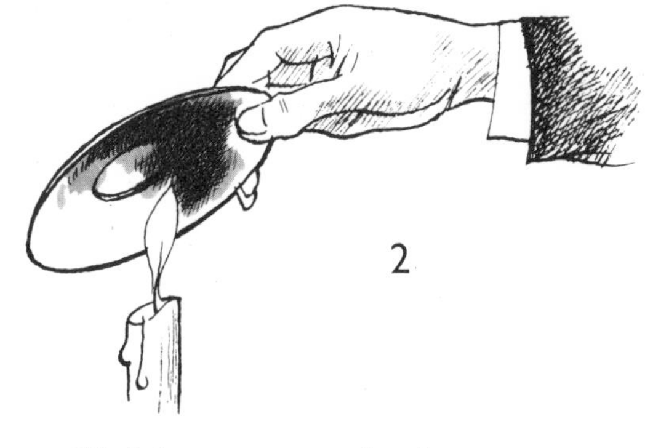

2. Hold an enamel plate or saucer over the flame.

3. Where the flame has touched the saucer, black soot will have collected.

4. Pour a little linseed oil on to the black candle soot and mix well.

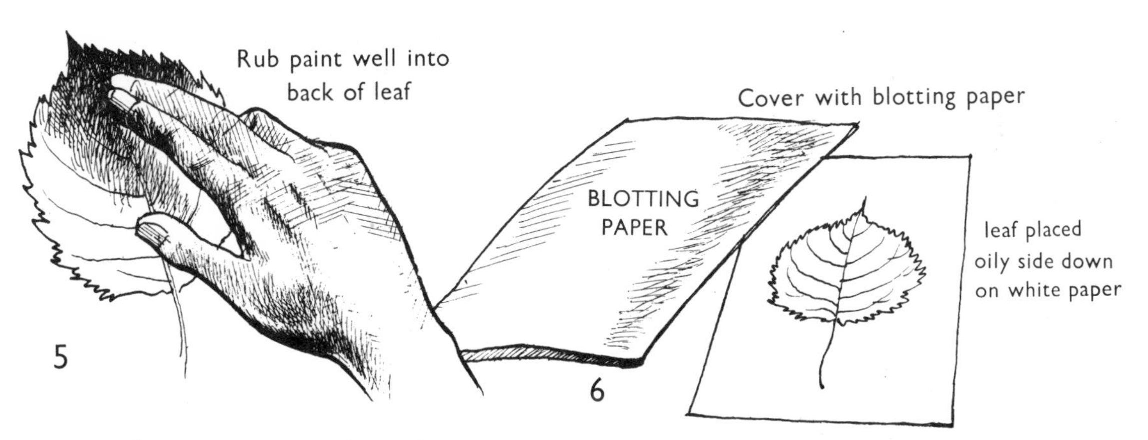

5. You have now made a black oil paint with candle soot and linseed oil.

Rub this black paint well into the *back* of the leaf with your finger or with a stiff paint brush, taking care to rub well over all the veins.

6. Place the leaf oily side downwards on a piece of white paper.

Cover with blotting paper and rub with the tips of the fingers on top of the blotting paper. When you lift up the blotting paper and the leaf, there will be a *print* on the paper.

COLOURED LEAF PRINTS

For these you can use oil paints or (more cheaply) shoe polish. Shoe polish can be obtained in various colours. Many of the brown, tan or reddish colours are very like the colours of autumn leaves. Choose carefully the colours which match your leaves.

Guelder rose

Here is the leaf of the guelder rose. You can see it in its autumn colours opposite page 160.

If you would like to make a colour print of the leaf of this plant, use oil paint or shoe polish.

Use fingertips or a small stencil brush to spread the colour on the back of the leaf. Apply the colour on to the leaf in the right places. You must put the red on to the red parts of the leaf and the yellow and green on to the yellow and green parts.

Place the leaf on a piece of white paper (oily side downwards) and cover with blotting paper. Rub gently but firmly on top of the blotting paper. Then take away the blotting paper.

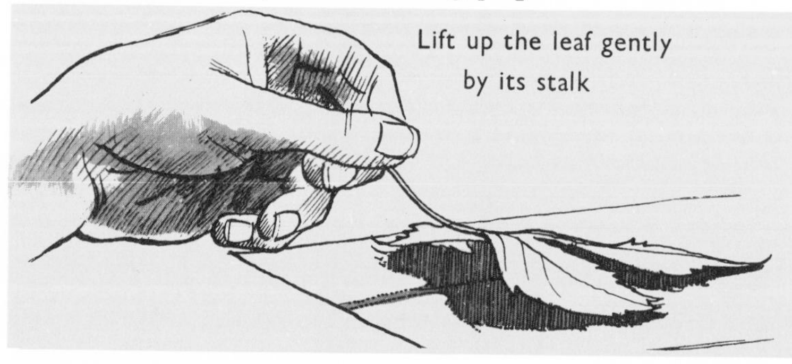
Lift up the leaf gently by its stalk

Lift up the leaf gently by its stalk leaving a colour print on the paper.

There are many beautifully coloured autumn leaves to use for your colour prints. Some are shown opposite page 160. Find others for yourself and discover their names.

BARK RUBBINGS

Compare these bark rubbing patterns.

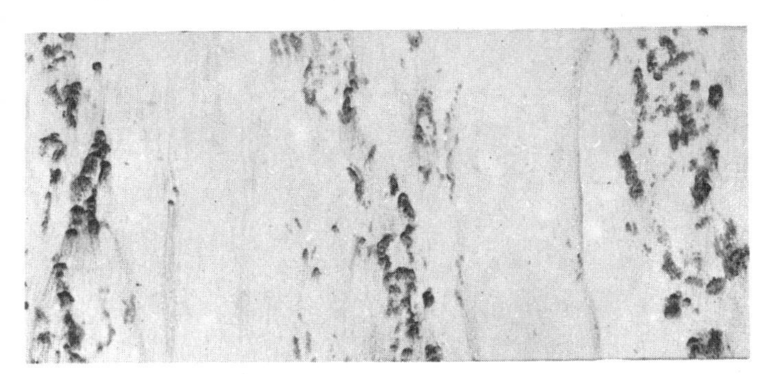

Oak

Oak has widely spaced ridges. The spaces between the ridges are called fissures. The fissures do not show on a rubbing so that only the ridges stand out.

Ash

In the ash-tree bark, the ridges are much closer together. We might say that the bark of ash is more closely grained than that of oak.

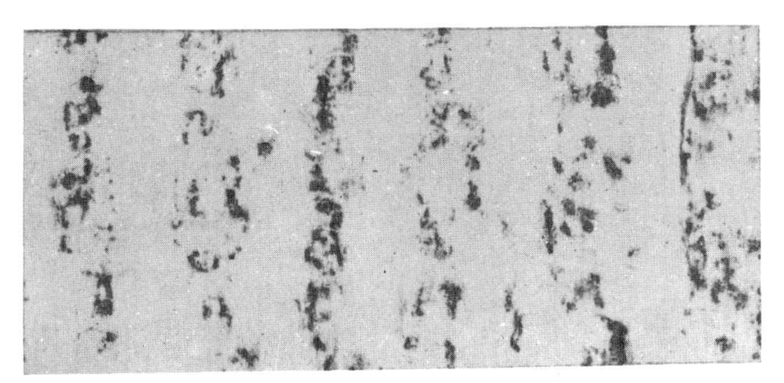

Holly

Holly makes an interesting bark rubbing. If you look at it carefully you will see that it has little dots and dashes all over it, rather like Morse code symbols.

Silver Birch

Another beautiful bark pattern is obtained from silver birch. This has lines going horizontally (across) as well as vertical (up and down) markings.

HOW TO TAKE A BARK RUBBING

1. Hold a piece of firm, tough paper (strong typing paper is good) against the bark of a tree and put a " T " at the top of the paper.

This " T " will show the right way up of your bark rubbing when you take it away from the tree.

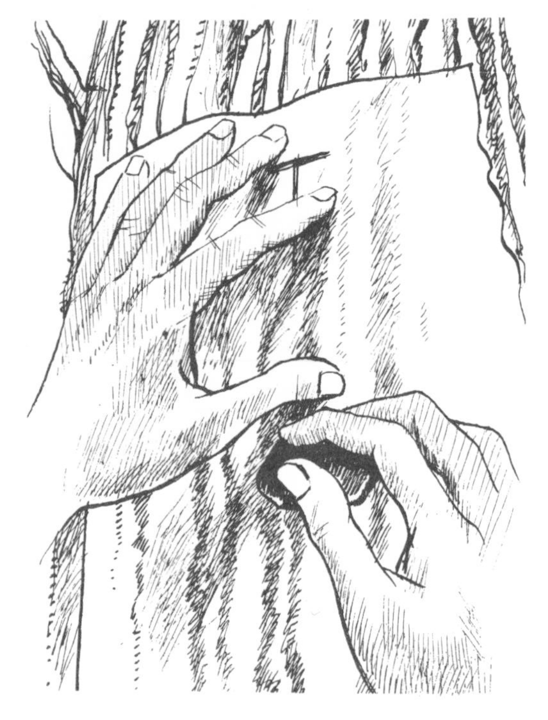

2. Next get a cobbler's heel-ball.

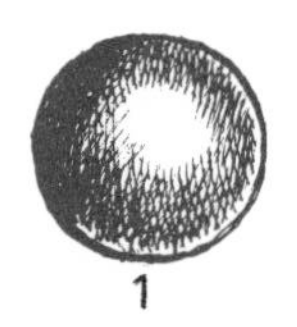

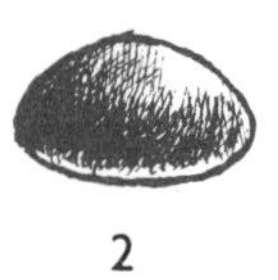

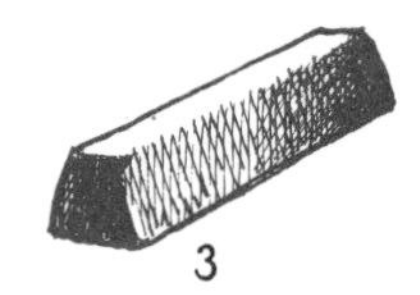

These are either spherical (Fig. 1) or hemi- (half) spherical (Fig. 2) or shaped like a stick of toffee (Fig. 3).

3. Holding the paper firmly against the tree, rub the heel-ball gently over it. Use the flat or the rounded surface of the heel-ball. If you use the stick do not poke into the hollows with it.

In the bark of the tree you will find ridges and you will find hollows, or fissures.

In a good bark rubbing you should see only the mark or pattern made by the ridges. The hollows, or fissures, should remain white.

If you use an oil crayon instead of a heel-ball, use the *side* not the *point* of the crayon.

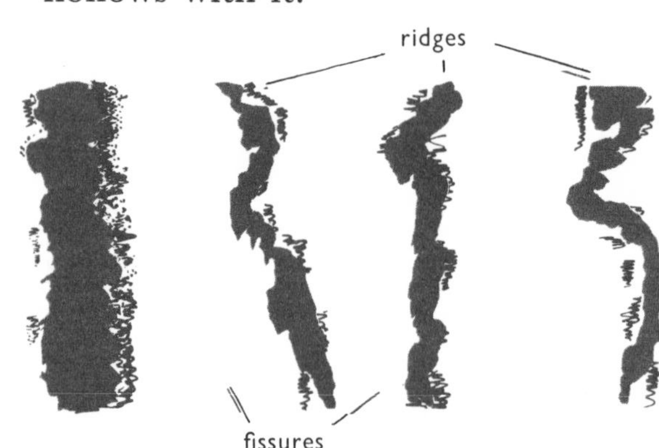

Diagram of bark rubbing showing ridges and hollows, or fissures

WINTER TWIGS AND BUDS

Here are five different twigs. The notes will tell you how to recognise them.

This is a node.

Ash

This is an internode.

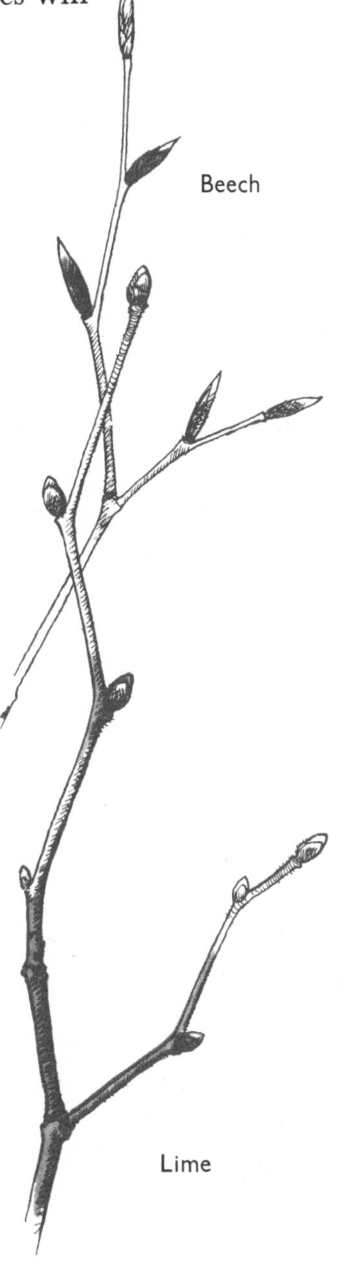

Ash buds are black.
Ash twigs have grey stems. The twigs are flattened at the nodes. The node is the part of the twig from which each pair of buds springs.

Beech buds are brown and cigar shaped.
The beech twig is thin and zig-zagged.

The oak twig has a crowd of egg-shaped, yellowish buds at the top; the "lateral", or side, buds are not as close together as the top buds.

Oak

Lime has zig-zagged twigs and red buds. The lower scale of each bud is very noticeable.

The horse-chestnut twig is thick and brown.
The buds are large and sticky. Underneath each lateral (side) bud is a horse-shoe-shaped scar. The terminal (end) bud has no leaf scar.

Horse chestnut

If you find winter twigs difficult to draw, try printing them in clay. You can see how to do this on the next page.

CLAY IMPRINTS OF WINTER TWIGS

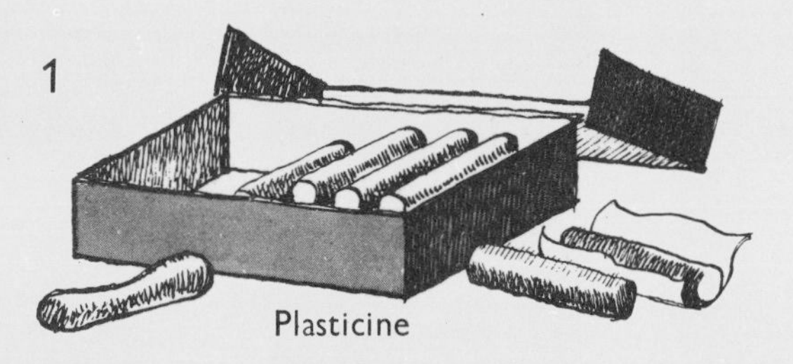

1. Dig up some clay, or use plasticine, or make clay by adding water to powdered clay.

2. " Work up " the clay with your hands until you have a soft, easily moulded ball.

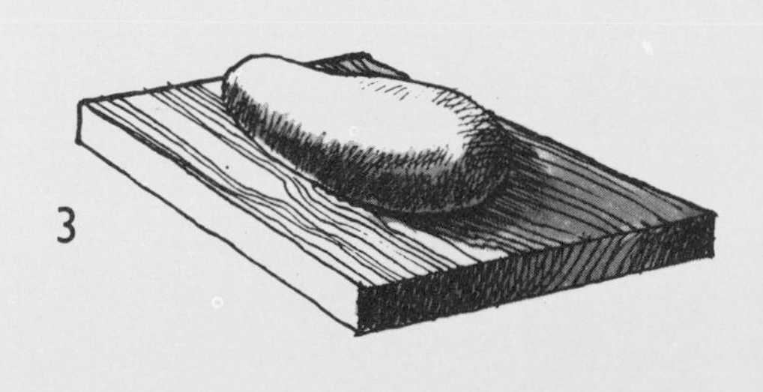

3. Put the ball of clay on a wooden board and flatten it.

4. Shape the clay into square or rectangular tablets (like tiles).

5

5. Here is the clay tablet. Lay the twig on it and press the twig firmly into the clay.

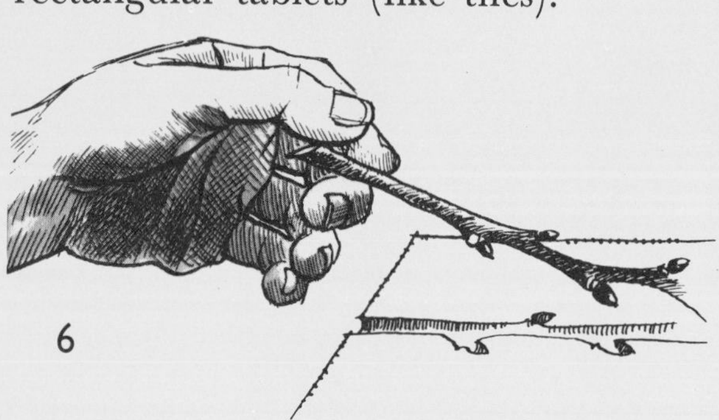

6. After pressing it in, lift the twig up from the clay, holding it by the lower end of the stalk. You will see that the shape of the twig has been made in the clay. This impression, as we call it, gives the exact size and perfect shape of the twig.

You can use this impression to make a more lasting record in plaster, as described on the next page.

PLASTER CASTS OF WINTER TWIGS

1. Find yourself some clay and "work" it as before until you have a round, soft ball.

2. Then make your clay into a tablet which is just the right size for the bottom of a cardboard box and make a twig imprint on the clay in the box (Fig. 1), as described on the opposite page.

Fig. 1

Fig. 3

Fig. 2

3. Now get some plaster of Paris and a bowl of water. (Fig. 2.)

4. Pour the powdered plaster of Paris into the water, a little at a time, stirring the mixture with a wooden spoon. Add enough plaster of Paris to make the mixture stiff enough (like thick cream).

Fig. 4

5. Spread the plaster of Paris over the clay in the box, after first moistening the clay surface with vaseline or liquid soap. (Fig. 4.)* Leave the plaster to set until the next day.

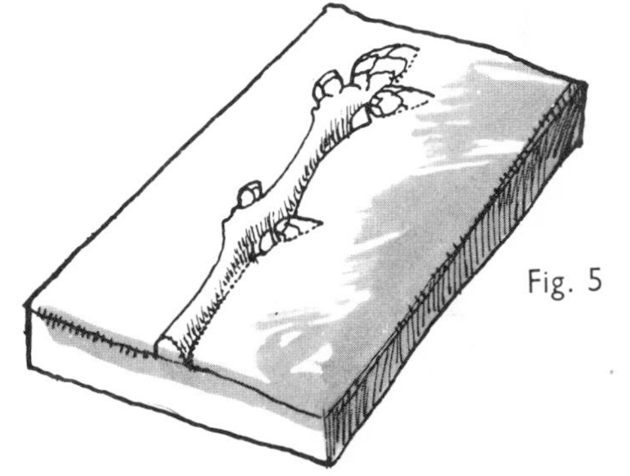

Fig. 5

6. Break down the sides of the cardboard box and take out the clay and the plaster. Separate the clay from the plaster and you will have a cast of the twig raised up on the plaster tablet. (Fig. 5.)

* If you wish to hang up your plaster tablet, make two holes in it by inserting match sticks in the plaster before it sets.

TREE OUTLINES IN WINTER

Look at trees against a winter sky. Try to name them from a distance. Below are some clues to help you.

Try cutting out tree shapes in black paper and mounting them on light coloured paper. These are called tree silhouettes.

Elm is a tall tree with crowds of small branches and twigs growing at the ends of the main branches. It has whisker-like twigs growing down the trunk.

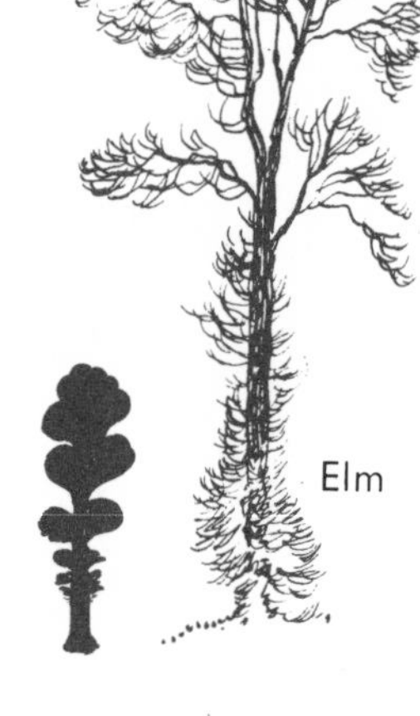

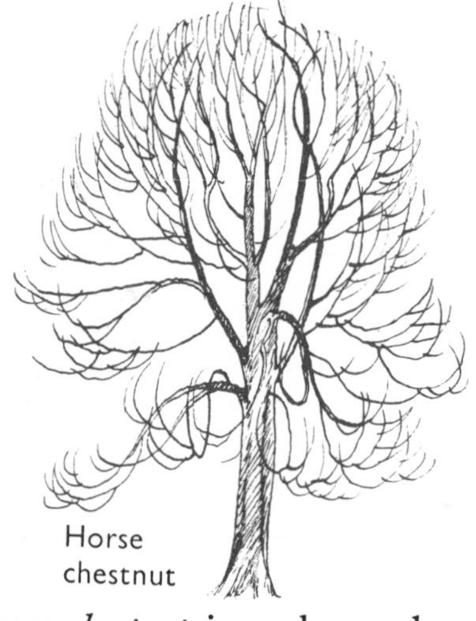

Horse-chestnut is a dome-shaped tree and all its branches turn upwards.

The picture on the right shows a *pollarded willow* by a river bank. A " pollarded " tree has had its top cut off and long thin branches sprout out from the crown. This tree makes a good silhouette.

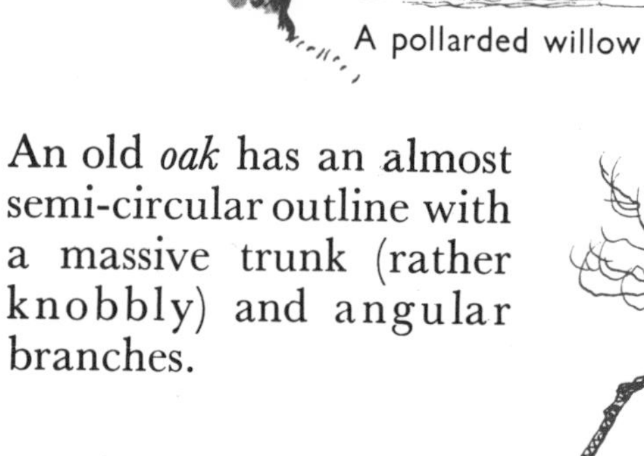
A pollarded willow

An old *oak* has an almost semi-circular outline with a massive trunk (rather knobbly) and angular branches.

An angular (sharply bent) branch

AUTUMN LEAVES

WEATHER STUDY

In autumn our weather is changing from summer conditions to winter conditions. In spring, it is changing from winter conditions to summer conditions. In spring and autumn, therefore, the weather is variable (often changing) and this makes it interesting to study.

By the words " weather conditions " we mean the state of the air around us, that is :—

whether it is warm or cold (temperature of air)

whether it presses heavily or lightly (pressure of air)

whether it is still or blowing about (calm or windy)

whether it blows quickly or slowly (wind speed)

whether you can see clearly through it (visibility)

whether it has any water vapour in it (humidity).

THE THERMOMETER

There are certain instruments which help you to find out about the conditions of the air. One of these is the thermometer which measures temperature.

There are different kinds of thermometers. The one used by a doctor to take the temperature of your blood when you are ill is called a *clinical* thermometer (see page 162).

The one shown on this page is a *room* thermometer to measure the temperature of the air in the room and tell you whether the room is warm enough or cool enough for you.

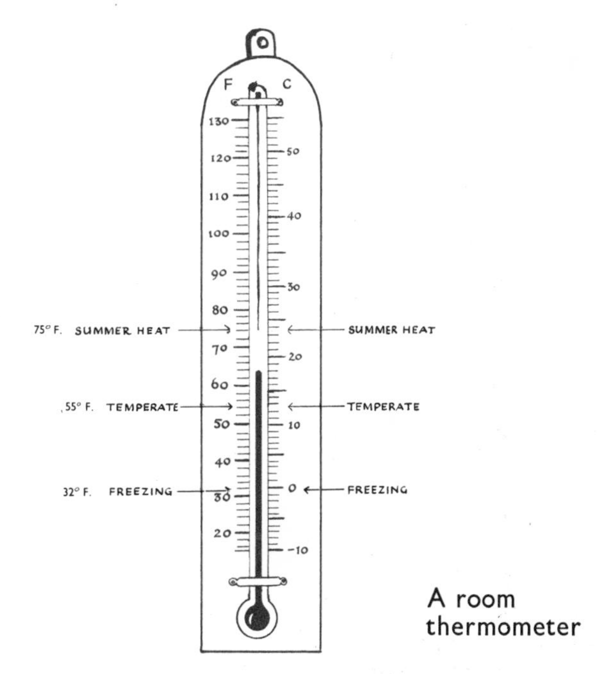

A room thermometer

Temperature is measured in *degrees*. The thermometer in the diagram shows both Centigrade (C.) and Fahrenheit (F.) scales. Temperature readings are now recorded mainly in degrees Centigrade. Try to compare the two scales in this thermometer. For instance : Freezing point which is 0° Centigrade is 32° Fahrenheit.

HOW A THERMOMETER WORKS

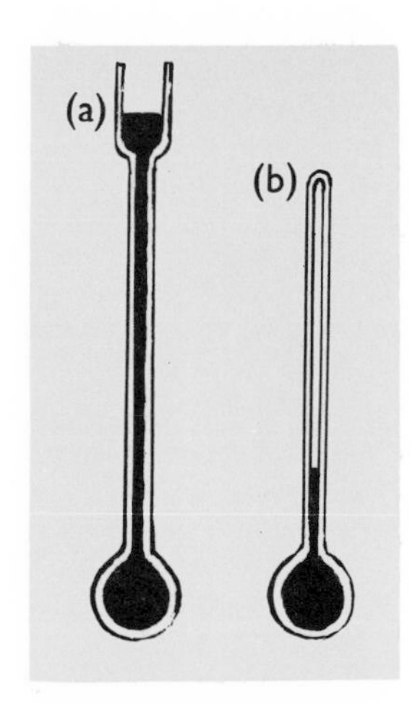

Thermometer bulbs usually contain mercury, a heavy metal which can flow like a liquid and is silver in colour. Pouring mercury into a thermometer tube is difficult because air pressure in the tube resists the downward flow. Bulb and tube are heated first to expel the air and then mercury will trickle down from (a) (see diagram). When bulb and tube are full the tube is sealed off at (b) while it is still hot. As it cools, the mercury contracts into the bulb but will rise again when the bulb is warmed. By means of a graduated scale marked in degrees, you can read the amount of this rise and fall of the mercury in the thermometer.

A. is a Fahrenheit thermometer, and B. is a Centigrade thermometer. You can see that they have different scales of measurement. You have already learnt that freezing point is 32° on the Fahrenheit scale and 0° on the Centigrade scale. Now you will see that boiling point is 212° Fahrenheit or 100° Centigrade.

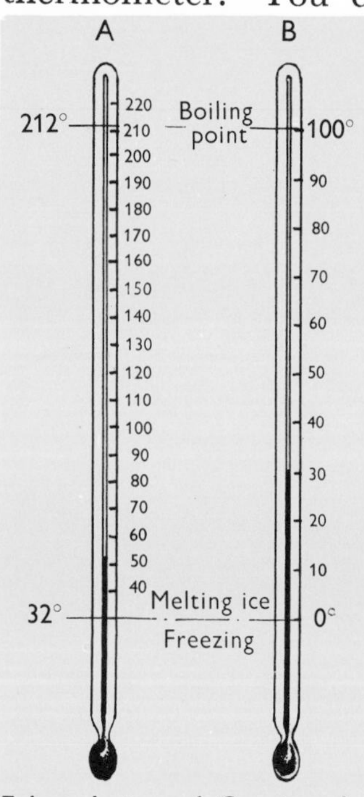

Fahrenheit and Centigrade thermometers

These thermometers are much longer than they appear here. They are usually about one foot or 30 centimetres in length

C. is a clinical thermometer. This is used by doctors and nurses to take human temperatures. It is a Fahrenheit thermometer measured like A, but you can see that it only includes readings between 95° F. and 110° F. You will see from the diagram that the normal human temperature is 98.4° F. When people are ill their temperature often rises to 99° F. or even to 100° F. or more, but it rarely becomes more than 103° F., which is very high indeed.

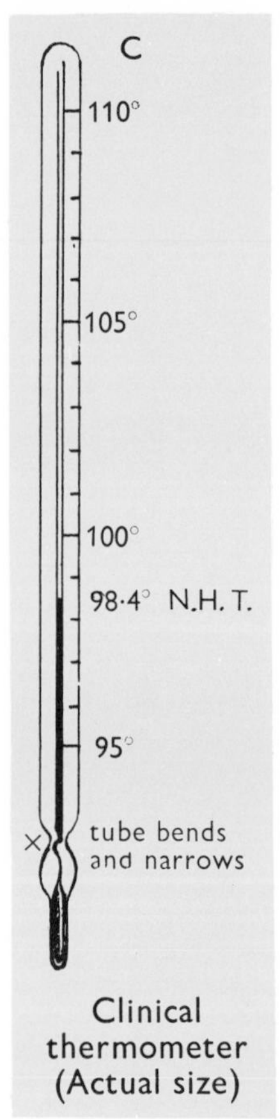

Clinical thermometer (Actual size)

The narrow bend in this thermometer tube (at X) stops the mercury from running back into the bulb after a patient's temperature has been taken.

MAKING A TEMPERATURE GRAPH

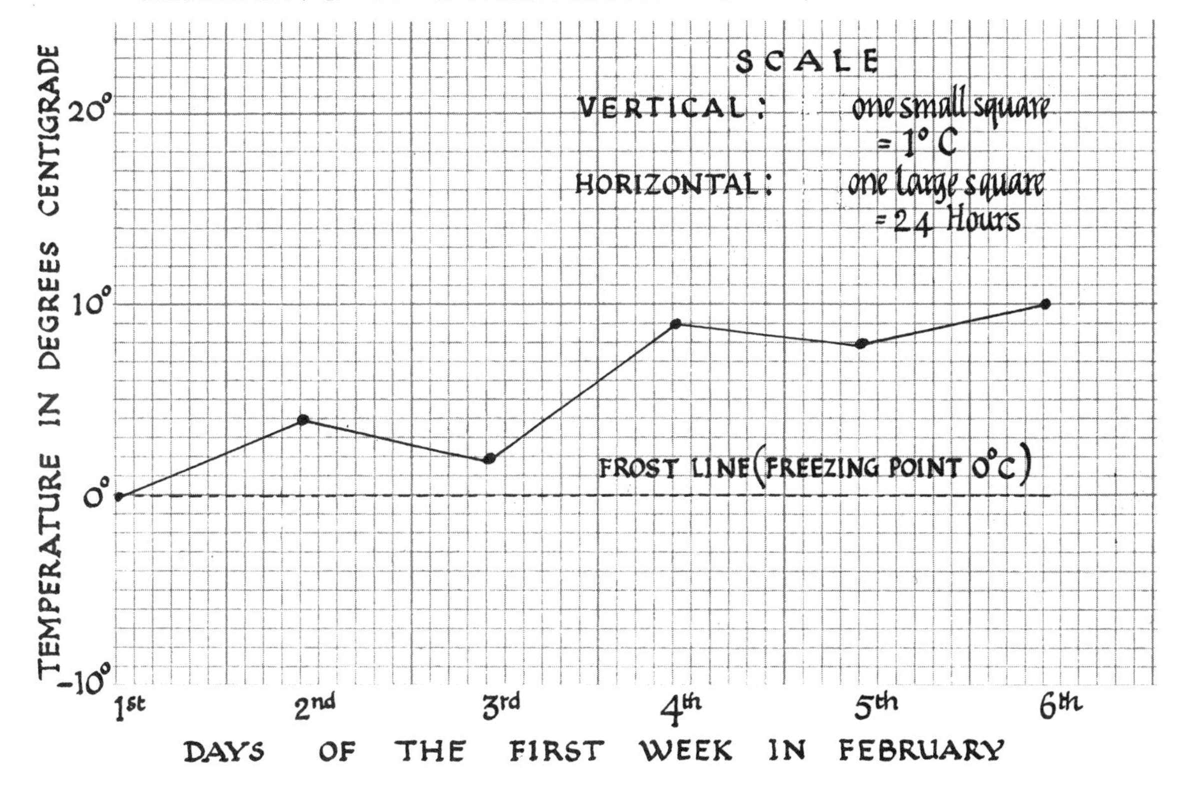

The temperature graph above is made on squared paper. This makes it easy to work out your scale and mark in each day's temperature.

There are two scales, or measuring lines, on a graph. One is vertical (upright) and one is horizontal (flat or along).

By the vertical arm of this graph is written "Temperature in degrees Centigrade", while along the horizontal arm is written "Days of the first week in February".

You must decide how many small squares are to represent your temperature readings along the vertical arm and how many will represent days or hours along the horizontal arm. In the graph above, the scale tells you that one small square represents one degree Centigrade vertically and one large square represents one whole day or twenty-four hours horizontally.

When you have drawn both arms of the graph and worked out your scale, you can "plot your points". Every day, at about the same time each day, you put a dot on the graph to show how high or how low the temperature reading is. Join your points by straight lines and look at your graph. Now you can see whether the weather is getting warmer or colder from day to day.

AIR PRESSURE

How do you know that air is all around you ?

Can you see it ?	No.
Can you smell it ?	No, only when it carries the vapour or tiny particles of a substance with a smell.
Can you feel it ?	Yes, when the wind blows.
Has air any weight ?	Yes. The weight of air, or " air pressure ", is approximately 1 kilo per square centimetre on *every* surface in *every* direction.
Does air push upwards as well as downwards ?	Yes.

EXPERIMENT TO SHOW THAT AIR CAN SUPPORT A COLUMN OF WATER

1

Air can support things. You can carry out an experiment to show that air can support a column of water. You will need a tumbler, a piece of cardboard and water.

Method

1. Fill a glass tumbler with water right up to the brim.

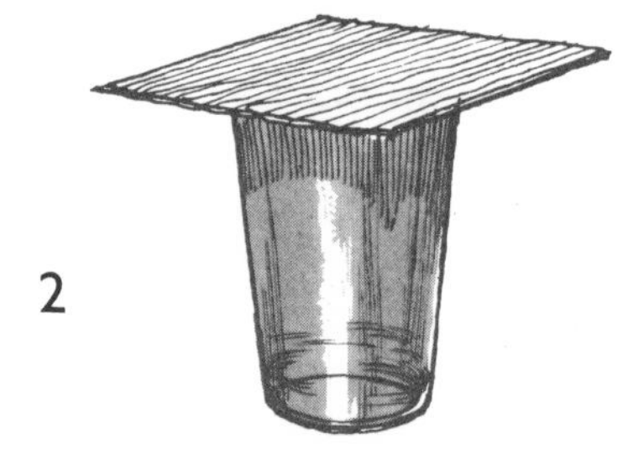
2

2. Slide a piece of cardboard over the top of the glass so that it touches the water. Do not allow any bubbles of air to creep in.

3

3. Turn the tumbler upside down holding the card against the glass. Take your hand away from the card. If you do this carefully, the water will remain in the tumbler. It will not fall out because it is supported by the air pressure below.

A VACUUM

A vacuum is a space where there is no air and therefore no air pressure.

It is difficult to find a complete vacuum in Nature. Under natural conditions, as soon as a vacuum occurs, one of two things may take place. Either air will rush into the vacuum, or, if this is not possible, the structure around the vacuum will give way, or collapse.

You can try making a vacuum for yourself. Boil a little water in an open tin (Fig. 1) and then seal up the tin tightly (Fig. 2). At first the tin is full of steam. Then, as the tin cools and the steam changes back into water again, a partial vacuum is formed inside the tin. The tin should then collapse (Fig. 3).

A large golden-syrup container (the type supplied to canteens and by school meals services) is excellent for this experiment.

Another interesting experiment is to drive the air out of an empty plastic detergent bottle by squeezing it hard. When you stop squeezing you will hear air rushing back into it. If you immersed it in water after squeezing, water would rush in in the same way.

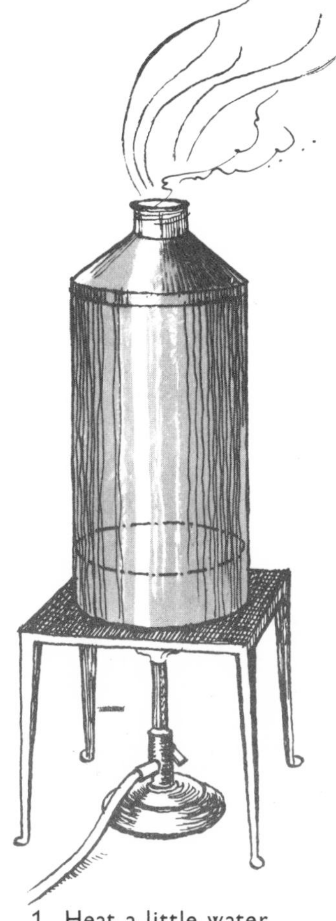

1. Heat a little water in a metal container.

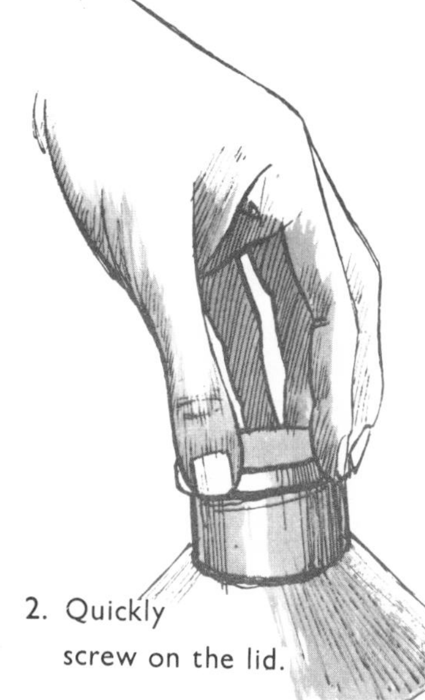

2. Quickly screw on the lid.

3. The tin will collapse.

A SIMPLE BAROMETER

Air pressure is measured by an instrument called a barometer. The earliest mercury barometer was invented by a man called *Torricelli* in the 17th century.

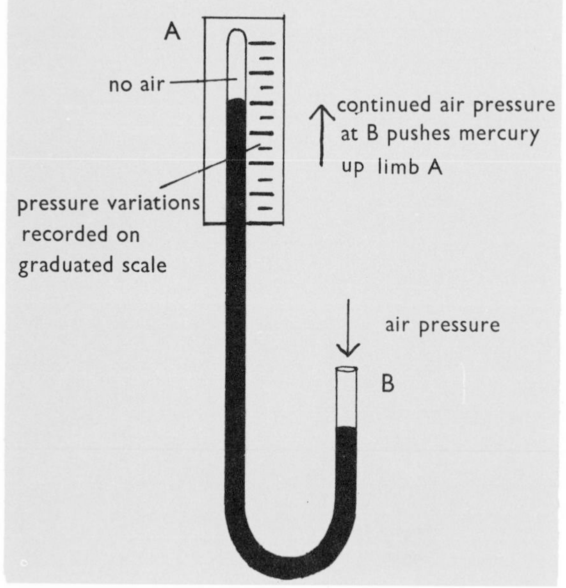

Above is shown a simple type of barometer. You will see that there is a short limb to this U tube, and a long one. The long limb is sealed at the top, but the short limb is open to the air. Mercury has been put into the tube and allowed to run back so that there is a vacuum at the end of the tube marked A. There is *no* air pressure inside the tube at A, but there is at the open end of the tube, marked B, and this is sufficient to push the column of mercury up the A tube. The normal pressure of air supports the mercury in the A tube to a height of about 76 centimetres above the level of mercury in the B tube.

Slight variations of this pressure take place when weather conditions change.
Dry air presses more heavily at B than moist air so that the mercury level at A may rise or fall.
A graduated scale is placed near the mercury level at A to record this rise or fall.

Another kind of barometer

The aneroid barometer has no mercury in it. The most important part of this barometer is a thin, flat, metal box from which air has been withdrawn so that the air pressure within it is reduced. The walls of this box can now be pressed inwards by the weight of the atmospheric air outside the box. Atmospheric pressure varies and the walls of the box will move inwards or outwards according to these pressure variations. These movements are recorded by a pointer moving over a dial and the pointer is worked by a pulley controlled by a spring.

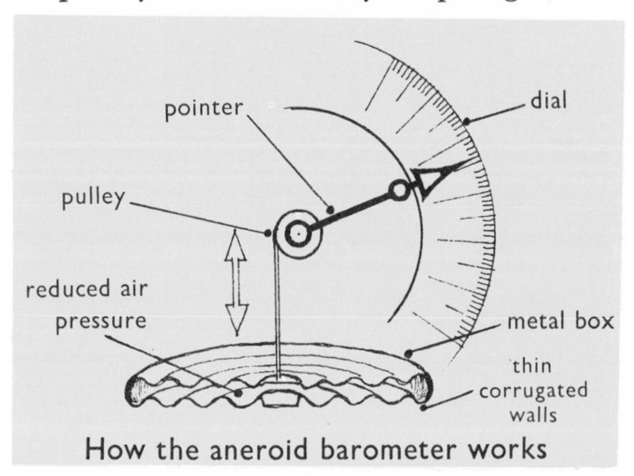

How the aneroid barometer works

RECORDING AND JUDGING WIND SPEEDS

The people who understand weather are called " meteorologists ". They have instruments called anemometers by which they can measure the speed of the wind.

The type of anemometer used most often in meteorological stations consists of four little cups attached to the ends of four arms of equal length, arranged in a cross (see diagram 1).

This is placed in a horizontal position on top of a vertical (↑) axis in such a way that it can spin round freely (see diagram 2). It is rather like a wind vane because it is moved round by the wind, but, as it turns, the speed of the wind is registered by a special instrument beneath it. The speed is shown on the face of a dial which looks rather like the dial you see on a gas or electric meter.

An admiral, Francis Beaufort, who lived in the 19th century, gave us an interesting way of judging wind speeds from various happenings in Nature. You do not need any special instruments for this. All you need is to be a good observer. For instance, when you see smoke rising vertically, the weather is said to be " calm " and the wind speed is less than 1 mile per hour.

When you hear leaves rustling gently, there is a " light breeze " and the speed of the wind is then 4 to 7 miles per hour.

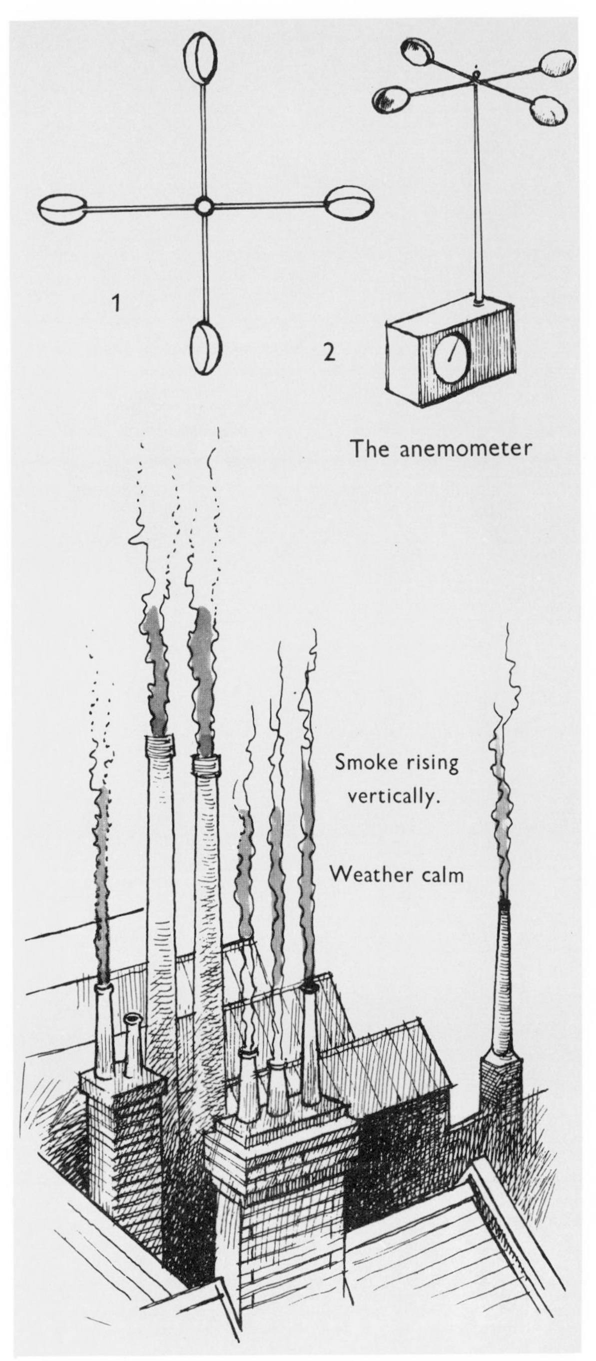

The Beaufort scale, which is built up from observations like these, is given on the next two pages.

THE BEAUFORT SCALE

No.	*Description*	*Effects*	*Speed of wind (in miles per hour)*
0	Calm	Smoke rises vertically.	under 1 m.p.h.
1	Light air	Smoke drifts gently. Wind vanes are not moving.	1—3 m.p.h.
2	Light breeze	Wind vanes begin to move ; leaves rustle. Wind noticeable on face.	4—7 m.p.h.

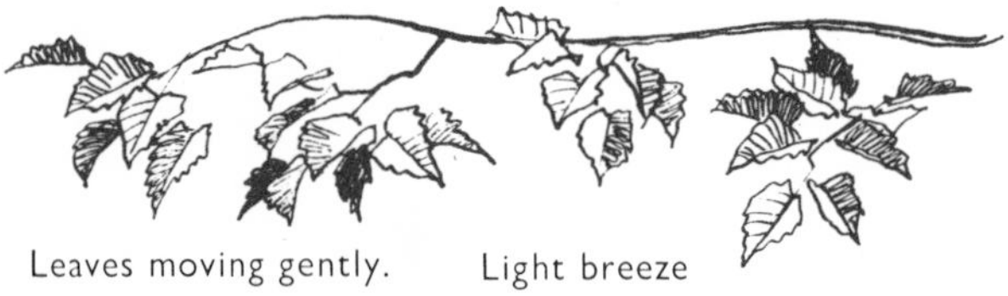

Leaves moving gently. Light breeze

No.	*Description*	*Effects*	*Speed of wind (in miles per hour)*
3	Gentle breeze	Leaves and small twigs constantly moving. Light flags extended.	8—12 m.p.h.
4	Moderate breeze	Dust and loose papers raised. Small branches are moving.	13—18 m.p.h.
5	Fresh breeze	Small trees in leaf seem to sway. Small waves form on inland waters.	19—24 m.p.h.
6	Strong breeze	Large branches sway. We can hear whistling in telegraph wires.	25—31 m.p.h.
7	Moderate gale	Whole trees in motion. We have to push against the wind when walking.	32—38 m.p.h.
8	Fresh gale	Twigs break off trees. Progress is difficult.	39—46 m.p.h.
9	Strong gale	Damage to chimney pots ; slates removed.	47—54 m.p.h.
10	Whole gale	Much damage. Trees uprooted. This sort of gale is more often experienced on the coast.	55—63 m.p.h.

No.	*Description*	*Effect*	*Speed of wind (in miles per hour)*
11	Storm	Usually experienced at sea. Widespread damage.	64—75 m.p.h.
12	Hurricane	Usually experienced in the tropics. Devastation !	more than 75 m.p.h.

WIND SPEEDS SHOWN BY SIGNS

The arrow signs you see below are often seen on weather maps. If you count the tails, you can discover at about how many miles per hour the wind is blowing.

This sign means that there is little wind and that its speed is less than 1 m.p.h. (mile per hour).

One tail means that the wind speed is 1—3 m.p.h.

Two tails mean that the wind speed is 4—7 m.p.h.

Three tails mean that the wind speed is 8—12 m.p.h.

Four tails mean that the wind speed is 13—18 m.p.h.

Five tails mean that the wind speed is 19—24 m.p.h.

Six tails mean that the wind speed is 25—31 m.p.h.

Seven tails mean that the wind speed is 32—38 m.p.h.

Eight tails mean that the wind speed is 39—46 m.p.h.

Nine tails mean that the wind speed is 47—54 m.p.h.

Ten tails mean that the wind speed is 55—63 m.p.h.

Eleven tails mean that the wind speed is 64—75 m.p.h.

Twelve tails mean that the wind speed is over 75 m.p.h.

MORE WEATHER SIGNS

Cattle and horses sheltering

Cattle and horses can often tell when rain is coming and gather together on the leeward side of trees and hedges. The leeward side is the side away from the wind.

A rookery

Rooks usually collect quite near to their rookeries when rain clouds gather.

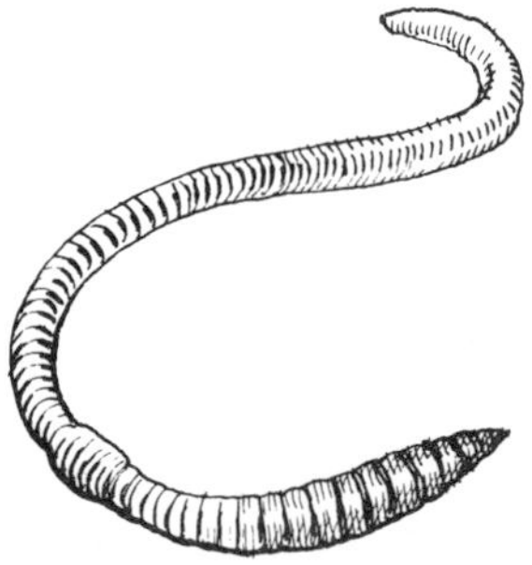

Earthworms come out of their burrows when the air is moist and may be seen travelling over the surface of the soil. Worms cannot live in a very dry atmosphere and often dry up and die if they are forced to be out in the sun's heat.

Earthworm

The black garden beetle or " rain beetle " can often be seen on top of the soil in the garden before a rain storm.

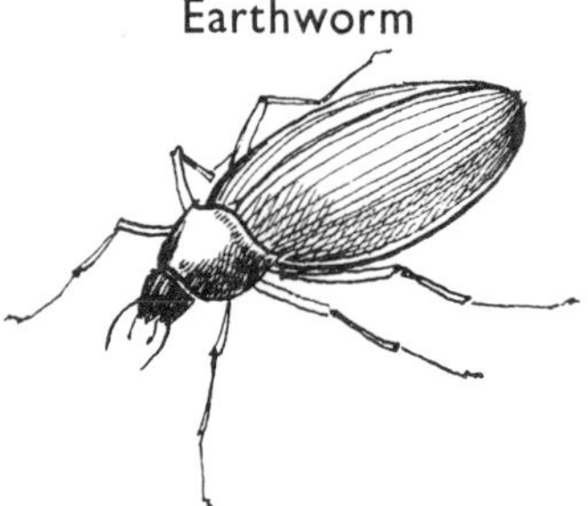

Rain beetle

It seems as if many creatures are able to " sense " the approach of rain or of moisture in the air.

Do you know any other signs which foretell bad weather ?

ANCIENT WEATHER LORE

Frogs croak loudly

Here are some more weather signs which country people know. You do not have to believe in them all, for some may be " old wives' tales ", but it would be fun to find out if any of them prove to be true.

Signs of rain (according to countrymen's lore)

It is said to be a sign of rain if crows make a noise in the evening, if geese are heard to gaggle loudly, and if oxen lie on their right sides. Frogs are said to croak loudly when rain is coming.

Weather forecasts

In the old days people tried to give weather forecasts although they had not much knowledge of depressions and wind behaviour as we have today. They had to look for ordinary, everyday signs. Here are some " old-fashioned" weather forecasts.

" If October and November be snow and frost, then—
January and February are likely to be open and mild."

The Greeks believed that this saying was true.

Weather changes were often forecast by studying the moon. For instance, it was said :

" A mist near full moon means fair weather ".

A ring, or halo, round the moon was important.

" Last night the sun went pale to bed,
The moon in haloes hid her head "—

wrote an old poet. Usually, a halo round the moon is taken to mean that wet weather is soon to follow, while there is another saying :

" Clear moon, frost soon".

You may have heard the saying, " Rain before seven, fine before eleven ". How many of these signs do you think are true ? Today, our weather forecasters use scientific methods. They study the atmosphere and speak of " depressions " and " warm or cold fronts ".

WATER IN THE AIR

The air is made up of several gases and one of these is water vapour. The amount of water vapour in the air varies from time to time. Warm air can absorb more water into itself than cool air can. The amount of water vapour in the air can be measured by meteorologists (weather experts) and the result is called the " relative humidity " of the air.

Humidity, as you probably know, is a kind of dampness and if the air becomes more and more damp, we say that the humidity is high. There is a point, however, at which no more water vapour can be absorbed into the air and we then say that the air has reached *saturation point.* If the air becomes warmer, the saturation point is raised, and if the air becomes cooler, it is lowered.

When air is cooled below its saturation point, then the extra water vapour in the air must be got rid of ; when this happens, the water vapour changes into water again and drops to the ground. There are several different forms of this water, however, as well as the most common form : RAIN. These are known as forms of PRECIPITATION. This long word means something which falls or separates out, and water, in some form or other, is often separating out of the air.

Here are some of the forms of precipitation that we know :

RAIN—the commonest form—drops of water falling upon the earth

SNOW—made of collections of tiny ice crystals

FROST—ice crystals which adhere (stick) to surfaces

HAIL—frozen rain falling to the earth in small, hard lumps. (On occasions hailstones can be very large)

SLEET—a mixture of rain and hail

FOG—tiny water droplets hanging in the air rather like a low cloud

MIST—a lighter sort of fog often seen in the mornings

DEW—water droplets which form on grass or other vegetation

Snowflake patterns

CLOUDS AND RAIN

1. The water here is exposed to the rays of the sun. The sky is clear and cloudless, but changes are taking place at the water surface.

2. As the sun warms the water and the air, some surface water changes into water vapour and rises into the air. In the higher part of the air, it is cooler and some of the water-vapour changes into very tiny droplets of water which hang together in the form of clouds.

3. The clouds are blown along in the main wind stream. In our country the main or " prevailing " winds are westerly, that is the wind often blows *from* the west or south-west. These south-westerly winds are usually rather warm moist winds as they have come over warm seas.

4. These winds blow the clouds towards our western shores. They may reach high mountains, such as those in Wales or Scotland, and as the moisture-laden clouds rise, they meet colder air which causes the moisture in them to condense and fall as rain. (See also page 136.)

WATER IN THE SOIL

What is the soil in your garden like? Rub it between your fingers. If it feels gritty and is fairly light-coloured, it is probably sandy. If it is smooth and slimy when wet, it is clay.

A good soil mixture of sand and clay is called a loam.

Sand and clay behave differently when water falls on them. You can prove this for yourself.

EXPERIMENT TO SHOW THE EFFECT OF WATER ON BOTH CLAY AND SAND

The plant pots are placed in glass beakers or jam jars

An equal amount of water is poured over the soil in each pot

Find two small porous plant pots of the same size. Put clay in one and sand in the other.

Stand each of the plant pots in a glass beaker or a jam jar. See that the sand and clay are packed very firmly into the pots. Now pour equal amounts of water on to the soil in each pot and watch carefully to see how quickly the water drains through each pot.

You will find that the water will drain more quickly through one of the soils than through the other.

Which soil is the more porous, that is, which soil allows the water to pass through it more quickly? You will find that water trickles more quickly through sand than through clay. We say that sand is more *permeable* to water than clay is. In fact the water may stay on top of the clay and form a puddle.

Water in the Soil

We have just seen that sand is more permeable to water than clay and that water can rest on clay without draining through it. The fact that clay can hold water to a certain extent can be proved if you make a clay vessel. The vessel can be made of natural clay or of plasticine, which is a prepared clay with plenty of oil in it.

Pour water into your clay vessel and you will find that it will hold water for a certain time, although the water will gradually seep through the clay and the whole vessel will become wet and slimy.

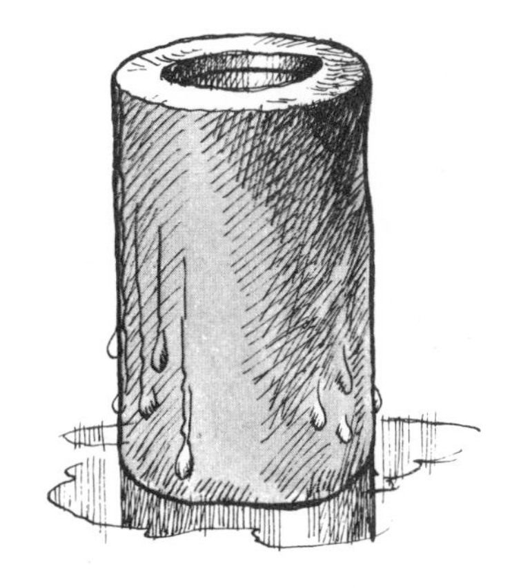

Clay vessel

ROOTS IN THE SOIL

Particles of sandy soil (enlarged)

Roots of plant in sandy soil — soil particles larger than those of clay soil—better drainage—good aeration of soil

If you look at some soil through a magnifying glass, you will see that it is made up of particles rather like the ones in the drawings. You can see that there are air spaces between the particles.

The roots of the plants push their way between the soil particles. Roots can breathe in the air spaces between the soil particles.

If it rains heavily on a clay soil, the air spaces in the soil often become filled up with water because the water cannot drain easily through clay soil.

Then the soil is said to be " water-logged ".

In a waterlogged soil the plant roots find it hard to grow because the soil is sticky and heavy. The air spaces are filled with water so that the roots cannot breathe.

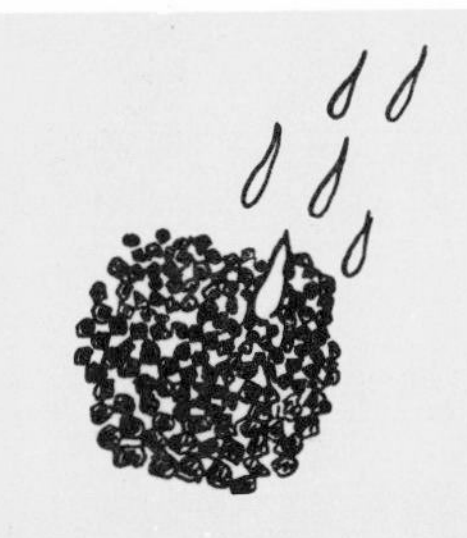

Particles of clay soil (enlarged) — heavy raindrops falling

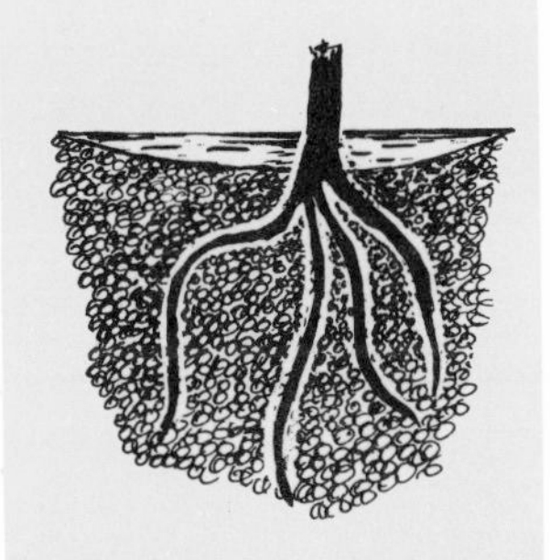

Roots in waterlogged clay soil — particles stick closely together—root growth difficult—bad drainage—poor soil aeration

WORMS IN THE SOIL

Fig. 2

You can prove that worms help to aerate (let air into) the soil by making a " wormery ".

For this you will need a transparent cylinder (Fig. 1). You can make one from an oblong piece of polythene, fastening the shorter edges together with sellotape. Or you can pour *boiling* water into a large jam jar which is standing in a little cold water in a bowl (Fig. 2). The bottom of the jar should break off evenly. (Fig. 3.)

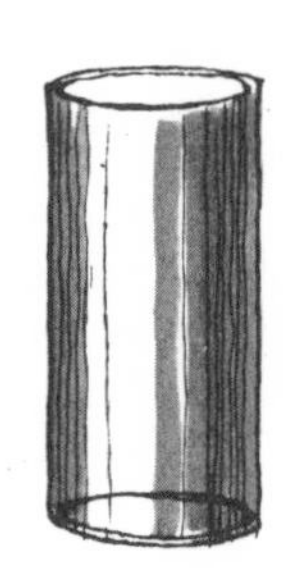

Fig. 1

Fig. 3

Stand your glass or polythene cylinder on top of a plant pot which has been filled with soil.

Put alternate layers of soil and sand in the cylinder, packing each layer down firmly. Plant grass seed on the top.

Gum a strip of white paper on the cylinder and mark on it the level lines of your sand and clay.

Then put in the worms. You need only put the worms (about six) on top of your soil layers. They will soon find their way down.

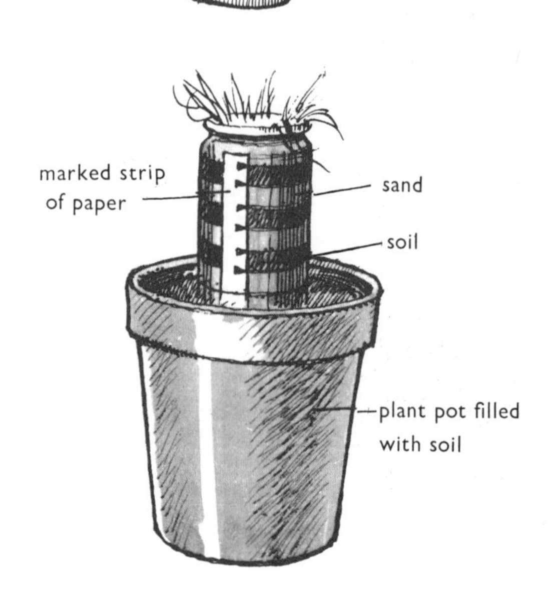

Here is the finished " wormery "

The earthworms will make burrows in the soil and sand and what do you think your neat sand and soil layers will look like in a few weeks time ? Try this experiment and see.

(Above) CALM — Beaufort scale 0.
(Below) A MODERATE GALE — Beaufort Scale

OXYGEN IN THE AIR—How green plants help us

Plants take in certain materials from the earth, from the air and from water, and with these materials they are able to make their own food. We often eat plants, thus making use of the food which plants have made for themselves.

While plants are making this special food, which they always do in *sunlight*, they give off the gas oxygen. In fact oxygen is a sort of waste material which comes out of the plant foodmaking " factory ".

Oxygen is, however, very valuable to all living things. We cannot breathe without it, nor can animals, nor can plants. Although plants give out oxygen, they need to take it in when they are *breathing*.

However, since plants give off this valuable gas, they are doing other living things a great service because, since all living things have to breathe, there could easily be a shortage of oxygen in the atmosphere. Green plants, in this way, help to maintain the balance of gases in the atmosphere.

It is impossible to see oxygen. Like most gases, it is invisible. We have to use special methods to prove that it is there. We have learnt that green plants give off oxygen in sunlight, when they are making food, and we can show, by an experiment, that this is true. In the experiment, we use a water-plant under water, so that we can see the bubbles of oxygen coming from the plant and rising in the water. The experiment is set up as shown in the diagram.

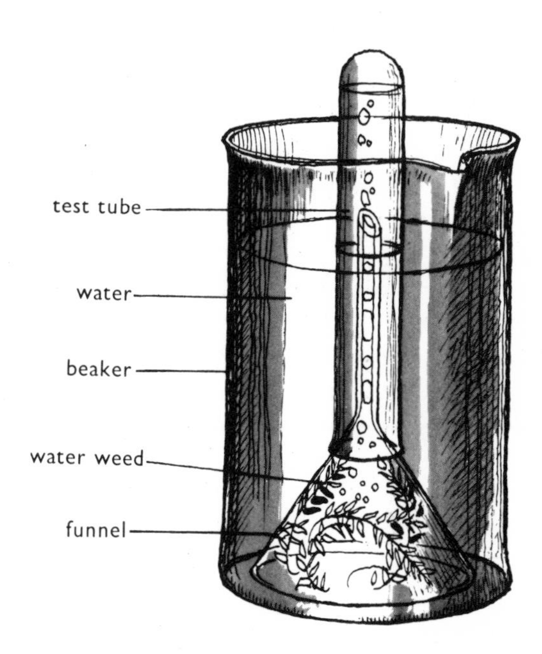

Set up your experiment like this and remember to place it in sunlight

The test tube must be quite full of water at the beginning of the experiment and you will find it easier to make sure of this if you set up the experiment completely under water.

If the apparatus is left in sunlight, you will see bubbles of gas arising in the test tube. This gas will gradually take the place of the water in the test tube. You will find out on the next page how to prove that this gas is oxygen.

A TEST FOR OXYGEN

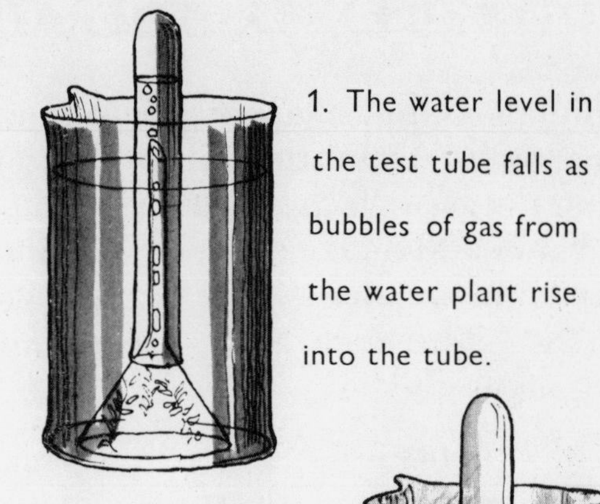

Oxygen is a "lively" gas. It is necessary to life. All living things have some method of taking it into their bodies. We usually describe this taking in of oxygen as "breathing".

Oxygen also makes fires burn, and a spark, plunged into oxygen, will immediately burst into flame. It is in this way that we can prove that the gas collected in the experiment described on page 177 is oxygen.

As the apparatus has been standing in sunlight, bubbles of gas have been rising from the plant and taking the place of the water in the test tube, until the test tube is almost full of the gas. (Diagrams 1 and 2.)

Now light a long splinter of wood and then blow out the flame, leaving a glowing end on the splinter. (Diagram 3.)

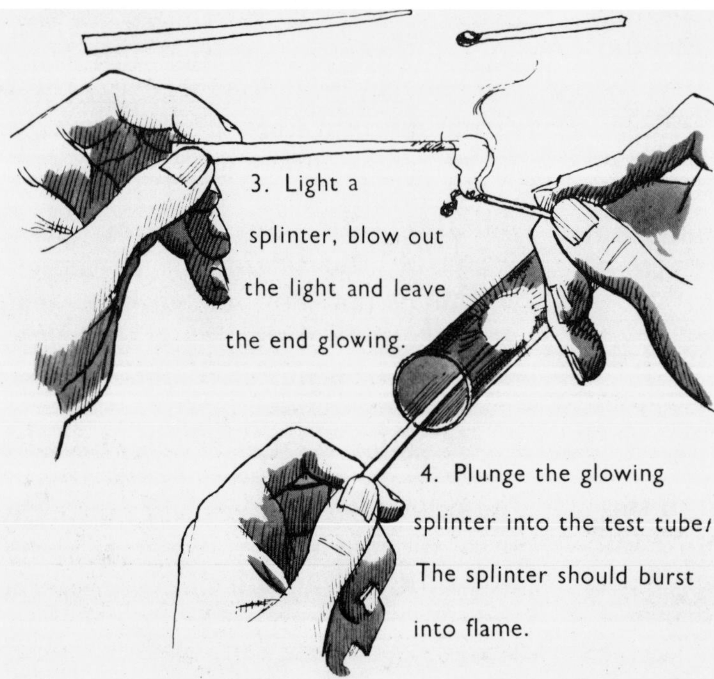

Carefully remove the test tube from the beaker.

Plunge the glowing splinter into the test tube immediately you take the tube away from the rest of the apparatus. The glowing end of the thin piece of wood should at once burst into flame. (Diagram 4.) If this happens, you have proved that the bubbles of gas, which you have collected in the test tube from the green water plant, are bubbles of oxygen.

AIR IN WATER

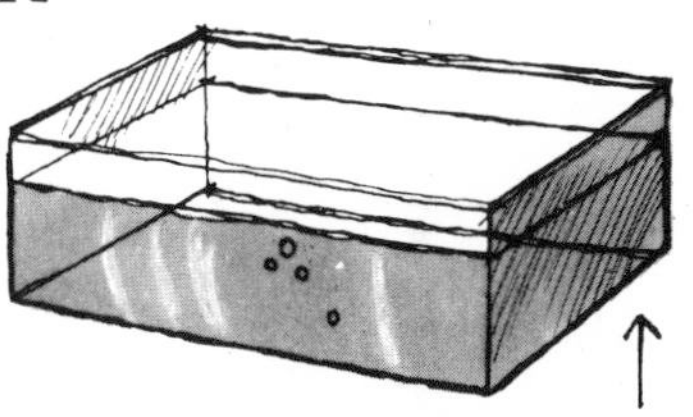

A shallow tank like the one above is better than a deep one like this one below

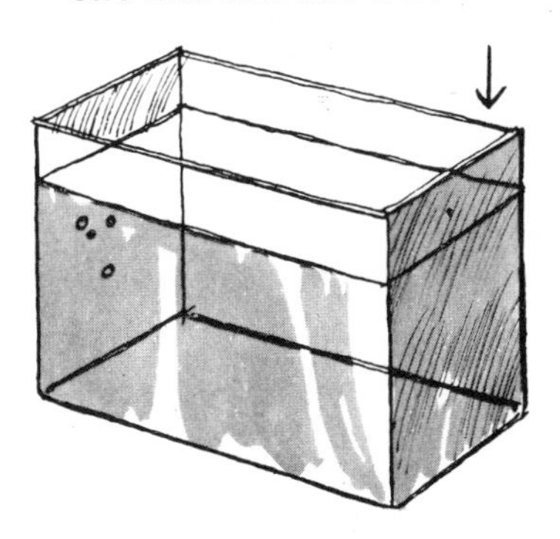

Animals and plants which live in water need air just as much as those living on land. There is air in water. Air is soluble, that is, it can dissolve in water. Some of the oxygen out of the air dissolves into the surface of the water. The dissolved air does not get very far below the surface. For this reason, it is better for aquarium tanks to be wide and shallow rather than deep. If the opening to the air is a small one, as it is in the old-fashioned type of goldfish bowl, the amount of air which can dissolve into the surface layer of water is not enough for the needs of fish and other water creatures.

The oxygen in water needs to be renewed constantly, if fishes are to live in it. As you read on pages 177 and 178 green plants, even water plants, give off oxygen in sunlight and this helps to replace the oxygen used up by animals and plants in breathing. It is necessary, therefore, to see that your aquarium tank is well stocked with green water plants, and you may also need to use an aerator to bubble air into the water.

The dissolved oxygen is washed over the leaves of plants which live in water and over the gills of fish, and so plants and fishes can take in the oxygen they need. These " water-breathers ", as we might call them, take in dissolved oxygen, while " air-breathers " take in " dry " oxygen.

Do not use a " goldfish bowl " like this, even for goldfish. The narrow opening does not allow enough air to reach the surface of the water

Stock your aquarium with good aerating water weeds such as : Water Starwort, Canadian Pondweed, Water Milfoil, Frog-bit (floating), Duck-weeds (floating), Watercress, etc.

ANIMALS AND THEIR NEED FOR AIR

Even in the egg, animals need air. If you examine a hen's egg when you have one for breakfast, you will see that the broad end where you crack the shell has an air space just below the curve of the shell. You can see this air space in the diagram, which shows a half-section of a hen's egg. The air space is between two thin skins, called integuments, and provides a developing chick with enough air for its life before birth. In the diagram, you can also see the positions of other parts of the egg.

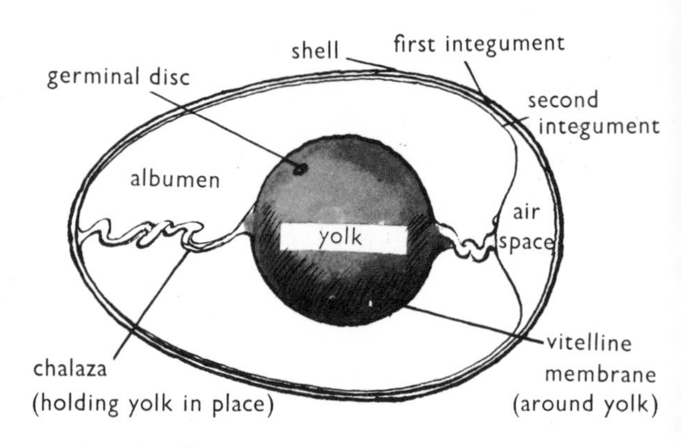

Look for the germinal disc from which a chick can grow in a fertile egg ; the yolk and albumen (white of egg) which supply the young chick with food (or supply you with food if you eat the egg) ; and the chalazae, which are thickened, twisted strands of albumen that help to hold the yolk in place.

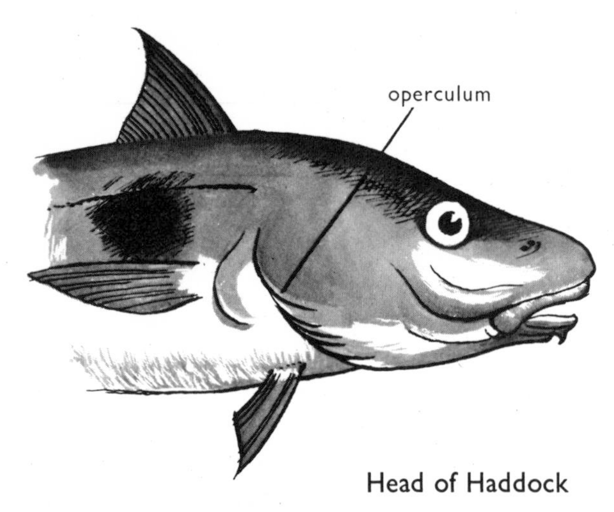

Head of Haddock

How a fish breathes

Fish, as you have read, are " water breathers " and use oxygen which is dissolved in the water. When a fish opens its mouth, it takes in water but does not swallow it. Instead, the water escapes by slits in the side of the fish's head. The flap called the operculum (see diagram) covers the gill opening on either side of the head of the fish.

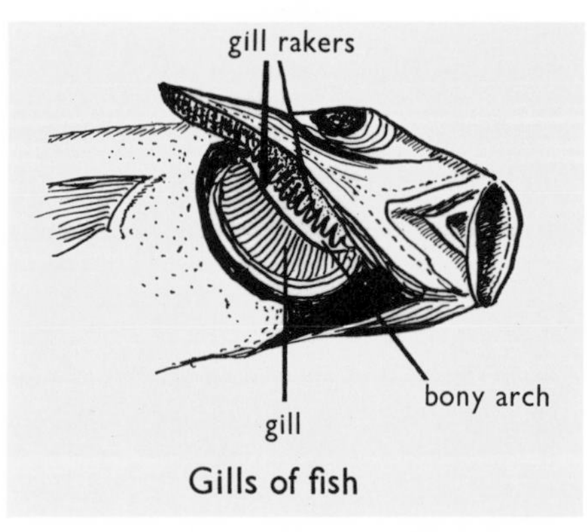

Gills of fish

If you were to lift up this flap you would see the gills, which are thin platelets full of tiny blood vessels. As fresh water passes from the mouth of the fish and out of the gill opening, it washes over these gills and oxygen is taken from the water and enters the blood of the fish.

How a frog breathes

Lungs of frog

You will now have realised that " breathing ", or " respiration ", is the way in which the body of an animal is supplied with fresh oxygen. Plants breathe, or respire, too, although they also give *out* oxygen.

In order to receive all the oxygen they need, animals have special organs. The organs concerned with the breathing process in fish are gills, which receive oxygen from the fresh water as it washes through them. Tadpoles, which grow up into frogs, also have gills when they are young, but as they gradually change from water animals into animals which live mainly on land, so their breathing apparatus changes for, on land, they will breathe *dry* air. The adult frog has *lungs* and its gills have disappeared. A frog, when swimming in water, must come to the surface to breathe. Sometimes you may see a frog sitting in the mud at the side of a pond, gulping in great mouthfuls of air. After it has taken this air into its mouth, it closes its mouth and nostrils and swallows the air into its lungs. This process is repeated and you can see the underside of its mouth cavity swell out and then collapse, as air is repeatedly taken in and swallowed.

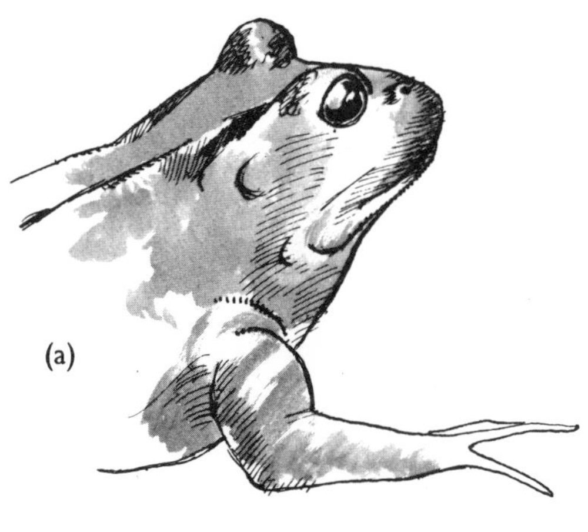

Diagram (b) shows a frog with air in its mouth

Compare this method with our *own* method of breathing. We do not need to *swallow* the air from our mouth into our lungs. The air we need is drawn through our nasal (nose) passages and windpipe because of movements of our rib muscles and diaphragm. These movements enlarge the cavity of our chest so that air can be drawn in. A frog can only enlarge the cavity of its mouth, which is quite a large one and can hold enough air for the small lungs of the frog.

HOW PLANTS BREATHE

Plants, as you now know, are valuable to other living things because they help to replace the oxygen which these living things take from the air in breathing, but we often forget that plants also must breathe and, just like other living things, they too take oxygen from the air. However, they *do* put some of it back and animals do *not*.

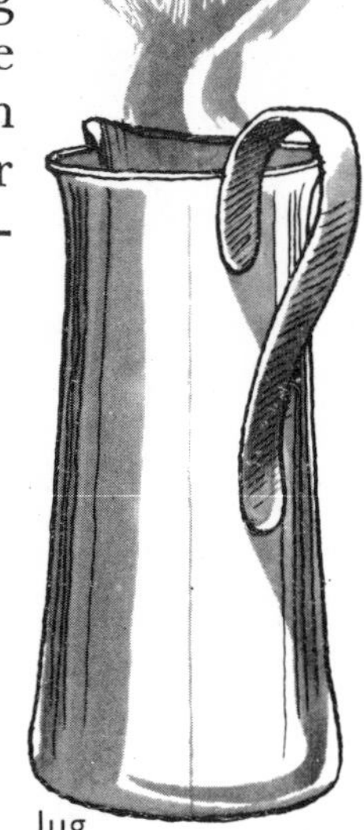

You can prove that plants have air in them. All you need is a beaker or a jam jar, some green leaves, and a jug of hot water (not boiling).

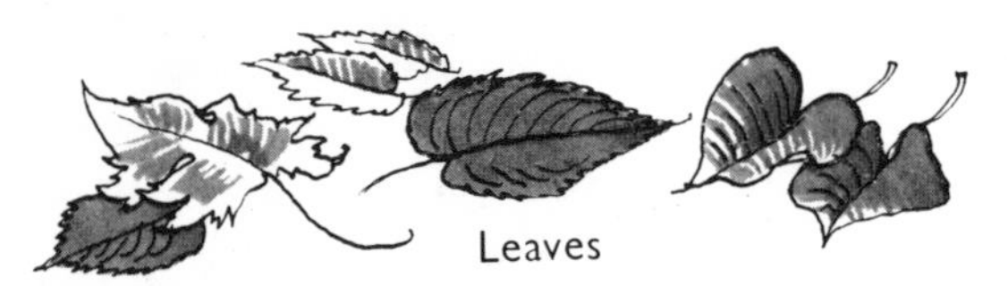

Leaves

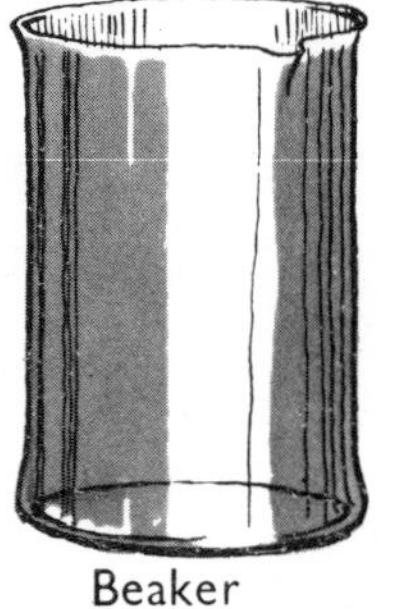

Beaker

Jug

Put the leaves in the beaker with some hot water and you will soon see bubbles of air rising from them. The bubbles of air have come from inside the tissue of the leaf. This air has been taken in by the plant.

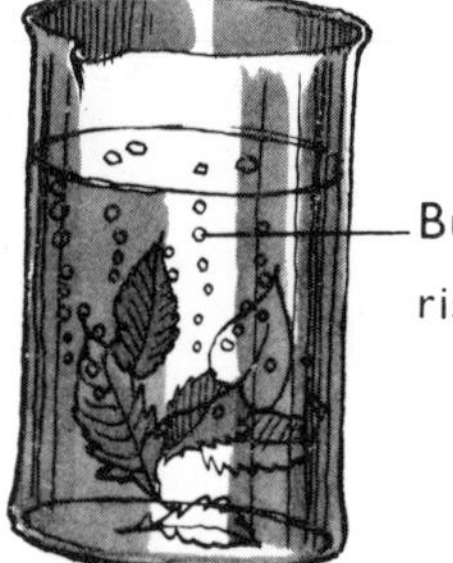

Put the leaves in the beaker and pour on hot water

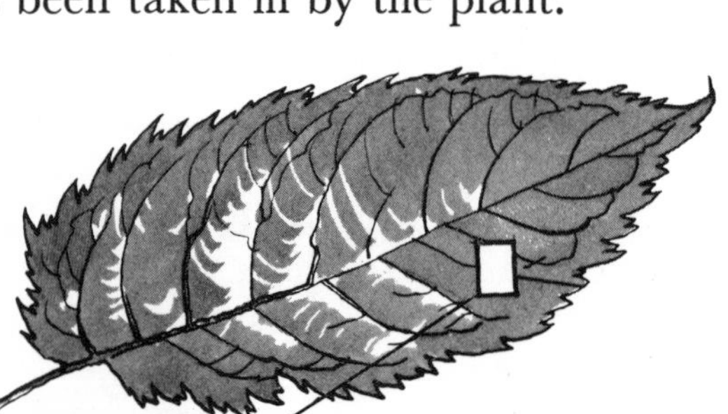

Just like all other living things, the plant has its own method and apparatus for taking in air.

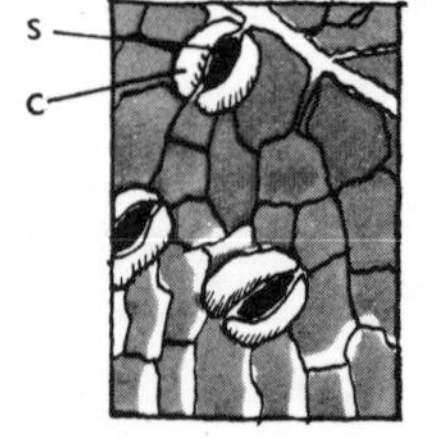

Part of leaf enlarged to show stomates

If you were to examine the under side of a leaf with a microscope, you would see tiny holes, or pores, called stomates (s). On either side of these little pores, you would see special sausage-shaped cells (c), which really *do* guard the pore, or stomate. By swelling out or collapsing they can open wide, or nearly close, the stomate. In this way the amount of air passing in or out of the leaf can be regulated.

HOW PLANTS TAKE IN WATER

The food of plants, which they make themselves, is manufactured out of the raw materials which the plants can get from the air, from the earth and from water. The plant makes use of the carbon dioxide from the air in making its food, and from the soil it takes water and the mineral salts which are dissolved in the soil water. Even without the mineral salts, a plant can make much of the food it needs from just carbon dioxide and water, and you will find that you can keep a green plant alive for quite a long time so long as you see that it has air and water.

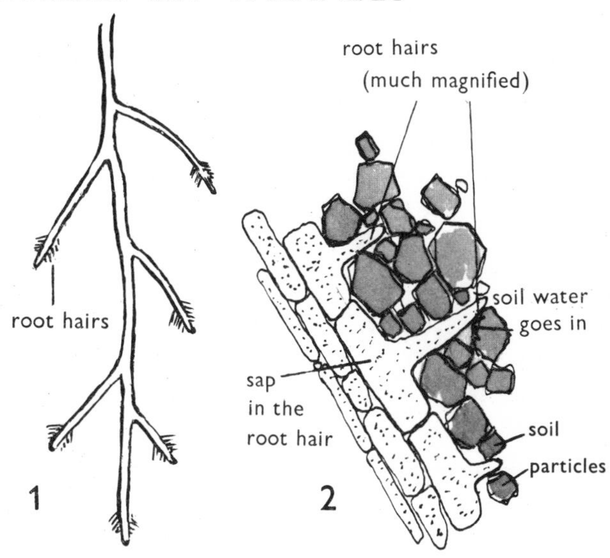

Plants take in the water they need from the soil by means of very small, thin hairs near the tips of their roots. If you wash the root of a plant and put it in a glass of clear water, you can often see these tiny hairs. (See Diagram 1.)

Diagram 2 shows you what the root-hairs look like if you look at them through a microscope. In this picture you will see that the root-hair comes into close touch with the particles of soil and also with the soil water, which clings around the soil particles. There is always liquid, known as " cell sap ", inside a plant, so that the root-hair is not empty but has some of the plant's sap in it. The soil water is thinner than this cell sap and, as thinner liquid tends to pass through living membranes and mix with thicker liquids, some of the soil water passes through the wall of the root hair and mixes with the cell sap. This is called osmosis.

Diagram 3 shows a model of a root-hair which you can make for yourself by scooping out half a potato and putting in a thick mixture of sugar and water. Stand this in a saucer of unsweetened water. If you have set up this experiment correctly, the water from the saucer will pass through the potato wall and mix with the sweetened liquid. After a time the potato will overflow.

Animals take air into their lungs, but they also give it out. The oxygen which is necessary to the life of an animal is taken from the air as it passes through the lungs, and other gases are given off.

Two gases which are given out " in the breath " are carbon-dioxide and water vapour.

It may take you a little time to get used to the idea that water vapour is a gas. Although it is still a form of water, it is invisible and it *is* a gas, as you read on page 172.

We can change the water vapour from our lungs back to water again very easily by breathing on to a cold mirror. Do this and you will see a mistiness on the mirror. (Diagram 1.) Touch the mirror and you will realise that the mistiness is moisture which you have breathed out.

You can also prove that you breathe out the gas, carbon dioxide. To do this, you will need a tumbler full of lime water and a glass tube or a drinking straw. Take a fairly deep breath and then bubble slowly through the straw into the lime water. (Diagram 2.) The lime water, which should at first be quite clear, will turn milky as you breathe into it. This is a scientific test which proves that the air which comes out of your lungs contains carbon dioxide, and this is also true of the outgoing breath from all animals.

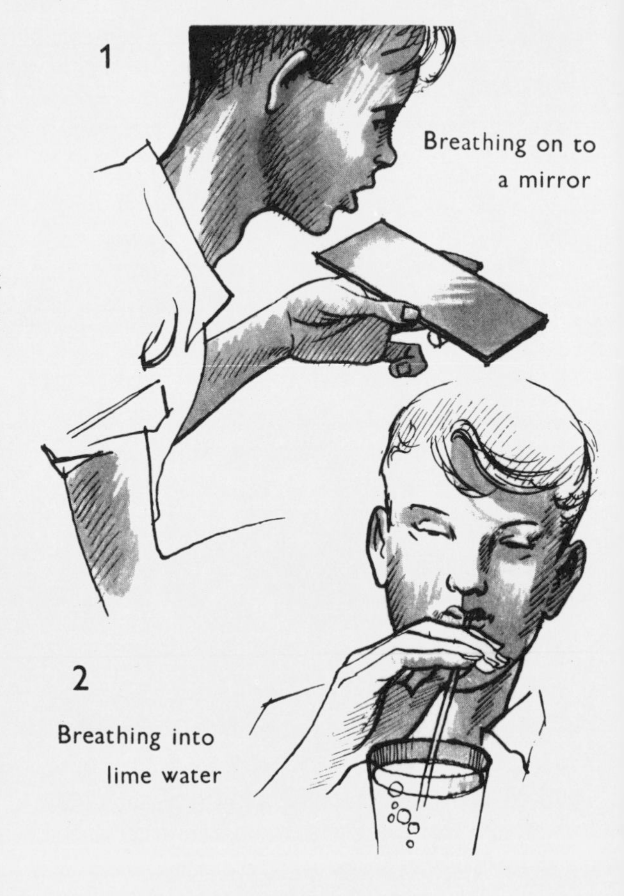

PLANTS

Like animals, plants also give off carbon dioxide in their breathing process. We can prove this by placing a green plant under a bell jar, as shown at the top of the next page, and hanging a small tube of lime water in the bell jar, catching the string between the stopper and the side of the jar. After a time the lime water should turn milky, thus showing the presence of carbon dioxide in the bell jar.

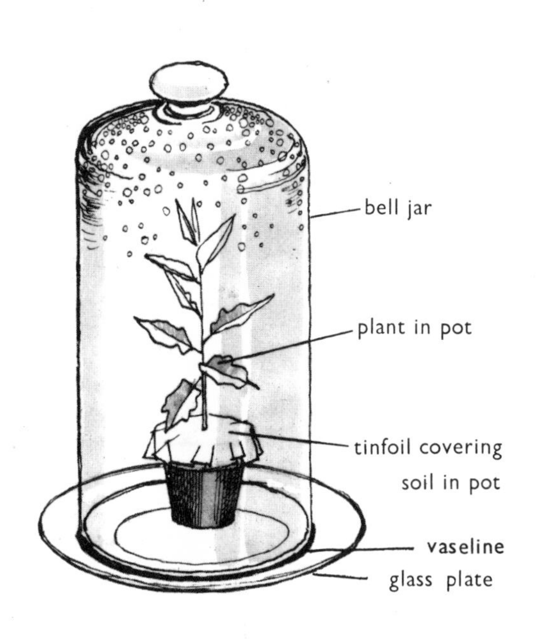

Plants also give out water vapour. In the picture at the top of this page you will see a plant under a bell jar. The rim of the bell jar has been coated with vaseline and the jar has then been placed on a plate to stop any moist air entering the jar. The soil in which the plant's roots are embedded is covered with tinfoil. If you leave the plant under the bell jar for only a few hours, you will soon be able to prove that water in the form of water vapour escapes from the plant, because this vapour condenses on the inside of the glass. At first there is a general mistiness on the inside of the glass, but soon quite large drops of water will appear.

This water has not come from outside or from the soil in the pot because, as you will remember, this was covered with tinfoil. Therefore the moisture must have come out of the plant itself. In fact it came, in the form of water vapour, from the small pores, or stomates, in the leaves. (Page 182.)

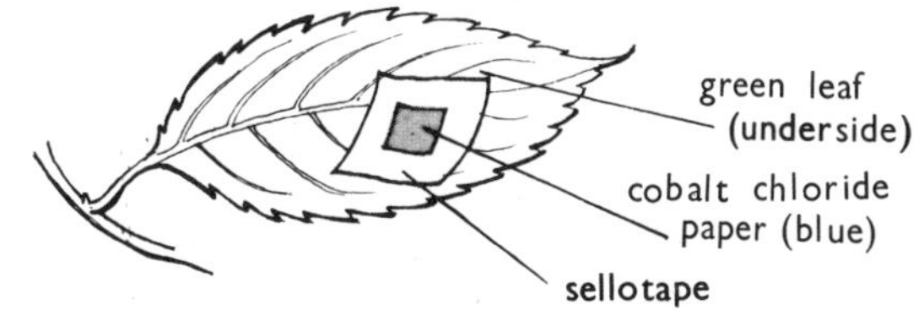

Gases enter and escape through the stomates. The blue cobalt chloride paper fixed on to the underside of this leaf will turn pink with the water vapour given off by the leaf. The leaf must be left on the plant.

The escape of water vapour from a plant is called *transpiration* and the changing of this water vapour into water drops on the cool surface of the glass is called *condensation*.

Condensation happens whenever moist warm air touches a cold surface or colder air. When a kettle boils (see diagram) we see this happening. The water vapour immediately emerging from the kettle spout is invisible until it reaches the colder air away from the warm kettle. Then it becomes visible as condensation takes place.

INVISIBLE THINGS IN THE AIR—SPORES OF MOULD

If you put a cork in an " empty " bottle, you are not really corking an empty bottle. The bottle is full of air and the air has thousands of invisible things in it. You can capture some of these invisible things and grow them.

Some of the invisible things in the air are mould spores.

AN EXPERIMENT TO GROW MOULD

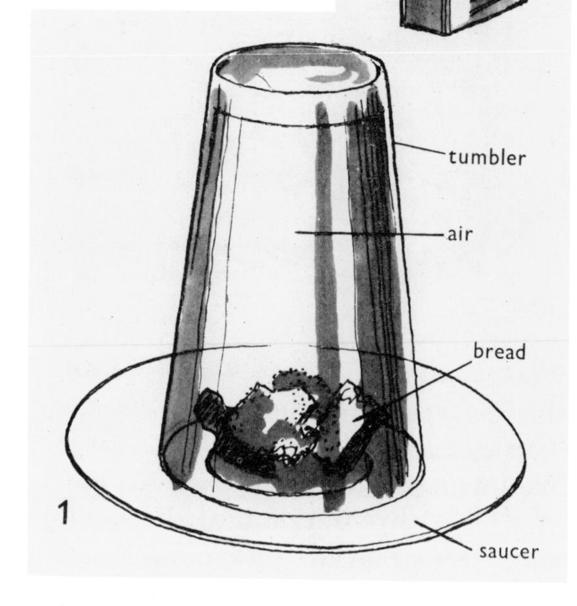

Put some pieces of bread in a saucer with a little water and place a glass tumbler upside down over the bread in the saucer. (Diagram 1.) The tumbler is full of air which contains many spores, and damp bread makes a good " growing " medium for mould spores.

You must now leave the experiment for a few days. Do not lift the tumbler during that time.

Soon the mould spores which are always present in air will begin to grow and feed on the damp bread.

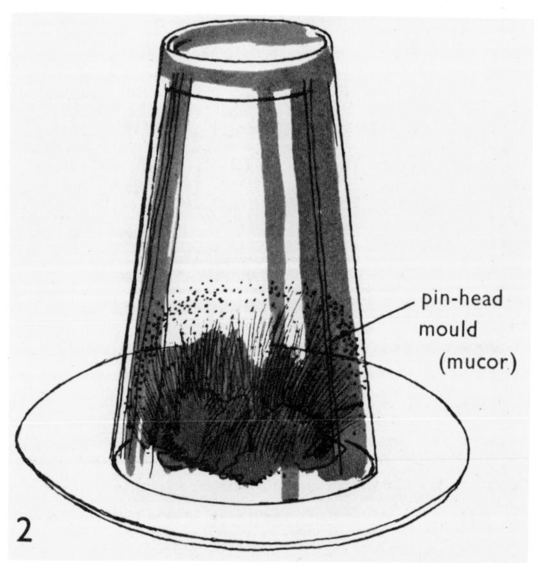

The bread will become covered with thin white threads and from these shoot up long white threads with little black heads, like tiny black-headed pins. (Diagram 2.) This " pin head " mould is called mucor.

If you carry out the experiment again using fruit instead of bread, you may find a growth of green mould called penicillium, from which the drug penicillin is made.

FUNGI

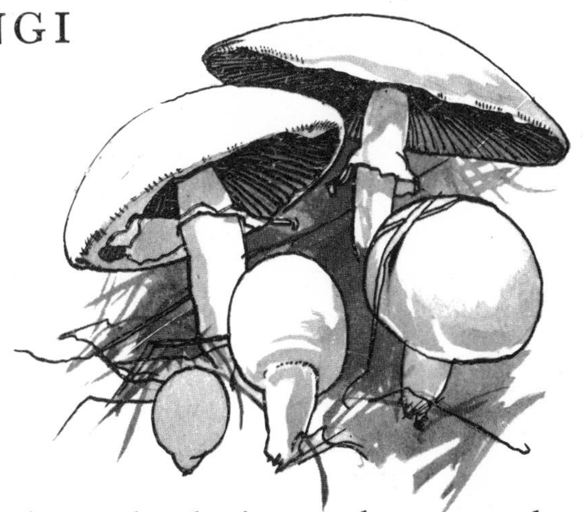

Fungi are different from other plants. They are not green. They have spores instead of seeds. They do not feed like green plants from air, water and mineral salts. They have to have specially prepared food which is, or has been, alive. This kind of food is called *organic*. The kind of food which other plants get from the soil and the air is called *inorganic* (not living). Animals must have organic (living or once living) food too, so that in their type of food, fungi resemble animals.

You saw that the pin-head mould, mucor, was taking its nourishment from bread made from wheat, which was once a living plant ; therefore this food is organic. The greenish mould, penicillium, also grows on a once-living food, preferably stale fruit or jam made from fruit. From one kind of penicillium comes the drug, penicillin, which kills the germs of pneumonia.

Fungi form an enormous group of plants which includes large things like mushrooms and other toadstools, as well as the moulds you have been developing under your glass tumblers. The toadstool types of fungi are known as *higher fungi* and the mould types as *lower fungi*. Among the higher fungi there is one type that you know very well. On this page is a picture of mushrooms. The mushroom is an edible (eatable) type of fungus. Many fungi are edible, but do not try to eat them unless you know a great deal about them, because some kinds are very poisonous.

Making a spore print

Mushrooms grow in grassy pastures, particularly where cattle have been ; they feed on the organic matter, or manure, in the soil. Mushrooms and toadstools reproduce by means of spores. Their spores are kept in special places, usually under the large " cap " or head of the toadstool. In the mushroom they grow on gills which you can find under the cap. It is difficult to see these tiny spores, but you can make a spore print of them by placing the cut-off head of a mushroom upside down on white paper. (See diagram.)

FOOD STORES IN GREEN PLANTS

Green plants can make their own food, using water, mineral salts and carbon dioxide as ingredients.

Some of this food they store in roots and underground stems, in bulbs, in corms and in seeds.

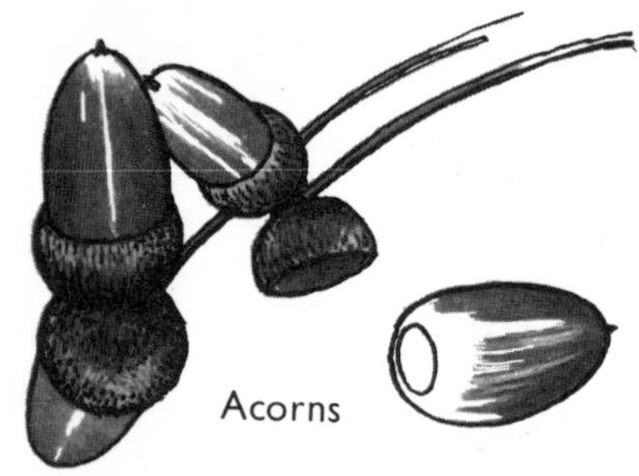

An acorn has food stored in it.

Beans and peas have food stores too.

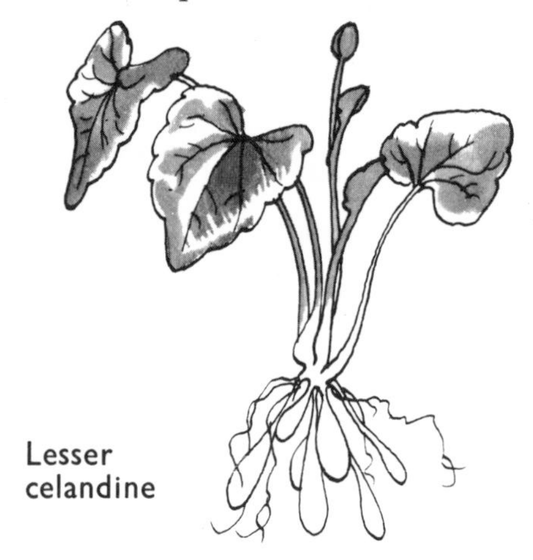

The lesser celandine plant has tuberous roots full of food.

The crocus has a corm full of food.

Daffodils, tulips and onions have bulbs full of food.

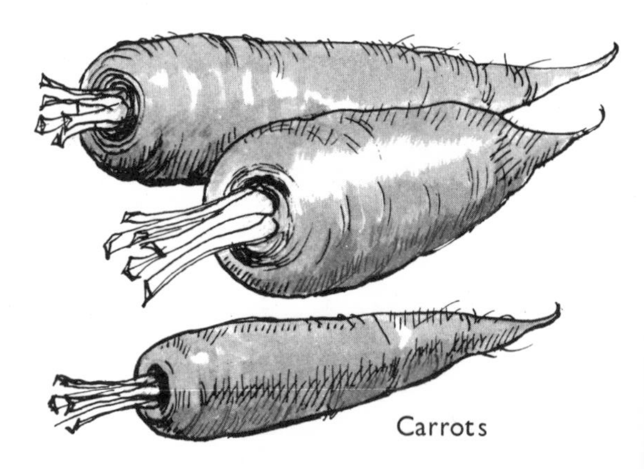

Carrots, parsnips and beetroots have large " tap " roots stored with food.

COLOUR REACTION TESTS FOR PLANT FOODS

Testing for starch

By means of simple chemical tests we can discover the presence of certain foods in these different plant stores. Some of the main types of foods found in plants are starches, sugars, oils and proteins.

Starches and sugars are similar to one another. They are both made up of carbon, hydrogen and oxygen in different proportions and, because of this, they are both called *carbohydrates*. We eat starches in bread, cereals and potatoes. We find sugar in fruits, and also in onions and carrots.

Testing for starch

A good test for starch is iodine. When added to anything which contains starch, it gives a blue-black colouration. Let us put iodine on pure starch first. Potato or rice starch consists of pure starch grains.

Put some drops of iodine solution on potato or rice starch grains and watch them change colour. They should turn deep blue or blue-black where the iodine has touched them.

Now that you have tried this colour test on pure starch, you can apply the test to many plant-food stores to see whether or not they contain starch. Try this same test on the plant-storage organs you see on the opposite page and enter the results in your notebook in table form.

Put a tick in the appropriate column according to the strength of the colour reaction.

Food store	Much starch	Little starch	A trace	None
Acorn				
Bean				
Pea				
Celandine roots				
Onion bulb				
Crocus corm				
Carrot				

Testing for sugar

A test for sugar can be made with Fehling's solution.

It is a good plan to try out this test first with pure grape sugar (glucose). You can buy glucose from a chemist. The sugar which we use for sweetening is usually cane sugar, which is different from grape sugar. Grape sugar responds better to the Fehling's test than cane sugar.

Warning: These tests should be made by a grown-up. Fehling's B is caustic and could burn your skin and clothes.

Using a spirit lamp or bunsen burner, heat a solution of grape sugar (glucose) in a test tube. When the liquid in the test tube is hot, add equal amounts of Fehling's solution A and Fehling's solution B. Take the test tube from the heat and watch the colour change. Brick-red precipitate of cuprous oxide forms when grape sugar (glucose) is present.

When you have proved that glucose (grape sugar) dissolved in water and heated with Fehling's solution gives a brick-red colouration, try out the same test with tiny pieces of different sweet fruits or vegetables. Some suitable ones are shown below. Put small pieces of the fruit or vegetable in a test tube containing water and repeat the test.

Testing for oil

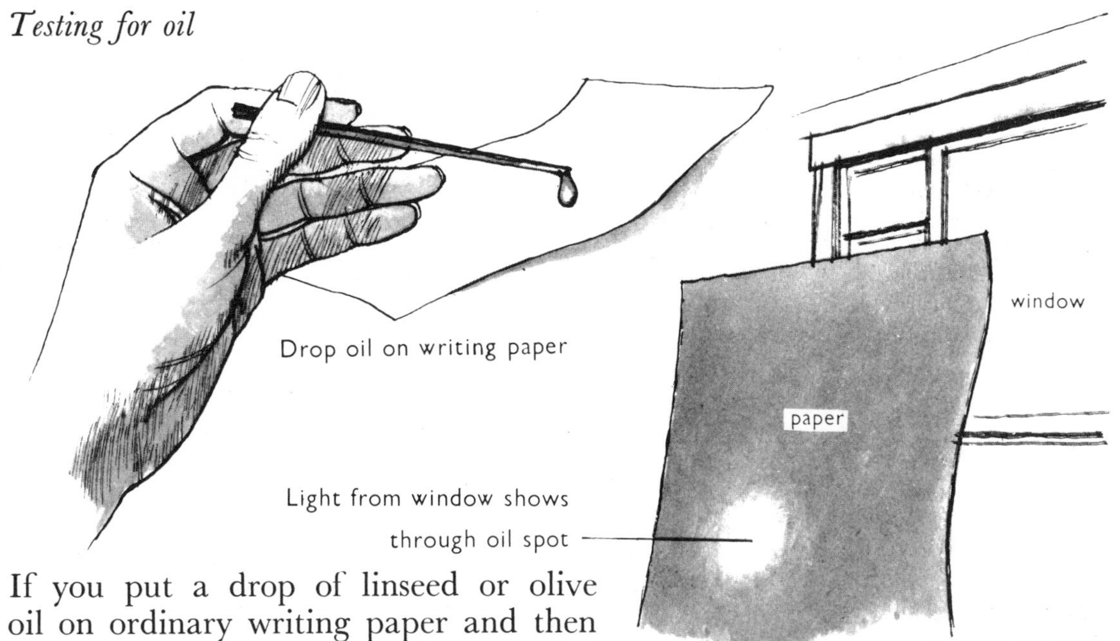

If you put a drop of linseed or olive oil on ordinary writing paper and then hold the paper up to the light, you will see that light comes through the oil spot. This mark is said to be translucent (" trans "—across ; " luce "—light). This is a good test for oil. If you cut open a nut and rub it on white writing paper, you will find that it makes a translucent mark, showing that nuts contain oil.

Testing for protein

Nuts also contain protein. Vegetarians, who do not take their protein food from meat, fish or eggs, often eat meals prepared from nuts and cheese, both of which contain protein. A colour test for protein can be carried out with two solutions which are very similar to the Fehling's solutions. These are :

(1) copper sulphate (the same as Fehling's A)
(2) caustic soda (which is contained in Fehling's B).

White of egg is probably the purest form of protein. It is called *albumen.*

Use egg albumen for your first test. Shake up some white of egg in a test tube with some distilled (very pure) water and a grain or two of salt. Add an equal amount (about three drops) of each solution (copper sulphate and caustic soda) and a violet or purple colour should appear in the test tube. This proves that there is protein in white of egg.

Now try this same test for protein with nuts and plant seeds. Take care, caustic soda can burn your skin.

Another good test is Millon's reagent which goes pink when protein is present.

THE CHEMISTRY OF AN EGG

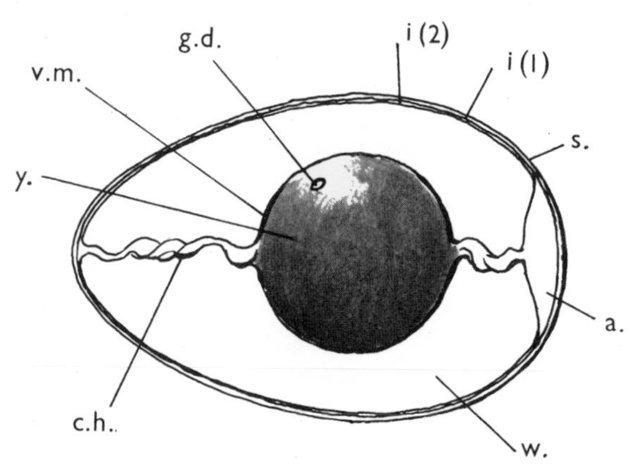

Section through hen's egg

Key to diagram :

- y. yolk
- v.m. vitelline membrane of yolk
- g.d. germinal disc from which chick can develop
- i (1). first integument (skin)
- i (2). second integument
- s. shell
- a. air space
- w. white of egg (or albumen)
- ch. chalaza—a twisted white cord which helps to suspend the yolk

As you have already learnt, white of egg is one of the purest forms of protein. It turns quite a deep purple if we test it with copper sulphate and caustic soda (see page 191). Do you remember the different parts of an egg? Look at the diagram above and its " key " at the side.

The yolk of an egg contains fat, but it would be difficult for us to prove this by a " spot on paper " test, as we did with nuts, because an egg is so sloppy. We can also test for fats with osmic acid, which gives a black colouration when added to fatty substances. If you separate out a little yolk of egg in a watch glass (or any small vessel) and put one drop of osmic acid into it, you will soon see this black colour.

Now, let us discover something about the chemistry of an egg shell.

Put some chippings of egg shell in a test tube with some water and add dilute hydrochloric acid (H.Cl.—see Diagram 1).

Bubbles of gas (carbon dioxide) arise from the egg shell almost as soon as the acid touches it and an " effervescence ", or " fizzing ", takes place (Diagram 2). The effervescence proves that the egg shell is made from some limy or chalky material.

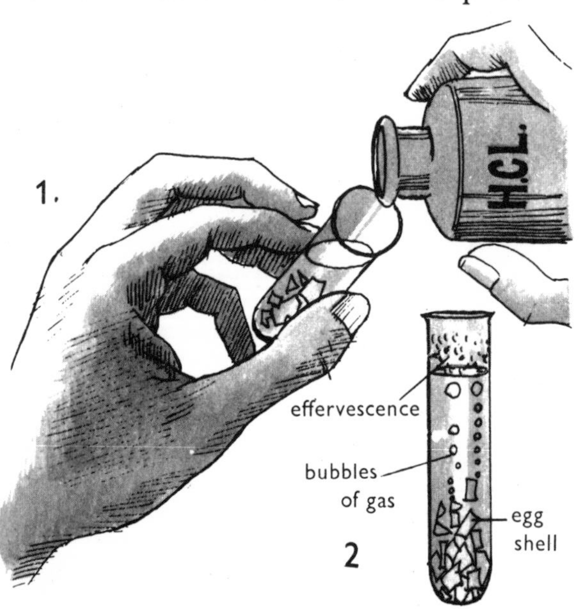

EGGS AND SEEDS

In one way, eggs and seeds are very much alike. Before they can grow into new individuals, they have to be *fertilised.*

Eggs

Eggs are fertilised by sperms.

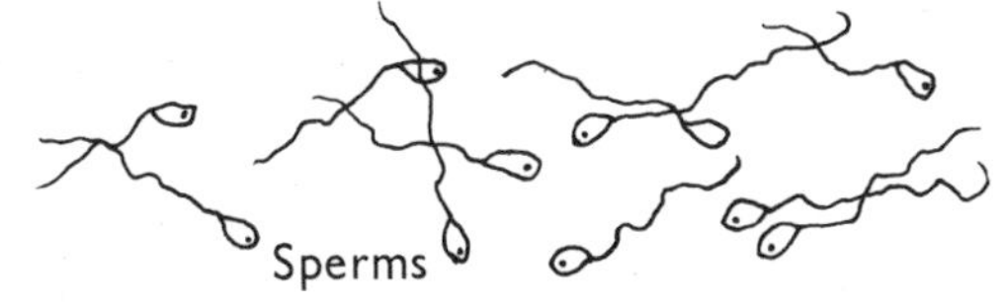

Sperms usually have little tails and can only be seen under a microscope.

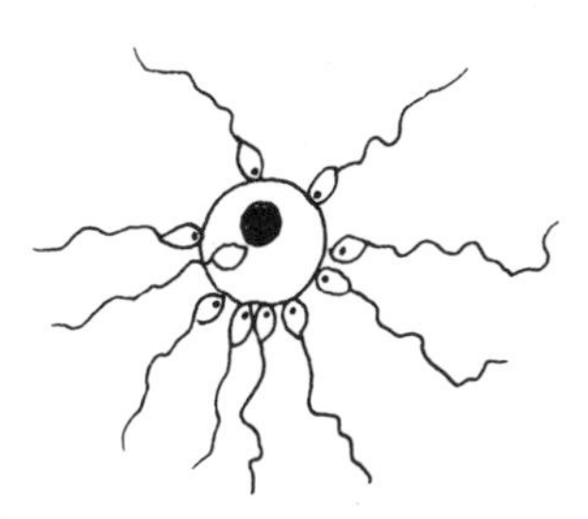

Here are some sperms swimming round an egg (not a hen's egg : it could be a starfish egg). *One* sperm penetrates into the egg.

When this happens, the egg is fertilised.

1 cell 2 cells 4 cells 8 cells

The egg now begins to divide into 2 cells, then 4 cells, then 8, 16, 32, 64 and so on until it begins to form a little body. This looks rather like a small mulberry at first. From this the young animal develops.

Seeds

Unripe seeds, known as ovules, are fertilised by pollen grains.

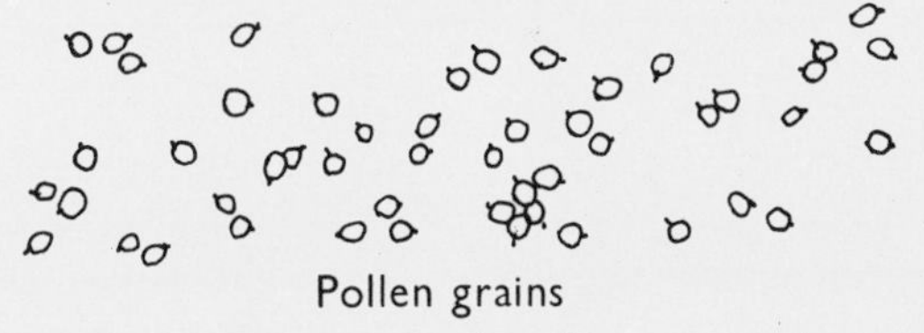

Pollen grains look like specks of yellow dust.

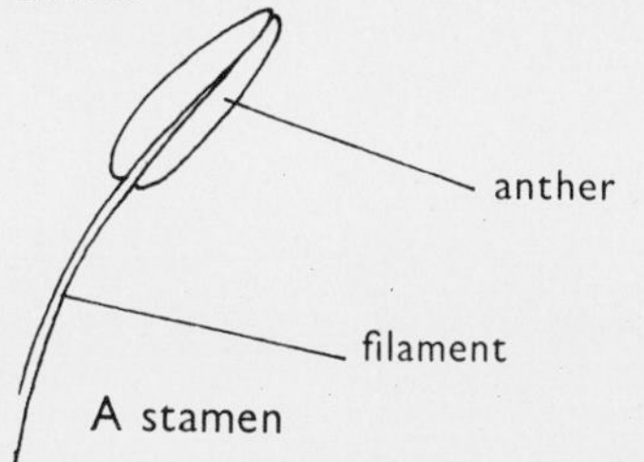

Pollen grains are developed inside pollen sacs in the anther of a flower stamen.

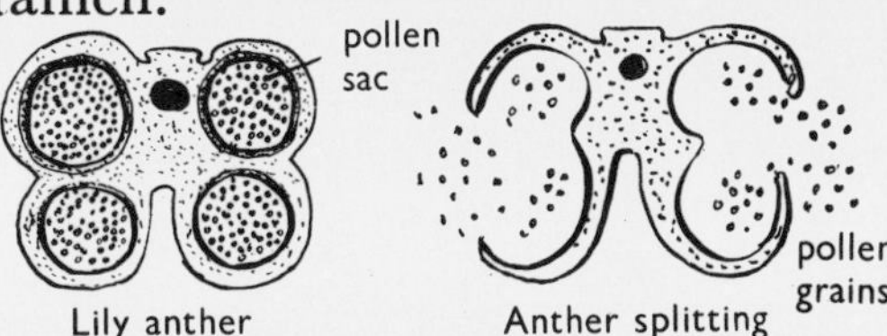

Section across lily anthers

If you cut across the anther of a stamen, you usually find four sacs inside it. The pollen grains are in these sacs.

You will see how pollen grains fertilise ovules on the next page.

Remember : All the drawings on this page are much larger than life size.

HOW SEEDS ARE FORMED

Some flowers, such as the yellow horned poppy, have one stigma. This stigma has two lobes. Other flowers, such as the buttercup, have more than one stigma.

Below is the primrose, which has one undivided stigma.

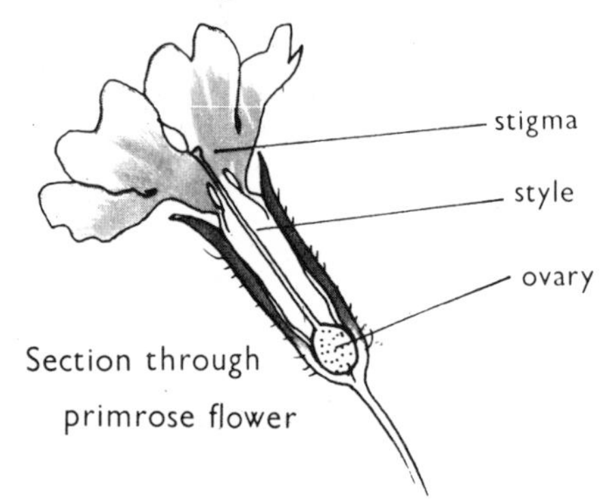

Section through primrose flower

Yellow horned poppy

Below the stigma is the long style and well down, at the base of the style, is the ovary, or seed box.

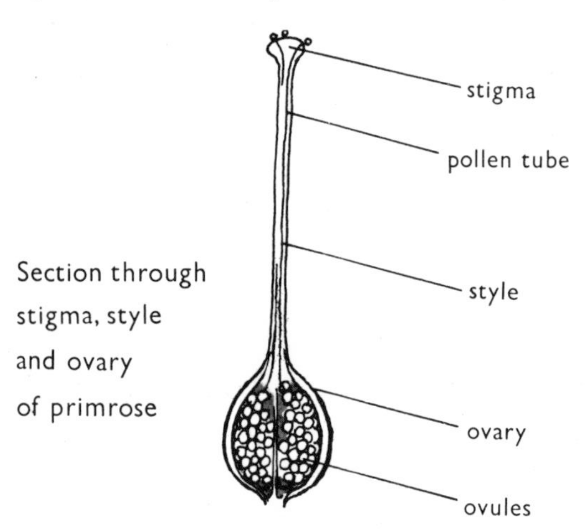

Section through stigma, style and ovary of primrose

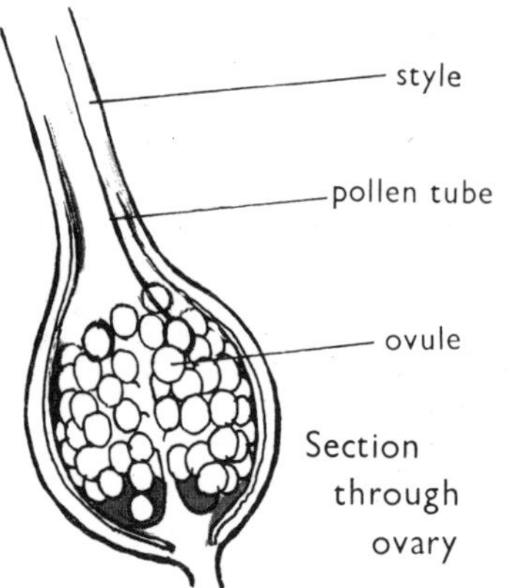

Section through ovary

In the drawing above, you can see that some pollen grains have landed on the stigma. One pollen grain has begun to grow down from the stigma towards the ovary (seed box).

The pollen tube grows until it reaches an ovule in the ovary. This ovule will now become a seed. The other ovules will also become seeds if pollen tubes reach them.

HOW SEEDS ARE SCATTERED

Here on the right is the fruit of the *field* poppy. This fruit has several stigmas which are joined.

All parts of this fruit are joined to form one big fruit, inside which are the seeds. As the long stem of the poppy head sways in the breeze, the seeds are shaken out of the pores at the top of the fruit. By this method (the "pepper pot" method) the seeds are dispersed (scattered).

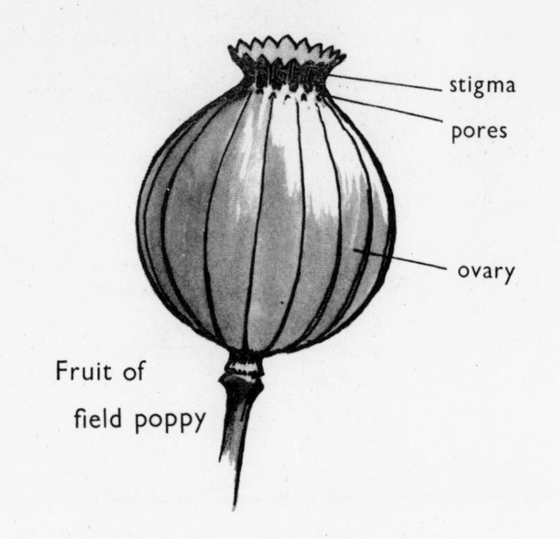

Fruit of field poppy

The scarlet pimpernel has fruits like those shown in the picture on the left. When the fruit is ripe, the top part cracks open and comes away like a little cap.

The shiny little seeds inside soon fall out of the cup which holds them.

Fruit of scarlet pimpernel

The willow herb fruit opens by splitting into five valves.

The seeds which escape all have tiny parachutes which carry them away on the breeze.

Fruit of willow herb with airborne seeds

THE DEVELOPMENT OF A BABY BIRD

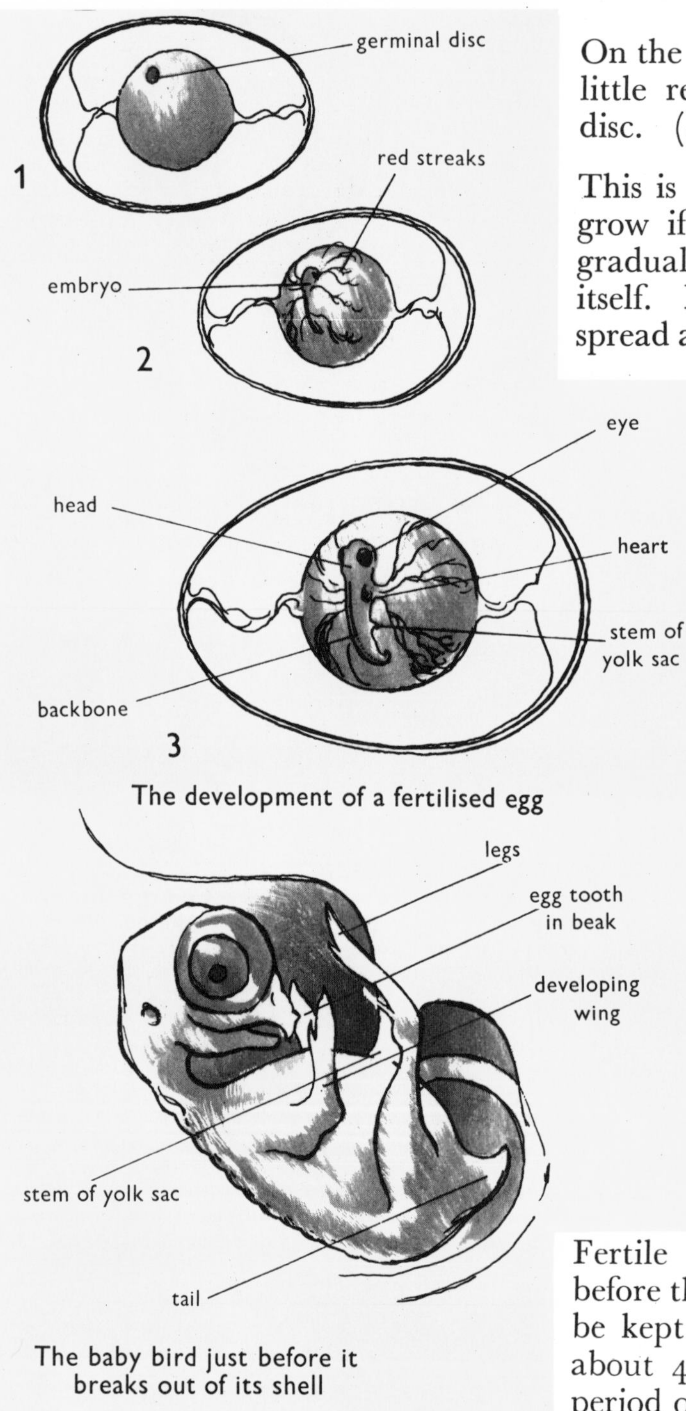

The development of a fertilised egg

The baby bird just before it breaks out of its shell

On the yolk of a fertile egg there is a little red spot called the germinal disc. (See Diagram 1.)

This is the embryo which begins to grow if the egg is kept warm. It gradually absorbs all the yolk into itself. Little red streaks, like veins, spread across the yolk. (Diagram 2.)

In Diagram 3 you will see that the embryo has developed further. This drawing has been made larger than the others so that you can see it clearly. The embryo is now larger and you can see its head with a developing eye. You can also see its backbone and its heart.

This is what the baby bird looks like just before it breaks out of the shell. On its beak it has a little lump called an egg-tooth with which it breaks open its shell. The yolk-sac is still fastened to its stomach and it has tiny wings (with no feathers yet) and legs and a tail.

Fertile eggs must be incubated before they can develop. They must be kept warm at a temperature of about 40 degrees Centigrade for a period of 21 days.

INCUBATING EGGS

Under natural conditions the mother hen keeps the eggs warm with her body until they hatch.

Poultry farmers use incubators. These are heated containers in which the air around the eggs can be kept at just the right temperature for hatching the eggs.

A mother hen with her chicks

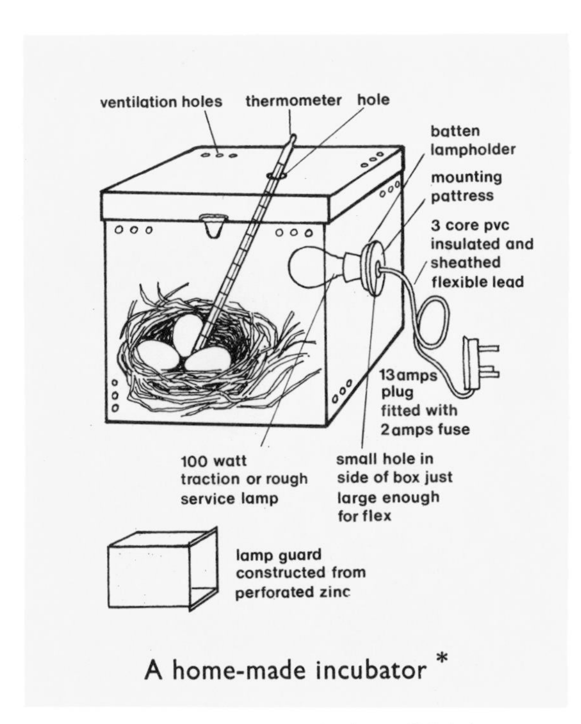

A home-made incubator *

You might find it interesting to try incubating some eggs for yourself.

The home-made incubator shown here is made from a strong wooden box with a lid. Small air holes allow the air to circulate gently. A 100 watt bulb should provide an adequate minimum temperature of approximately 100° F for egg development. Heat will also be retained by the straw around the eggs. The bulb should be well above the straw and protected by a lamp guard (see lower diagram) which can be screwed to the box over the lamp.

The eggs will need to be turned twice a day.

* The lampholder should be a " high temperature " type batten holder mounted on a pattress of British Standard Specification. The 3-core pvc insulated and sheathed lead should be connected to the terminal block mounted on the pattress and thence to the lampholder via heat resistant or insulated leads. The lamp guard should be connected to the earth (green/yellow) wire of the flex.

WATCHING BIRDS—A BIRD "HIDE"

Bird watchers usually have a "hide". A hide is a canvas tent with a "peep hole", a little slit in the canvas through which you can watch birds without the birds seeing you.

You can watch birds quite well from the shelter of some bushes. Remember that you must keep very still.

A bird "hide"

The time and place must be right for bird watching. Spring is a good time when birds are busy nest-building and rearing young.

Begin by watching birds in the garden. Put food out first to attract them.

Watch birds which come to collect food in your garden and use field glasses, if possible, to watch flying birds

RECOGNISING BIRDS

Before you can describe a bird that you have seen, you must know the names of all the parts. Look at the drawing below and learn the names of the different parts of a bird's body.

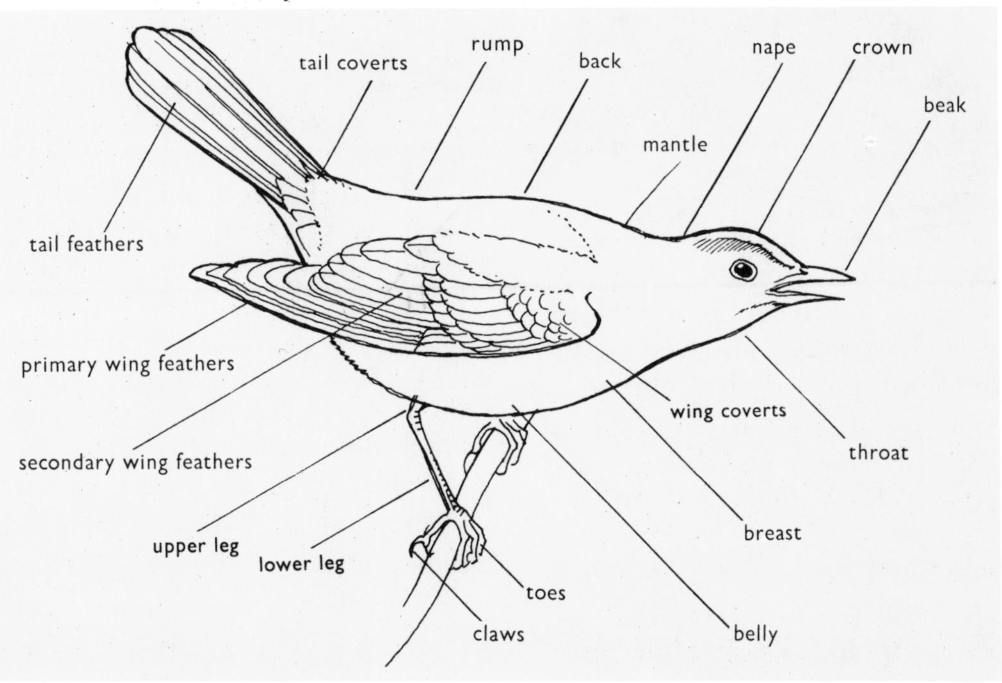

Now look at the descriptions below. Compare the parts mentioned with the picture above. Study your bird identification books and discover what each bird is called.

Bird 1
Slate-blue head
Chestnut mantle
White flecks on wings

Bird 2
All black plumage
Yellow beak
Dark brown legs

Bird 3
Olive-brown back
Orange-red breast
White belly

Bird 4 (water bird)
Bottle-green head
Curled black tail feathers
Chestnut breast
Blue streaks under wings

Bird 5
Blue-black back
and breast band
Forked tail
Chestnut throat
White underparts

Answers
1. Chaffinch
2. Blackbird (cock)
3. Robin
4. Mallard (drake)
5. Swallow

LOOKING AFTER PET ANIMALS

Golden hamsters like these make good pets

If you are a really good naturalist, you need to observe and record the habits of as many living things as possible. Some animals can be kept as pets and then you have a good opportunity for watching them.

Golden hamsters

Golden hamsters are excellent animals to keep as pets. They store food in a corner of their cage so that they do not need feeding every day. Feed your hamster mainly on cereals, such as soft oats, and fresh green vegetables. Female hamsters with babies should be given milk.

The golden hamster lives for about $2\frac{1}{2}$ years. This does not seem very long, but during that time you can see a baby hamster grow up and perhaps have babies if you keep a male and a female together for a short time.

There is a diagram below of a suitable type of cage hutch to make for your hamster.

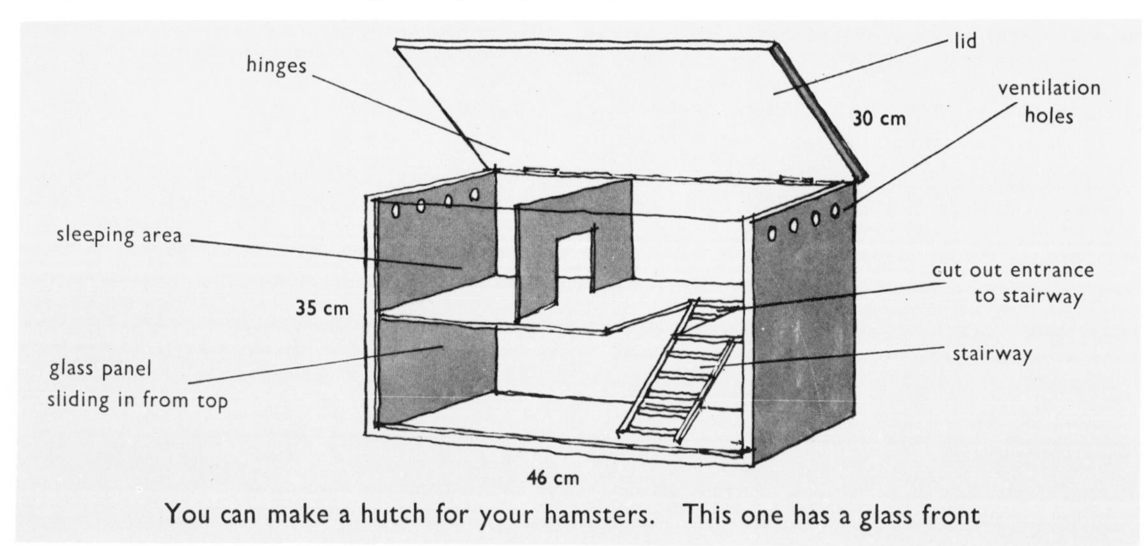

You can make a hutch for your hamsters. This one has a glass front

This photograph shows Barbara exercising on the lawn. On the right is a rain gauge

Rabbits make nests for their babies and Barbara made a " baby nest ", with straw which we gave her, every spring, whether she was due to have babies or not.

Rabbits and guinea pigs

Rabbits make good pets. It is wise to let them out of their hutches occasionally, particularly if you have an enclosed piece of grass on which they can run about, nibble the grass and take exercise.

Rabbits and guinea pigs can be exercised together and in fact they can be reared together in the same pen, with separate hutches. The rabbit and guinea pig pictured together here represent actual animals which became great friends. The rabbit even helped the guinea pig to rear her babies, and, although rabbits are silent animals, this particular rabbit (named Barbara) used to utter soft grunts in answer to her guinea pig friend's constant chatter.

Rabbits, guinea pigs and most animals with gnawing teeth are vegetable feeders and must have their ration of greens and cereals each day. They do not drink much water, except in hot weather. When they have young to feed, they should be given some warm milk. Rabbits appreciate a warm bran mash in winter and in wintertime too you must see that they have extra straw bedding and that this is kept dry and clean.

Rabbits and guinea pigs can be good friends

INSECT-HUNTING

Look for insects and other small creatures in the wood. You might find some of the ones shown above. These are :

1. Dor beetle (burying beetle)
2. Moth pupa
3. Devil's coach horse
4. Wood ant
5. Click beetle
6. Spider
7. Caterpillars

By digging at the roots of trees or beating the bushes with a stick so that caterpillars and other creatures can fall on to a white sheet of paper or cloth (see diagram), you can collect a number of insects to study.

Put your specimens in a polythene bag, with a little leaf mould, and bring them home for further study.

The caterpillars could be collected in an aerated cardboard box together with some of their food plant.

The spider, which you can see in the middle of the web, is *not* a true insect. It has eight legs. True insects all have six legs. The caterpillars and the pupa are early stages of insects. The caterpillars are the larvae of moths or butterflies. These will turn into pupae before they become insects. Several insect pupae and larvae are found in soil.

You will not be able to recognise all the insects you find because there are thousands of different ones, but you may be able to recognise *types*.

SOME OF THE MAIN INSECT TYPES

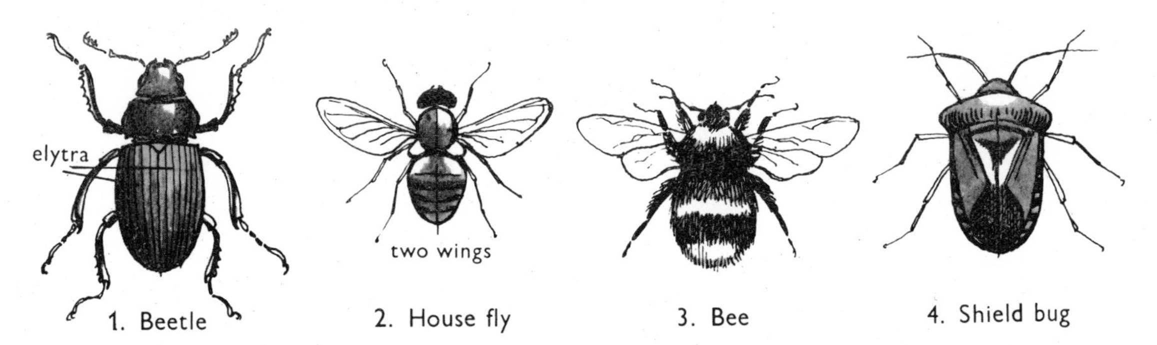

1. Beetle 2. House fly 3. Bee 4. Shield bug

1. Beetles always have a straight line down the middle of their back. This is made by the meeting of the two hard wings (elytra) which come together in a long straight line. Under these hard wings are two membranous ones.

2. A house-fly has only *two* wings. Bluebottles also have only two wings.

3. Bees, ants and wasps are the highest forms of insect and are fairly intelligent and live in communities (groups). Bees have large hairy bodies and four delicate wings; they take nectar from flowers.

4. Bugs are often found as pests on plants, and sometimes in houses. Their wings appear to be crossed over and in fact half of each outer wing is horny and the other half is membranous.

5. Grasshoppers, cockroaches and earwigs belong to related families. Their bodies and wings are long and straight. Grasshoppers have long hind legs for jumping. Female cockroaches have no wings.

6. Dragonflies spend the first part of their life in water, but then leave the water and change into flying insects with beautiful membranous wings. They have long bodies and look rather like aeroplanes in flight. They are always found near water.

7. Moths and butterflies are very much alike, but you can recognise a moth by its soft furry body and by its feathery antennae. When asleep it folds its wings flatly across its back. A butterfly holds its wings upright.

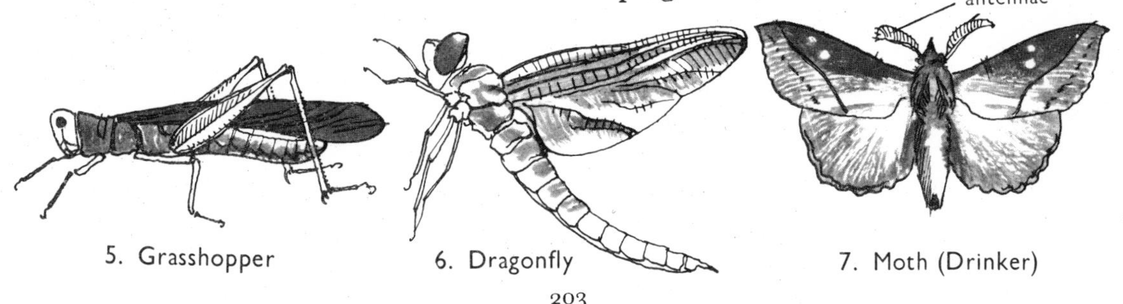

5. Grasshopper 6. Dragonfly 7. Moth (Drinker)

STILL DISCOVERING INSECTS

Here are some more ways of looking for them.

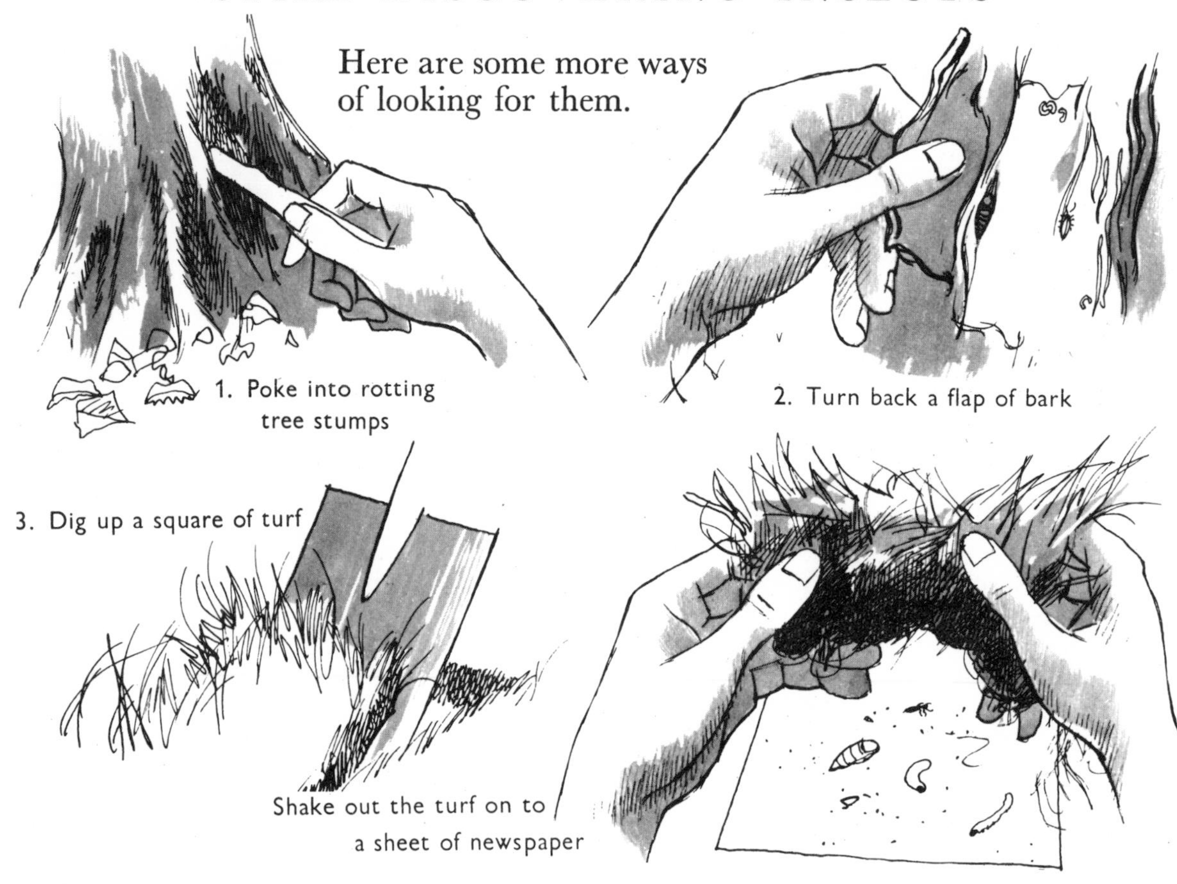

You may find larvae while you are looking for insects.

On the left is the larva of the wood tiger beetle which lives underground in a burrow.

Wood tiger beetle larva

Below it is the stag beetle chrysalis. Its larva feeds on rotting wood. You may find this creature as you poke about in rotting tree stumps.

Stag beetle chrysalis

On the right is a " subterrarium " which you could make for your larvae. It is made of two pieces of glass slotted into a wooden frame and is filled with soil.

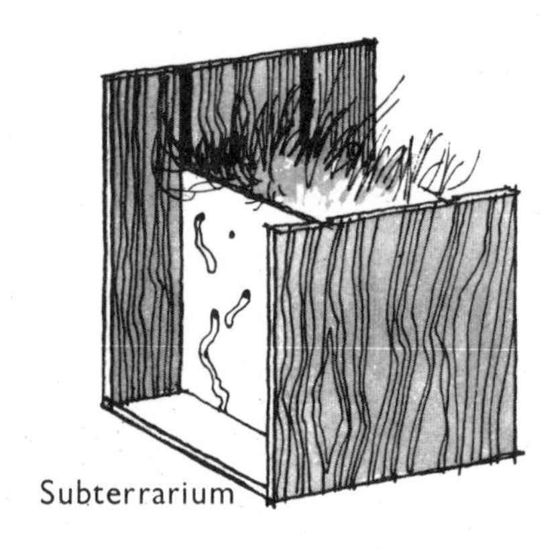
Subterrarium

HOW TO MAKE A CATERPILLAR-REARING CAGE

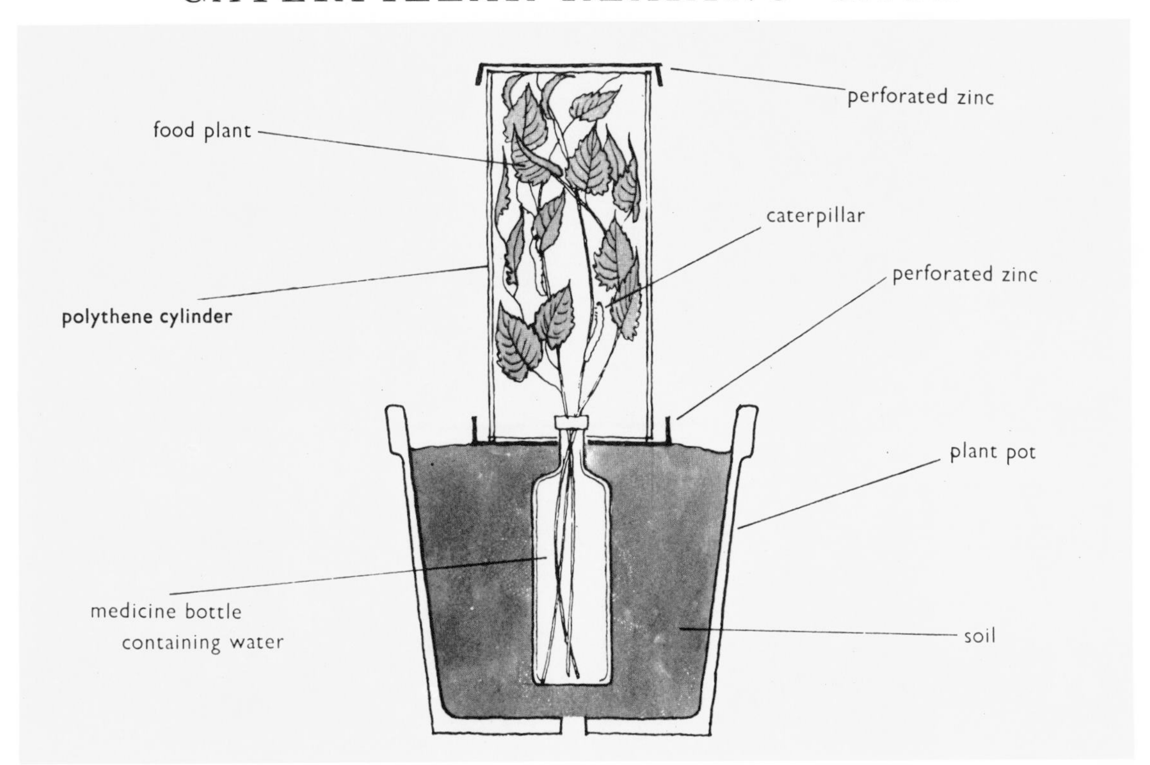

You can make this observation cylinder for watching the development of your caterpillars until they pupate or become chrysalids.

Roll a piece of polythene, about 20 cm by 35 cm, into a cylinder and fasten the join with sellotape.

The polythene cylinder stands on top of a plant pot of soil. The food-plant, on which the caterpillars feed, stands in a medicine bottle of water embedded in the soil with its neck just above the surface (see diagram). Perforated zinc is used to cap the cylinder. You can also place a sheet of perforated zinc at the base of the cylinder as long as you cut a hole for the stem of the plant, but this is not essential and, as some caterpillars burrow into the soil when they are about to pupate (go into a sleeping stage), it may be as well to leave them a little soil surface.

Remember that the food plant you put with your caterpillars must be the same as the one on which you first find them feeding, because caterpillars will usually feed only on the kind of leaves upon which they develop from the egg. For instance, silkworms which began life on mulberry leaves will not eat lettuce.

THE EQUIPMENT YOU WILL NEED FOR FIELD-WORK

This boy has a small haversack with him. You will find a haversack very useful. Some of the things you may need to carry in your haversack are a sandwich tin for your lunch, a bottle of fruit juice (you can get very thirsty when out collecting), a torch, a knife, some string, some polythene bags, a pocket compass, a hand lens, some little tubes, bottles and matchboxes, your field note-book and pencil, and a trowel. If you can get it in, a long tin box is useful, but your polythene bags will be a good alternative.

Now why do you want these things ?

Remember that you may be searching beneath undergrowth or looking into holes or caves, so you will need a torch.

You may have to dig for ferns or insects, so you will need a trowel.

You need to keep your plant specimens fresh, and a tin would be very useful for this. You could use your sandwich tin after lunch, or you could put your specimens in the polythene bags.

For small specimens of any kind, you will find that your specimen tubes and matchboxes are very useful.

You need a compass to check upon your direction, for you can easily get lost in a wood. It is a good plan to have a map with you too.

You may have to go into damp or rough places in search of your quarry so wear good watertight shoes or boots. If you decide to wear wellingtons like the boy in the picture, put on some thick socks first so that your feet will not blister.

HOW TO MAKE YOUR OWN POND NET

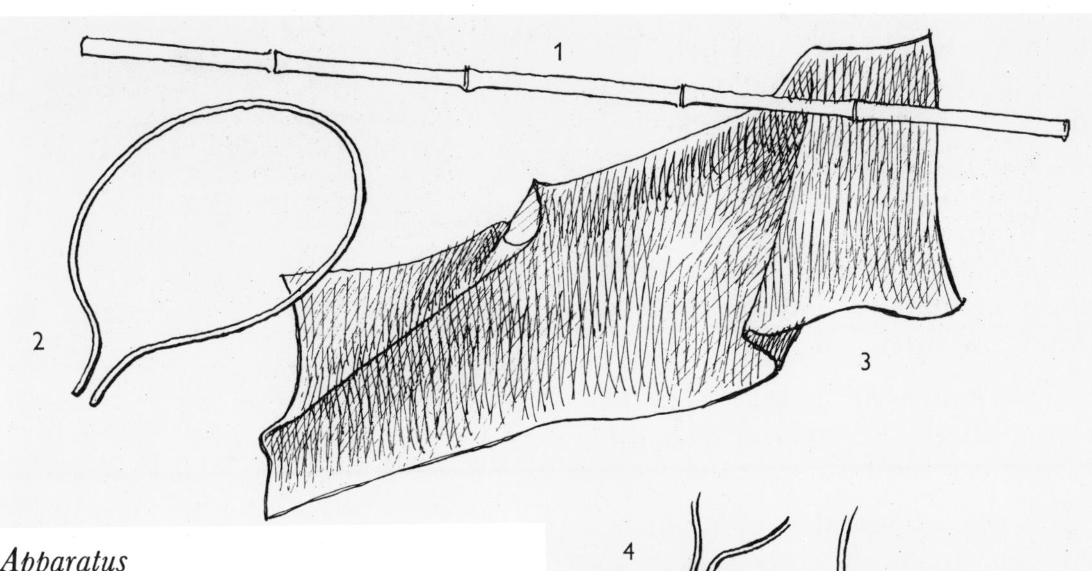

Apparatus

You need a strong stick of bamboo (1), a strong piece of wire which you bend into a shape like that shown in Diagram 2, a piece of closely woven muslin (3), two used match sticks, some fine binding wire and some strong linen thread.

Directions

When you have made the wire framework of the net (see Diagram 2), push the two ends into the hole in the end of the bamboo rod (Diagram 4). Push the ends farther in still (Diagram 5) and wedge them with the two match sticks (m).

Hammer the two ends and the match wedges well down into the hole and do not worry if the bamboo splits a little because you are now going to bind it with binding wire (w).

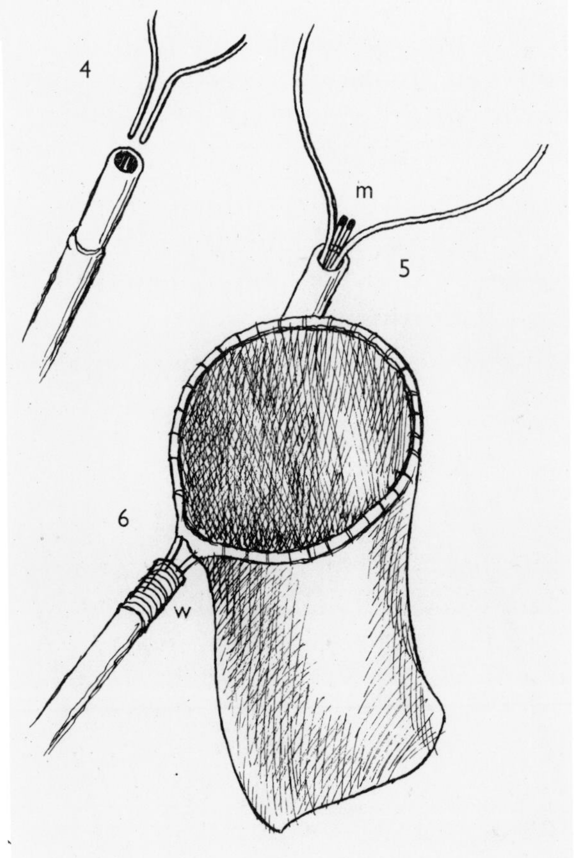

Lastly, make the cloth rectangle into a bag and sew it on to the wire framework with strong linen thread.

The next three pages will help you to name the flowers that you find. You will then be able to keep a record of these flowers. There are several ways of doing this. You can press your flower specimens and mount them in a book. Each flower should be named and the date of finding it written beside it.

You can enter the names of plants in your diary, giving the date and place of finding for each one.

If you like sketching and painting, you can make a record of your flowers by drawing them and colouring the drawings.

You can make a scrap-book record, sticking pictures of flowers you know, or have discovered, in a notebook with unlined pages.

You can keep a card index (page 215).

A book for mounting flowers

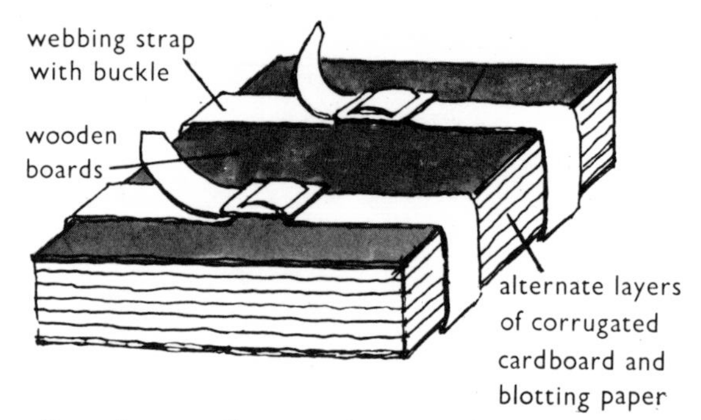

Pressing and mounting specimens

You can make a good flower press with two wooden boards strapped together with webbing or leather straps, or with a discarded buckle belt. Use two straps if possible because this helps to make sure that the flowers are being evenly pressed.

Between the two wooden boards put alternate layers of corrugated cardboard and either newspaper or botanical drying paper, both of which are better than blotting paper. If you can get nothing else, buy the very thin, cheap kind of blotting paper, coloured rather than white. Thin pressing paper is more porous than thick and allows air to circulate better, while darker colours help to preserve the flower colourings.

The book for mounting your flowers and grasses after they are pressed can be made from sheets of drawing paper or it can be a drawing book. Mount only *one* specimen on each page. Use a paint brush to arrange the sprays and stick with diluted gum arabic or with strips of transparent sticky tape.

WAYSIDE & WOODLAND FLOWERS

Naming your specimens

The best way to name your flower specimens correctly is to look through some good flower identification books. There is a list of identification books on page 239.

Coloured illustrations are helpful and you may be able to identify your flower first merely by turning over the pages until you come to a picture which is like your flower.

Before reading about it, make a good description of the flower for yourself and then check up with the descriptions in the book. Count the petals and sepals, note the colour of the petals and look at the shape of the stalk and leaves.

It will be easier to identify the flower if you know what family it belongs to. In the next few pages you will find descriptions of some flower families.

THE WHITE DEAD-NETTLE FAMILY—LABIATAE

All plants in this family have *square* stems. They also have irregular flowers with a lip, or labium. The fruit (see base of flower in diagram) is made up of four little nutlets. There is a long forked stigma (see diagram). The stalk of the stigma is called the style. The stamens are attached to the inside of the flower tube. Find these in the diagram. Remember, too, that you are looking at a half-section of this flower. The rest of the stamens are in the other half. How many stamens will there be all together?

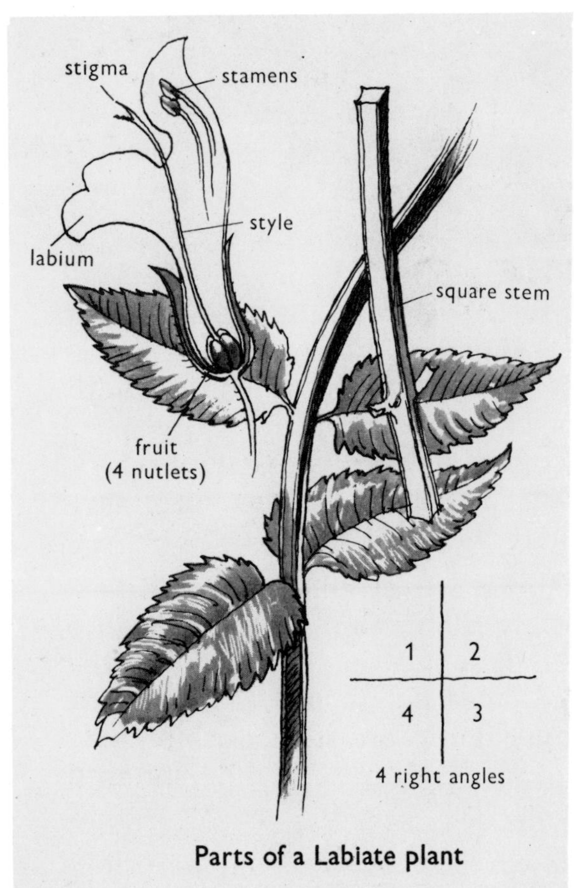

Parts of a Labiate plant

The leaves in a Labiate, or white dead-nettle type of plant, are arranged in pairs and each pair is at right-angles to the next pair. Do you know what a right-angle is? Look at the little drawing marked 1 2 3 4 and then find out why we say these leaves are at right-angles.

THE DAISY AND DANDELION FAMILY—COMPOSITAE

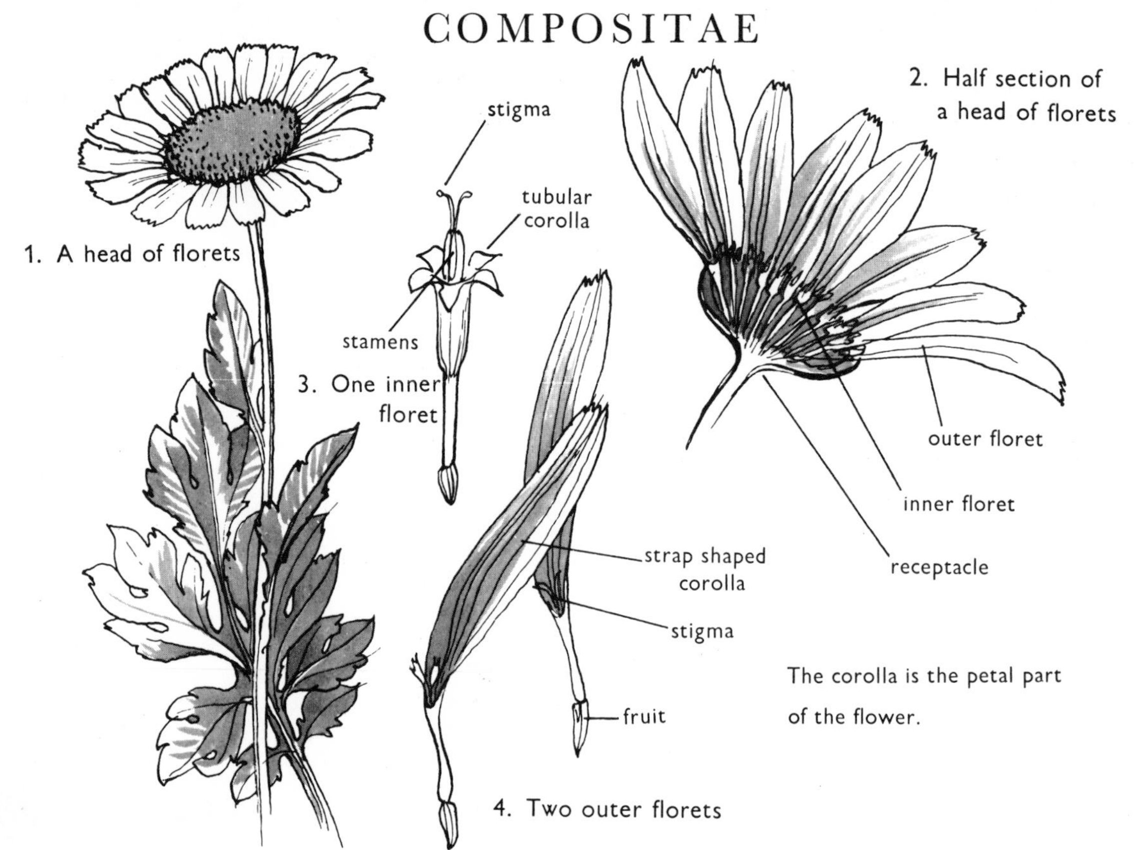

Flowers of this family are *composite*, which means that the head of flowers (1) is made up of many tiny flowers or *florets*. If you cut through the head of a marigold flower, you will see that it looks like diagram 2 and that the florets are arranged on a platform called a receptacle.

Daisies are arranged like this too. Many people think that the head of the daisy is a single flower, and that the outer florets are petals and the inner florets are stamens, but this is not so. They are two different kinds of florets.

Now is the time to get out your hand lens and to examine the florets through it. You will see that the inner florets are like Diagram 3 and the outer florets are like Diagram 4. The inner florets are said to be " tubular " and the outer ones are described as " strap shaped ".

Do you think you could recognise a flower from the Compositae family? Here are some: common daisy, dandelion, cornflower, marigold, goat's beard, ox-eye daisy and coltsfoot. You will find that the dandelion has only strap-shaped florets and the cornflower has only tubular florets.

THE SWEET PEA FAMILY—LEGUMINOSAE

A large number of plants in this family have flowers rather like butterflies. These flowers have one large petal called the " standard " petal which stands erect, like a standard, at the back of the flower (Diagrams 1 and 2). There are also two side petals called the " wing " petals, and there are still two more petals in the front, covered by the wing petals. These two inner petals are joined together in the form of a boat and are called the " keel ". Look right inside your flower and find this joined keel which is made of two petals. How many petals has the flower all together ?

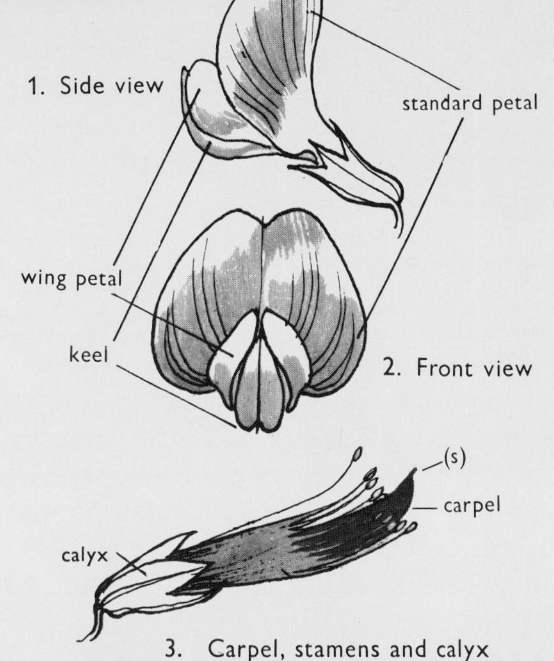

3. Carpel, stamens and calyx

Inside the flower is the fruit, which is a pod. When the fruit is young it is called the carpel. The stamens are grouped together in flowers of this family and group round the carpel as though they are protecting it. You can see the stigma (s) of the carpel standing up among the stamens in Diagram 3.

THE SNAPDRAGON FAMILY—SCROPHULARIACEAE

What a long name this family has—but do not worry. You may remember the first part of the name which sounds like " scroff ". Flowers of this family often have a " lipped " appearance. Can you see the lower lip in the drawings below ?

1. Flower of snapdragon

2. Flowers of foxglove

3. Flower of figwort

RECORDING PLANT DISTRIBUTION

You must learn to record plants in their own homes, or " habitats ". It is a good plan to begin by making a record of all the plants you find growing in one small area. Make your record in the form of a map.

On the right is a map—called a distribution map—which shows, by means of symbols, the plants found on a small area of woodland floor. Direction is shown by the arrow pointing to the north. The trees act as landmarks. Between these trees, on the woodland floor, are seven different kinds of ground plants and shrubs. These are dogwood, honeysuckle, blackberry, ground ivy, primrose, wood anemone and dog's mercury.

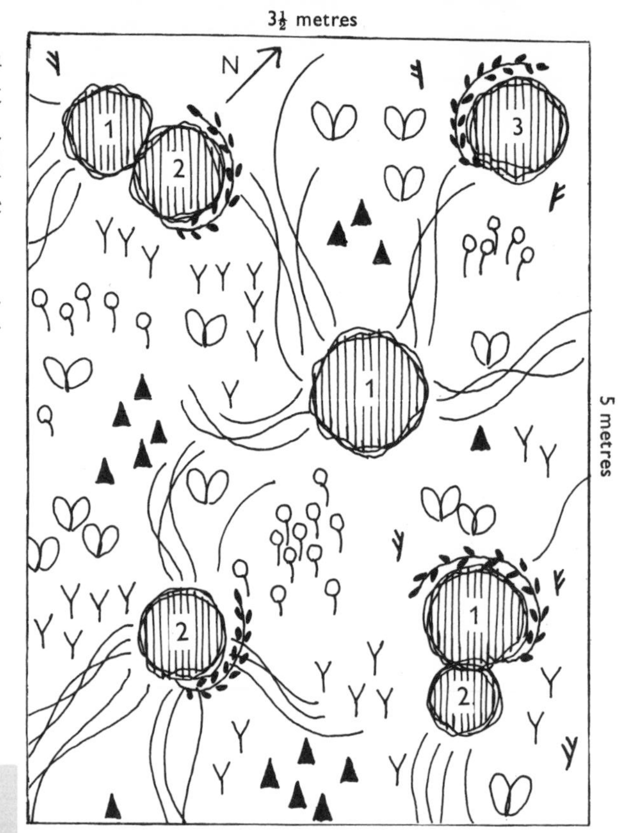

Symbols	Key
	Trees 1. Oak 2. Ash 3. Sycamore
	Shrub, dogwood
	Honeysuckle
	Trailing bramble
	Ground ivy
	Primrose
	Anemone
	Dog's mercury

Do not worry about drawing your map exactly to scale when you first start field-work. Mark off your territory with rope or string, and measure the length and breadth of it. Draw a rectangle on paper the same shape and write in the measurements along the sides as shown. The trees or large shrubs will serve as landmarks. Mark these first.

Then note the positions of other *groups* of plants, such as a clump of nettles. You cannot mark every *separate* plant on your map. Mark the plants by means of special signs, or symbols, and make a " key " to show what the symbols mean.

TAKING A PLANT PROFILE

Two or three bamboo poles are driven into the ground and a rope is stretched tightly between them.

A spirit level is used to make sure that the rope is horizontal.

You will see that the rope A D in the drawing above is stretched between poles A, B, C and D, and covers 3 metres of ground. At each of the points 0 to 15, which are 20 centimetres apart, the distance from line to ground is measured and recorded on a graph (see below). The plant growing immediately below each point is identified (named) and is recorded on the graph by means of a symbol. The graph of the plant profile will look like this:

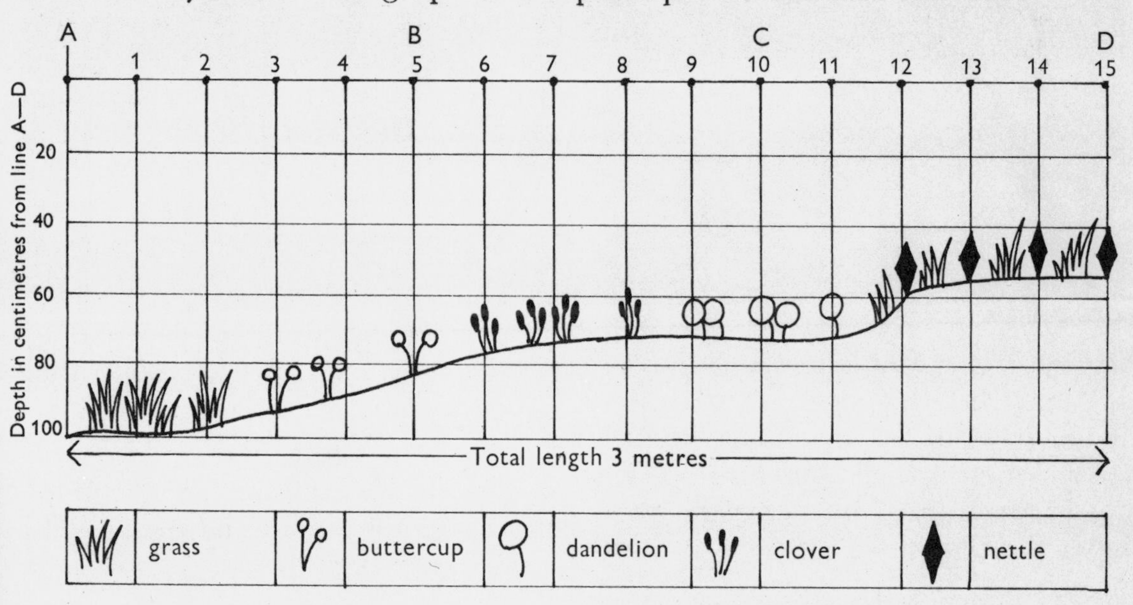

LOOKING AFTER YOUR PLANT SPECIMENS

When you plan to go out to collect flowers, you must remember to prepare some vessels in which to place your specimens. The bigger these are, the better. Find some buckets, tin baths, washing bowls or large jars. See that every specimen you bring home has its stalk in water. The plants can stay in these large metal or earthenware vessels all night and you will find them revived and fresh next day when you come to sort them out. It will help in sorting out if you keep plants from the same locality together.

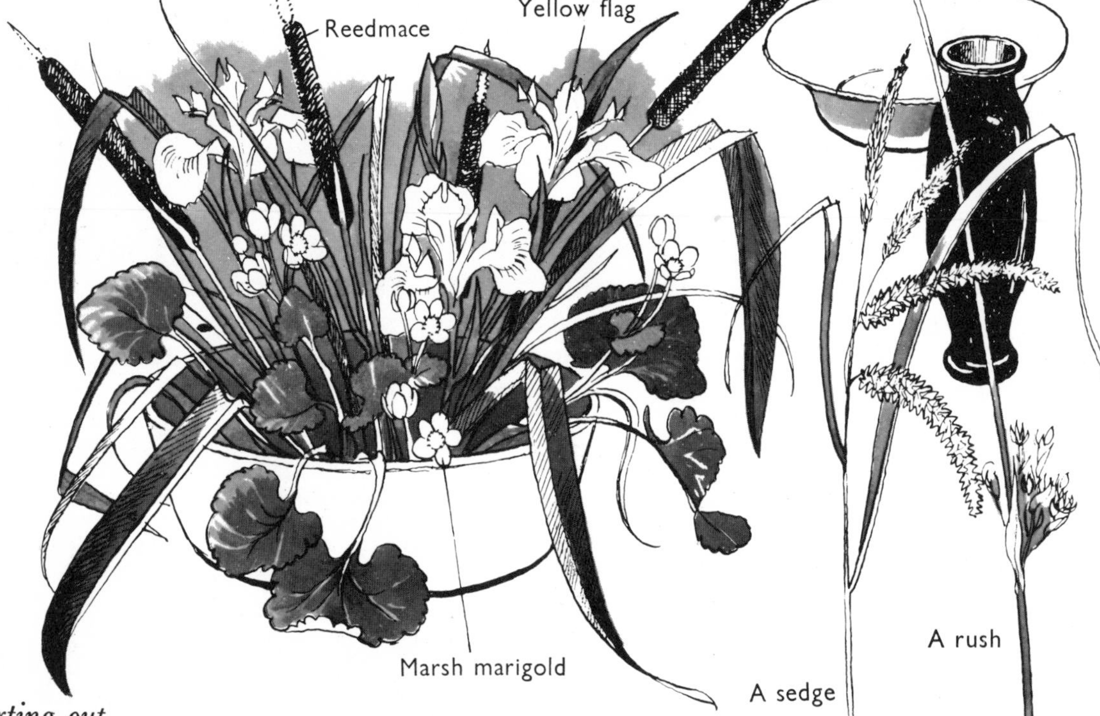

Sorting out

Above are various plants which have all come from the edge of a stream. In the bowl are yellow flags, reedmace (often called bullrush) and marsh marigold. By the side are shown a sedge and a rush.

We can sort out the sedges and the rushes first from the other plants. Sedges and rushes are grass-like plants with green or brown flowers. Both live in moist places. The rush shown in the picture has a tuft of flowers at the side of the stem. The sedge has long drooping heads of flowers.

After we have separated out the rushes and sedges, let us look at the plants with coloured flowers, such as (a) the marsh marigold and (b) the water iris, or yellow flag.

The marsh marigold belongs to the same family as the buttercup. You will see that it has five petals and numerous stamens. Its leaves are large and spreading, with a network of veins.

The iris, or yellow flag, belongs to a different group altogether. Its flower parts are in *threes*, not in *fives* as in the marsh marigold, and its leaf is long and sword-like, with parallel veins. These two flowering plants represent two big groups of flowering plants.

Labelling

If you have been using your flower books, you will probably have got to know, by now, the names of many different flowers. Remember to keep records of these plants whose names you have tracked down. One way is to make a card label for each plant. On the label you will write :

1. the common name
2. the Latin name
3. the date when flowering
4. the place where found,

so that you will have this sort of record :

Marsh marigold
Caltha palustris
April 5th 1970
Local pond.

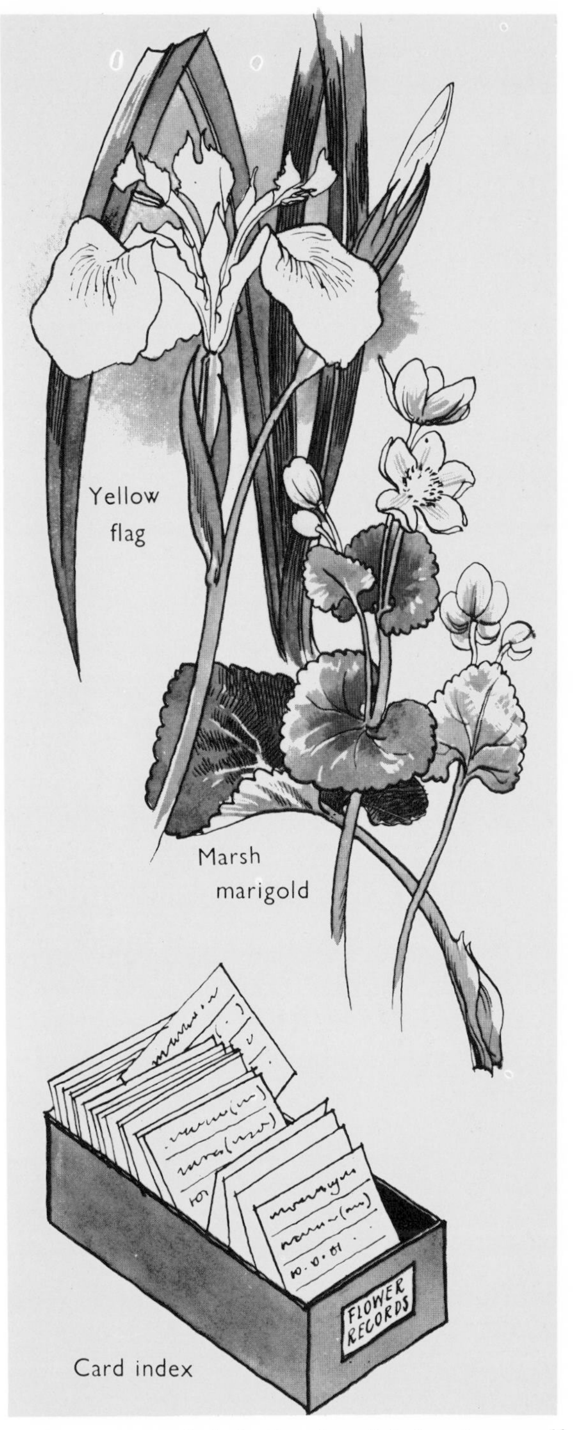

Your card labels should be kept all together in a box in alphabetical order.

You must now learn to recognise some plants which have no flowers at all.

You may wonder how plants which have no flowers and therefore no seeds can reproduce themselves. The answer is that these plants produce spores, sometimes on their leaves, sometimes in a special container which we call a capsule. You will remember that fungi have spores.

Let us examine three types of non-flowering plants.

A liverwort

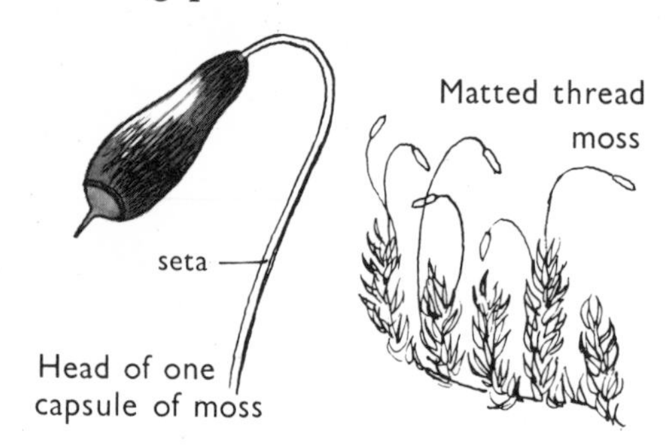

LIVERWORTS. Liverworts are simple green plants. They grow flat on moist surfaces. They were called " liverworts " because they look rather like the lobes of a liver. They have midribs, but no stalks, and never stand upright. They have fruiting structures which look like small umbrellas.

MOSSES are very small green plants with short stems and small, delicate leaves. They have spores which are kept in special capsules which stand up on stalks, or *setae* (singular, *seta*), away from the plant.

FERNS. The leaves of ferns are called *fronds*. Instead of seeds, ferns have tiny spores. These are usually found on the back of the fronds in small brown spots, known as *sori* (singular, *sorus*). The frond of a fern is usually very much divided up, except in one or two kinds such as the hart's tongue fern shown in the illustration.

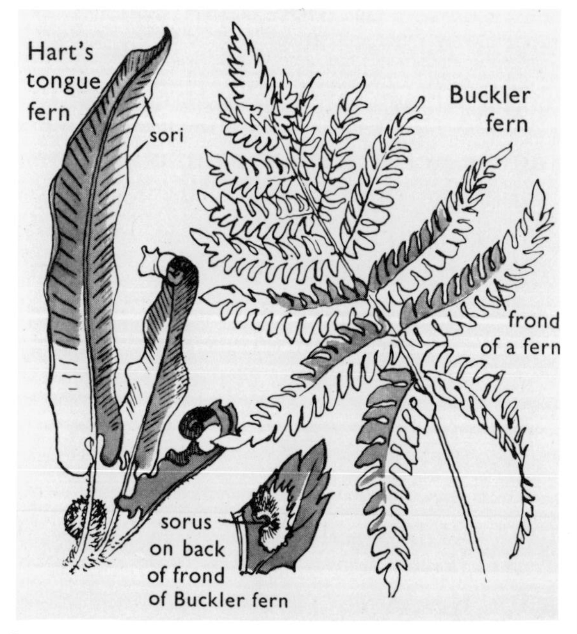

A MOSS WITH A "WORLD RECORD"

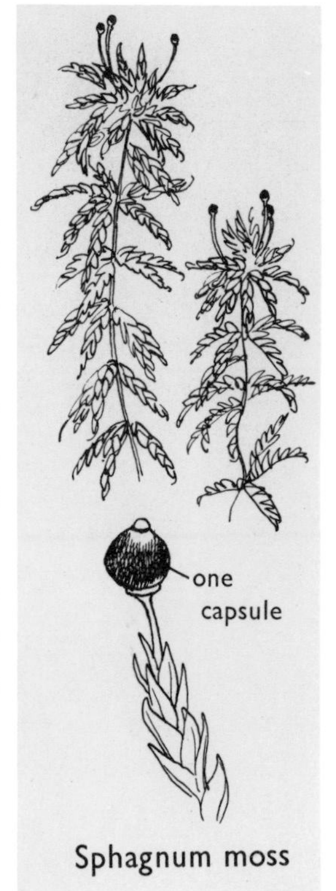

Sphagnum moss

You may have seen moss in hanging flower baskets or placed around the base of flowering bulbs in a pot. Used in this way, moss helps to keep other plants moist, because most mosses can absorb, or take in, many times their own weight in water. One, sphagnum moss, can absorb a great deal of water. We can think of it as holding a sort of " world record " among the mosses for the amount of water it can absorb.

On the right is a diagram of sphagnum moss. It is sometimes found in very damp woods, but more often you will find it growing in a bog, in the same kind of area that you will find cotton grass, another bog plant.

You can try an experiment to see how much water sphagnum moss can absorb. Let some pieces of moss dry and then put their stems in water. You will see that the moss will soon absorb a small glassful of water.

A MOSS GARDEN

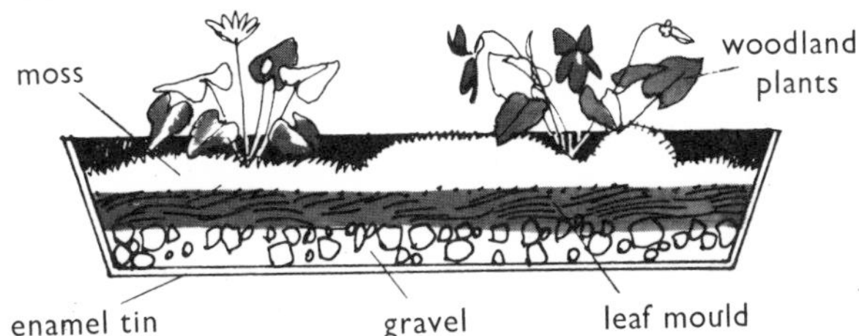

Perhaps you would like to make a moss garden. This could either be to cultivate some of your favourite mosses, or it could be a base for growing other little plants (particularly woodland plants) which like moist surroundings. You can keep many small plants alive and fresh for weeks in a moss garden. For this you will need a flat enamel tin such as a toffee tin (painted if you do not want it to rust), a bagful of woodland leaf mould, some gravel and the moss.

Mosses should be kept in a moist atmosphere and sheltered. An old aquarium tank would make a splendid place in which to grow them. Plant your mosses in a base of leaf mould. Cover the tank with a sheet of glass. This will keep the atmosphere moist so that you will only need to spray with water at rare intervals.

Growing mosses in an old aquarium tank

EXPERIMENTS WITH GROWING SEEDS

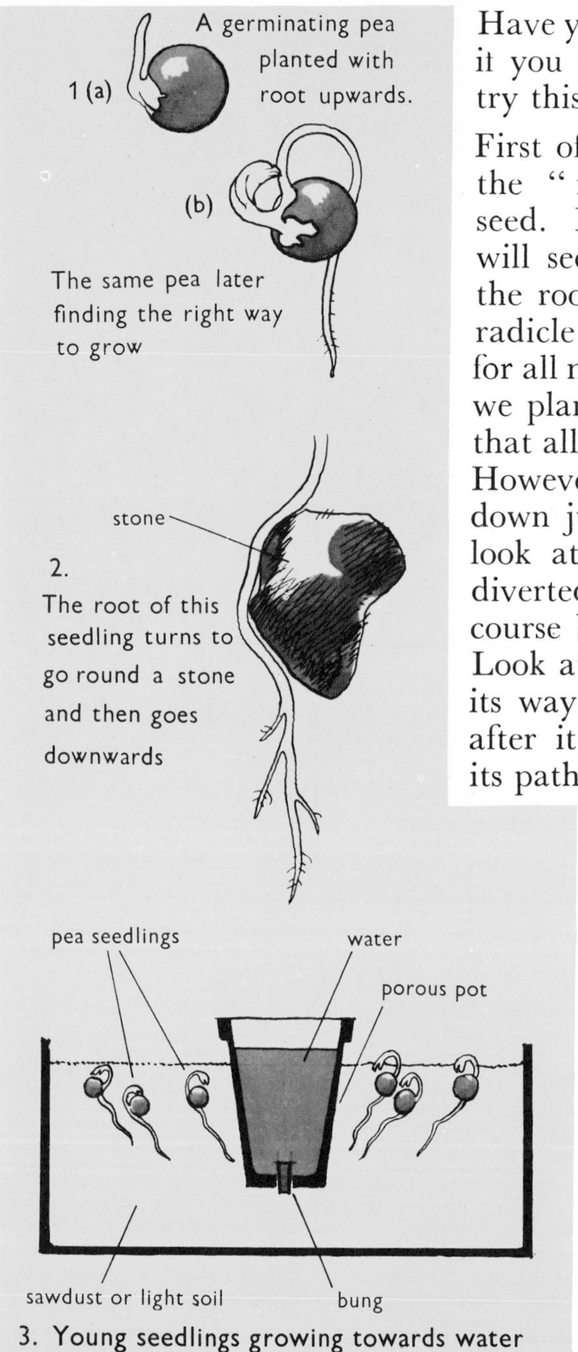

3. Young seedlings growing towards water

Have you ever wondered what would happen if you planted seeds upside down? Let us try this.

First of all you will have to know which is the "right way up" of the seed. Look at a pea seed. You will see a little lump which is the root, or radicle (r). This radicle should point downwards, for all main roots grow downwards, but when we plant pea seeds we do not bother to see that all the radicles are pointing downwards. However, the radicle manages to find its way down just the same, as you will see if you look at Diagram 1. Roots are sometimes diverted (turned) from their downward course but they always go back to it again. Look at the drawing of a main root finding its way round a stone. You will see that, after it has grown around the obstacle in its path, it continues to grow downwards.

Roots are sometimes diverted from their path by water. Roots need water and cannot resist growing towards it, even if it means growing away from their normal downward path. Diagram 3 shows some young seedlings growing in a box. Their water supply is in a small porous pot which allows a little water to seep through it. A small plant pot with its hole bunged up with a cork will do for this. You can see that the roots of the seedlings are diverted off their course towards the damp area around the porous pot.

We say that roots respond to gravity unless they are diverted. They also respond to darkness and grow away from light.

TO SHOW THAT SHOOTS SEEK THE LIGHT

Shoots grow upwards when light comes to them from above. If the light comes from another direction, the shoots will turn towards it. This shows that shoots can be deflected (turned aside) towards light, just as roots are deflected by water.

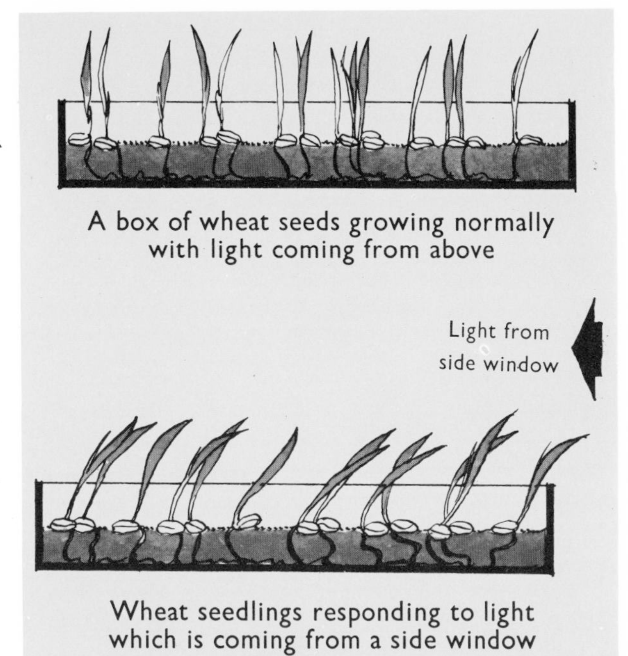

A box of wheat seeds growing normally with light coming from above

Wheat seedlings responding to light which is coming from a side window

Experiment 1

Plant some seeds in two different seed boxes; when the shoots begin to come through, place one box in a position where light comes from above or from two different directions and the other where it receives light from only one window. See whether you get results similar to those in the diagrams.

Experiment 2

Here is an experiment you can set up to show that shoots of germinating seeds will find their way to the light even when this comes through a little hole.

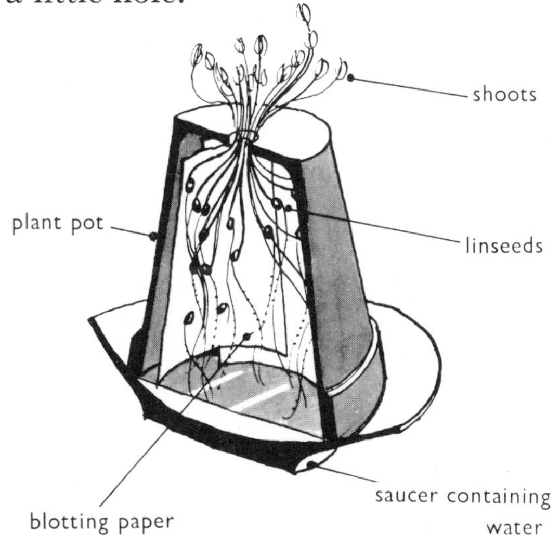

Section through plant pot in Experiment 2

Line a plant pot with blotting paper and dampen this thoroughly.

Sprinkle some linseeds (seeds of flax) on the blotting paper. You can buy these seeds at a chemist's. Linseeds are sticky and will stick at once to the damp blotting paper.

Stand the plant pot upside down in a saucer, or a shallow bowl of water, so that the blotting paper is touching the water.

Record what happens.

Record the results of these experiments in your notebook. Say *why* you think your results " happened ".

CAN PLANTS MOVE?

Parts of plants often *do* move, but very slowly. Some of these movements are growth movements, as when the shoots of a plant grow towards light.

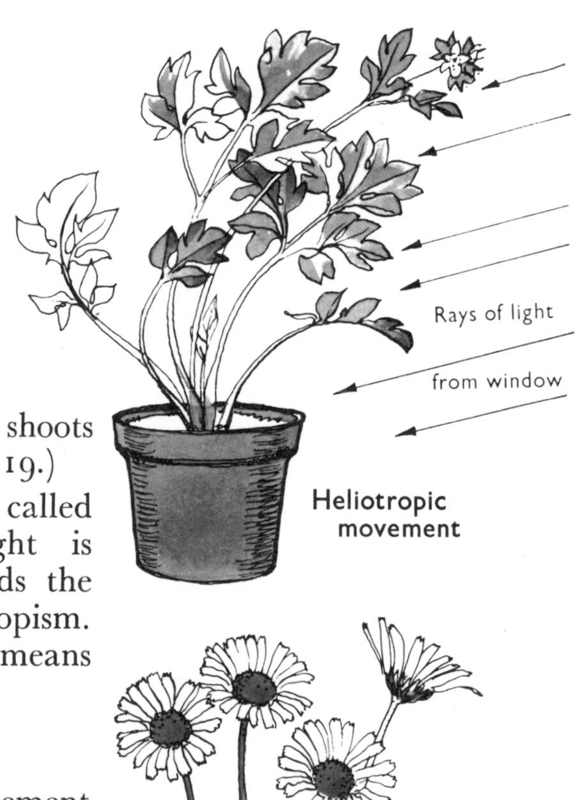

Heliotropic movement

1. Heliotropic movements

If plants are growing in a room with light coming from one window, their stems, leaves and flowers will turn towards the light. You saw that the shoots of seedlings behaved like this. (Page 219.)

These growth movements of plants are called *tropisms*. Growing towards the light is called heliotropism. Growing towards the earth, as roots do, is called geotropism. " Helio " means " light ", " geo " means " earth ".

2. Sleep movements

There are other kinds of plant movement known as " sleep movements ". Look at the two pictures of daisies. The first shows daisies in the daytime with their flowers open to sunlight. The second shows the daisies after sunset. They have closed up so that the outer florets (the white ones) protect the inner (yellow) florets from the cold air.

1.

2.

Sleep movements of daisy

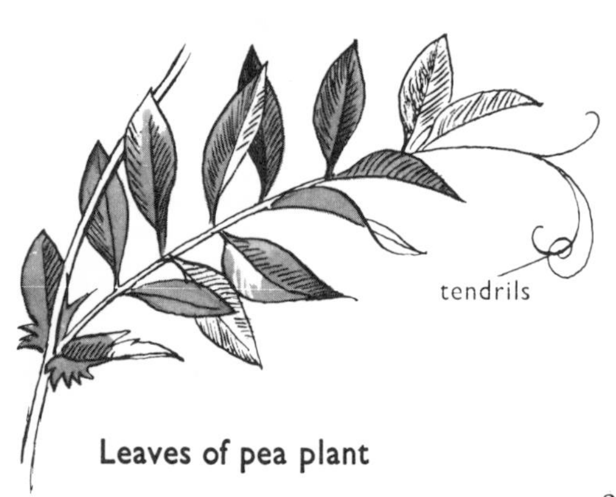

Leaves of pea plant

3. Climbing movements

The pea plant has sensitive parts called tendrils which develop in the place of leaflets at the ends of the leaves. You can see these tendrils in the picture. When the tendrils come up against something to which they can cling, they curl round it. This is how the plant climbs up to seek the light and warmth.

Many weak-stemmed plants must reach the light by climbing. They often use stronger plants as supports and sometimes they twine themselves round so tightly that the supporting plant is strangled.

The plant shown here, the honeysuckle, has a twining stem. The tip of its stem is constantly moving round in a circular fashion (round and round) as it grows. This movement of the tip of the stem is called *circumnutation.* In the honeysuckle this movement is in a clockwise direction, that is, in the same way that the hands of a clock move round.

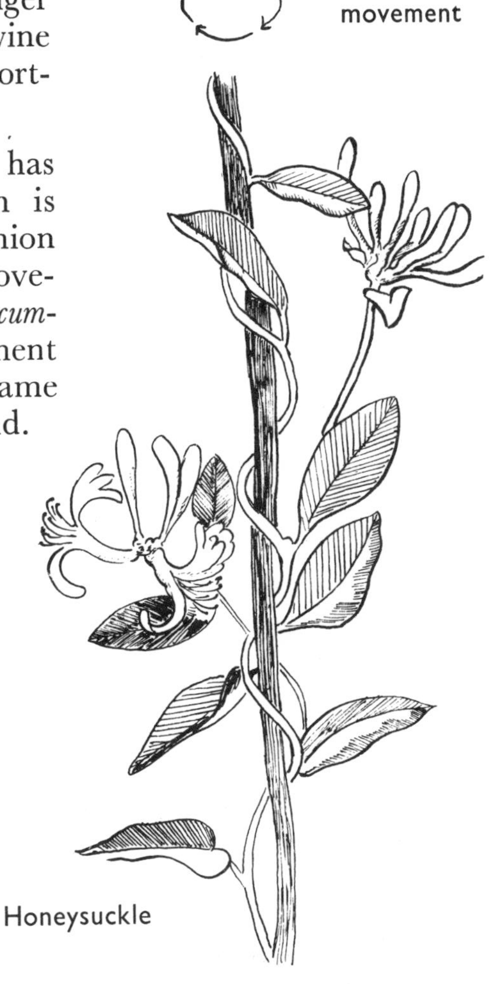

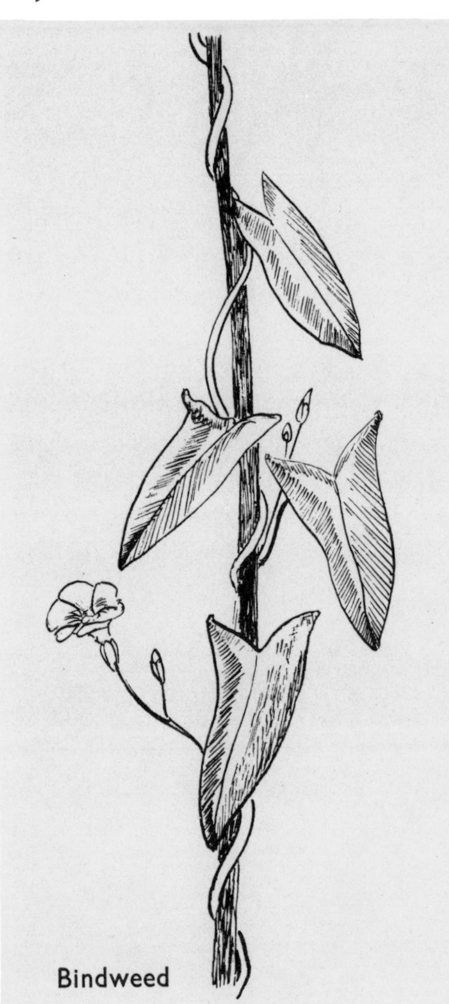

Another twining plant, known as bindweed, also climbs up the stems of stronger plants in a similar way, but the stems of this plant twine in an anti-clockwise direction, that is in the opposite direction to the movement of clock hands.

Below is a diagram to show anti-clockwise movement.

Anti-clockwise movement

MORE PLANT MOVEMENTS

A plant which sleeps in the daytime

Among the " sleep " movements of plants (see page 220) is an interesting one which you will see in a yellow flower called, " John go to bed at noon ". It has another name: goat's-beard. This plant grows on waste land, roadsides and railway embankments. Unlike the daisy, which closes up at night, this flower closes at noon-day.

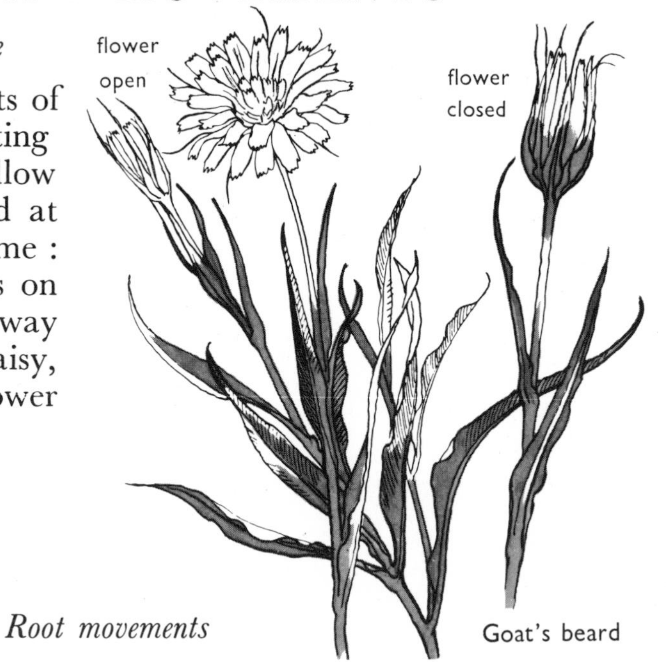

Goat's beard

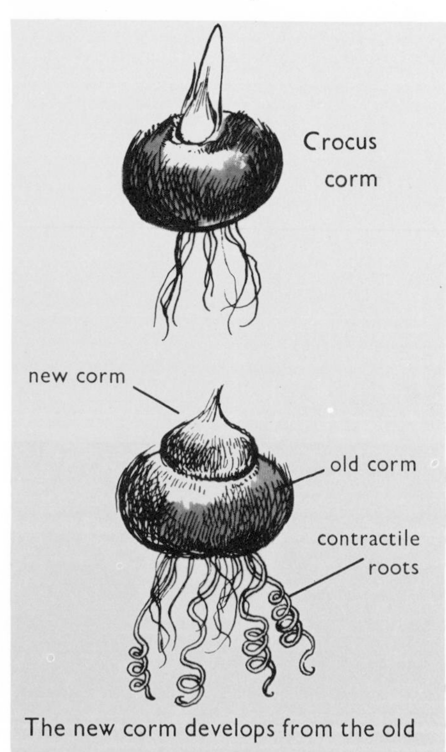

The new corm develops from the old

Root movements

The crocus is a plant which can move its roots. You will notice that some of the roots of this crocus corm are thicker than others. These thick roots act in a special way. Each year the new crocus corm forms on top of the old one. If this went on happening year after year, the corms would be pushed above the ground. However, the thick roots stop this from happening. They contract (draw up closer together) in spirals (see diagram) and pull the corm down into the soil.

Explosive movements

Although most plant movements are slow, some plants can make sudden explosive movements. One of these is the pansy whose fruits suddenly split open and throw out their seeds, when ripe.

Pansy

ANIMAL MOVEMENTS

We are used to seeing animals using legs and feet for moving around, but some animals have other methods of movement.

Earthworms

The earthworm's muscular body is made up of numbers of segments (see Diagram 1). On the lower part of each of these segments are tiny bristles. It is by the movement of its muscles and the contact of these bristles on a rough surface that the earthworm is able to travel.

In Diagram 2 you can see the bristles on one side of the worm's body. It has the same number of bristles on the other side too, so if you work out a sum you will find that there are eight bristles on each segment.

These bristles are called *chaetae* or *setae*. If you put a worm on some brown paper and listen carefully, you will hear the bristles scratching on the paper as the worm travels along.

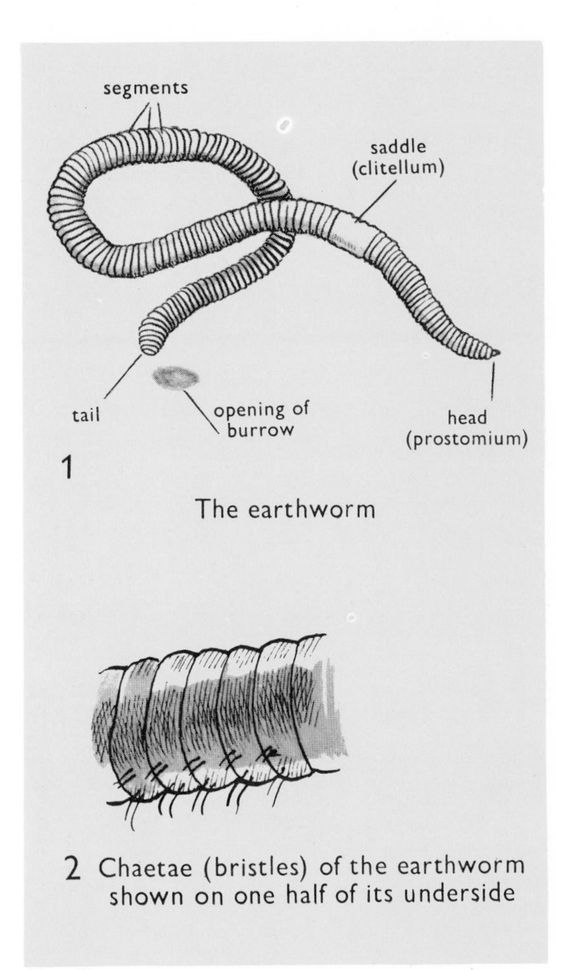

The earthworm

2 Chaetae (bristles) of the earthworm shown on one half of its underside

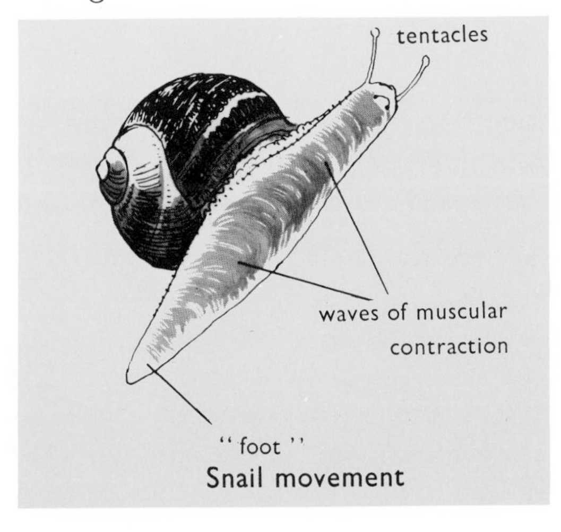

Snail movement

Snails

A snail moves along by means of a muscular " foot " underneath its body. If you place the snail on a piece of glass and look at the " foot " from under the glass, you will see rhythmic waves of movement (we call them waves of muscular contraction) passing along the base of the " foot ".

ANOTHER KIND OF ANIMAL MOVEMENT

Swimming in fish

A fish is well adapted for a swimming life. It has a streamlined body which allows it to move smoothly through the water. Notice too that all its scales point backwards so that they do not ruffle as the fish swims forwards.

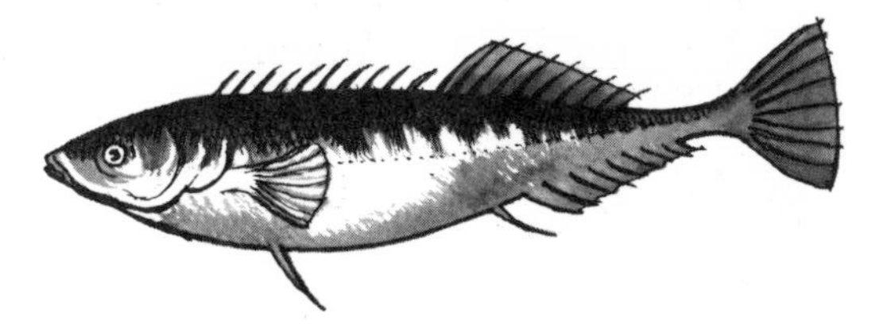

A ten-spined stickleback

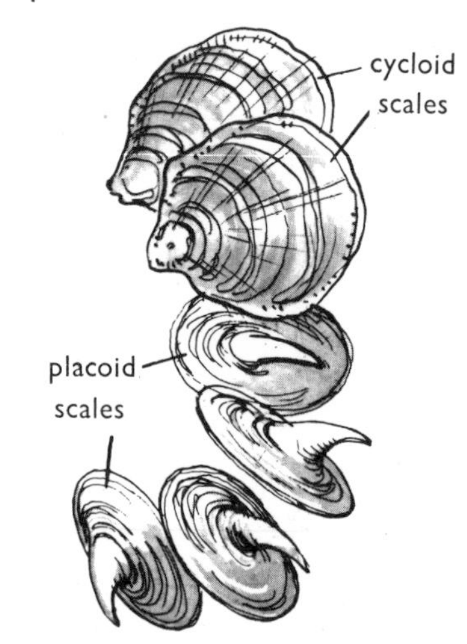

In the largest group of fishes—known as bony fishes, or *teleosts*—the scales are roughly circular with ring-like markings on them. You can tell the age of a fish by counting these groups of rings. Herrings, salmon, trout and many other fishes have scales like these. They are called *cycloid* scales.

Non-bony, or *cartilaginous*, fishes (such as dog fish, skates and rays) have different scales which look like tiny spines and are like simple teeth in structure. These are called *placoid* scales.

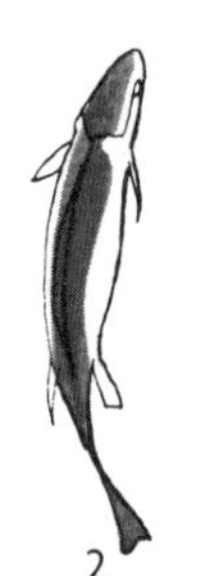

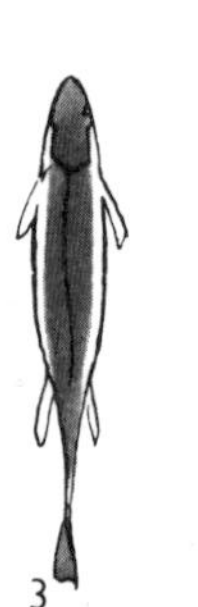

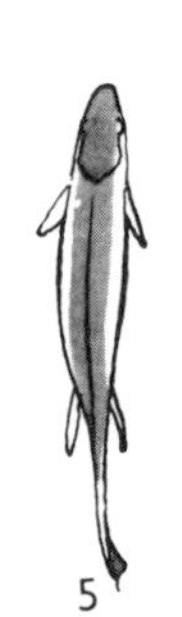

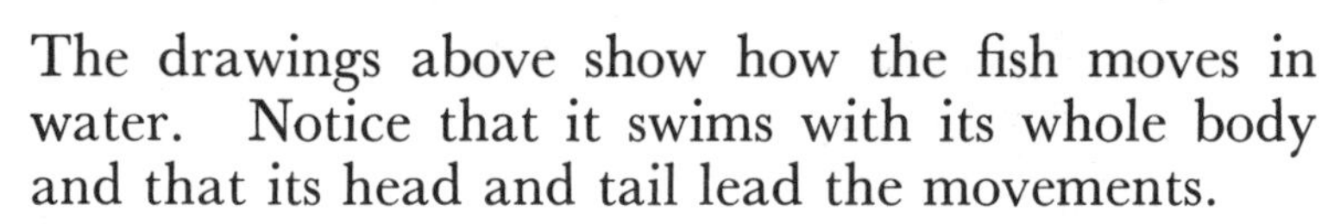

The drawings above show how the fish moves in water. Notice that it swims with its whole body and that its head and tail lead the movements.

The drawing on the left shows the path of a fish as it moves through the water. If you follow its progress from A to B, you will see that a fish does not swim in a forward fashion but zigzags from side to side. This movement begins at the tail, which the fish lashes from side to side, and the wave-like motion is carried through the body.

VERTEBRATE ANIMALS IN MOTION

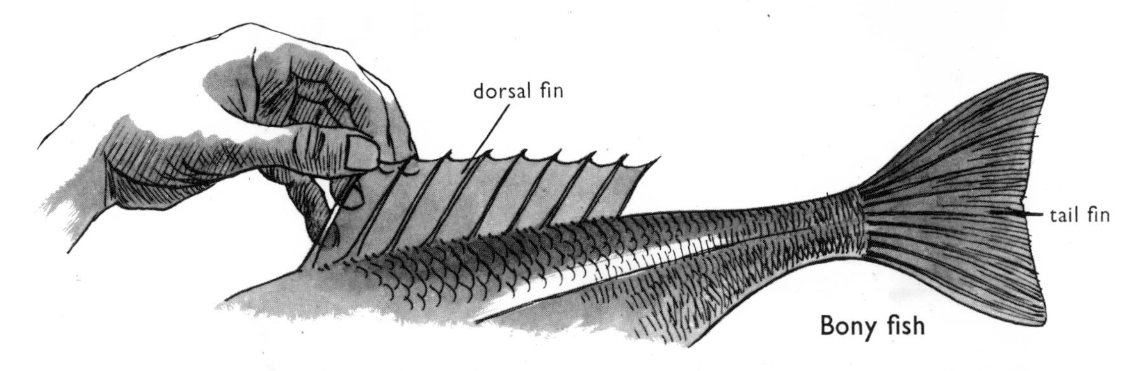

Bony fish

The fins of fish are made of small bones with skin stretched between them. Pull up the dorsal (back) fin of a fish with finger and thumb, and see this for yourself.

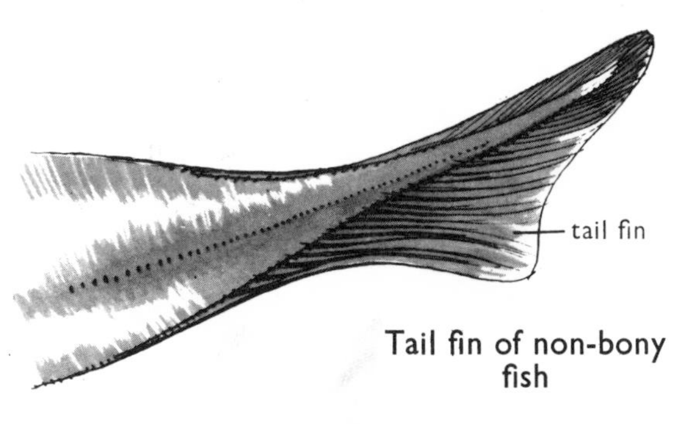

Tail fin of non-bony fish

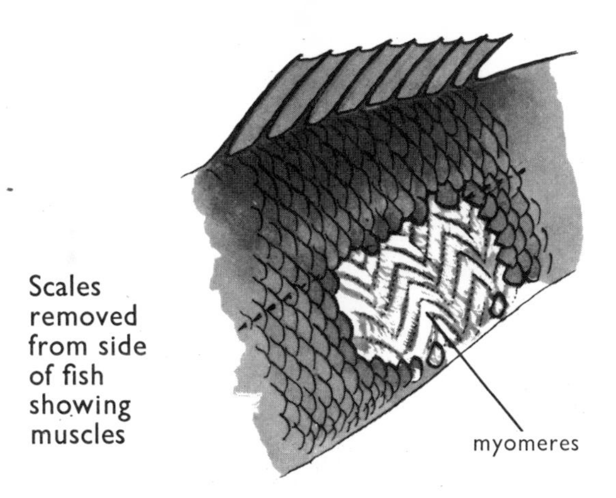

Scales removed from side of fish showing muscles

Fish usually have two sets of paired fins, as well as their tail fin, their back (or dorsal) fins and their under, or " ventral ", fin.

The tail fin of a bony fish is similar in construction. It is made up of fin rays (little bones with skin stretched between them) and the sides of the fin are *symmetrical* (the same shape on both sides).

The tails of cartilaginous, or non-bony, fishes are different from those of bony fishes. Instead of being symmetrical (evenly spaced above and below the centre), the tails of cartilaginous fishes (dogfish, skates and rays) are uneven or asymmetrical (see diagram).

As you saw on page 224, a fish swims by lashing its tail and by the strong movements of its body. You can see the strong muscles which the fish uses to make these movements. If you scrape away the scales from the side of the body of a fish, you will see zigzag muscles which are called *myomeres*. A fish uses its fins to balance and steer itself in the water.

THE FIVE-FINGERED LIMB

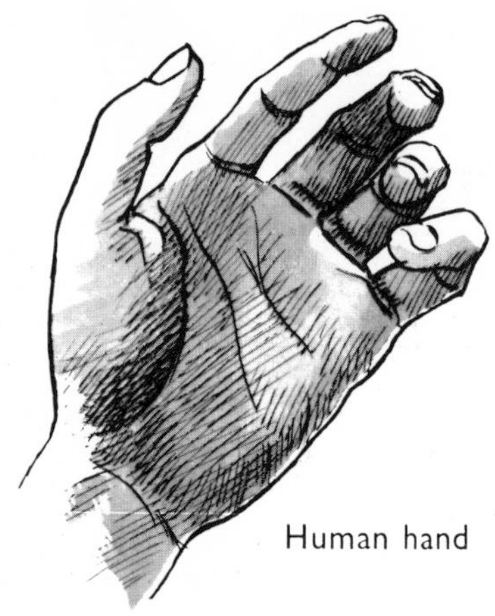
Human hand

Have you noticed that most vertebrates (backboned animals) have a five-fingered limb ? Perhaps the most perfect five-fingered limb is our own hand. Count how many joints there are in each finger, and in your thumb. Bend your thumb and first finger to touch one another, and then bend your thumbs to touch your other fingers. We say that the thumb and fingers are " opposable ". Because of this, men can do many things with their hands which other animals cannot.

Because of his clever hands, and his active mind, man is much more skilled than are other animals. The animals which most nearly resemble man are the apes. They have opposable big toes as well as opposable thumbs and use their feet as well as their hands for climbing and clinging.

Gorilla

Orang-outang

THE FEET OF BIRDS

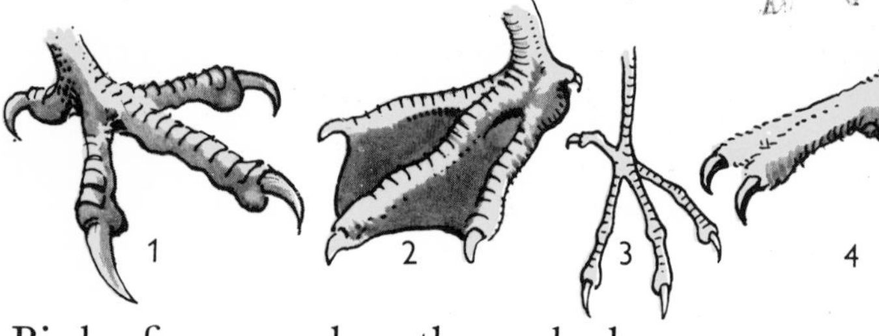

The drawings show four types—

(1) a bird of prey
(2) a swimming bird
(3) a perching bird
(4) a climbing bird

Birds of prey, such as the eagle, have curved claws (talons) on their four toes. Swimmers, such as ducks, have webs between their toes. Perching birds, such as robins and starlings, have slender little toes, three in front and one behind. Climbing birds, such as owls and woodpeckers, have two toes in front and two behind.

You will notice that birds have a modified (slightly altered) kind of limb—they have only four toes instead of five.

A BIRD'S WING

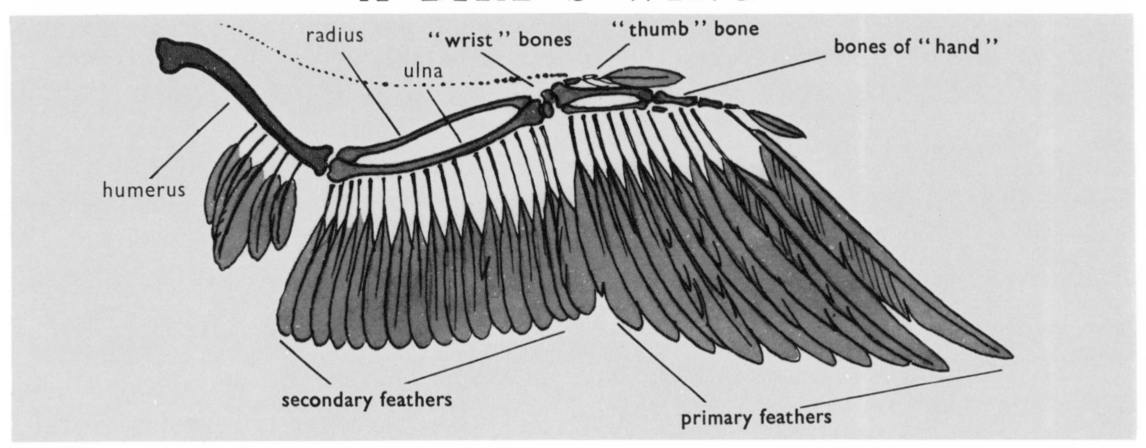

Birds have special organs to help them in moving from place to place. A bird's special organs of locomotion are its wings. These are modified front limbs and have some of the same bones that we have in our arms and hands, as you can see in the diagram above.

The upper bone is usually buried in the body and the wing feathers are supported by the arm, hand and finger bones below the " elbow " joint. There is also a little tuft of feathers on the " thumb " bone.

Examine the wing of a bird. You can often get the wing of a chicken from a poulterer's. Scrape off the little " cover " feathers from the top of the wing until you can see the big quills of the flight feathers and scrape the flesh off the bones to which they are attached.

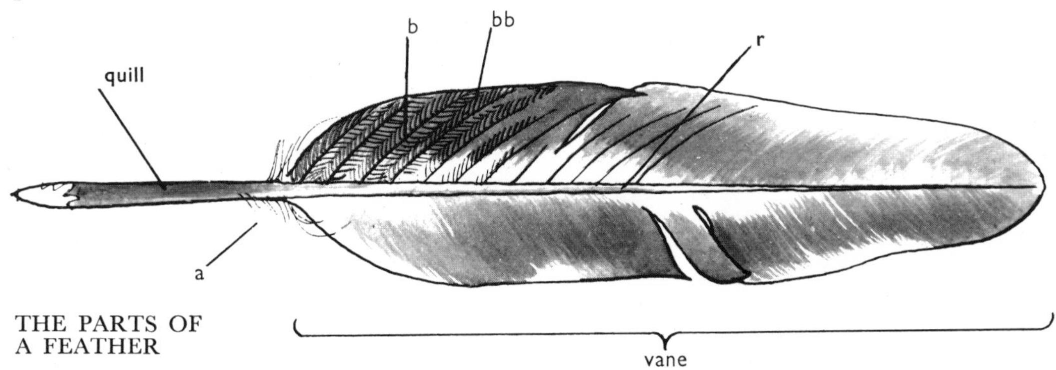

THE PARTS OF A FEATHER

The " stem " of a bird's feather is called the *rachis* (r). The flat part is the *vane*. From the rachis branch out small bony rays called *barbs* (b). These are linked by tiny hooked structures called *barbules* (bb). The part of the rachis from which the barbs come is often known as the shaft. The lower part is known as the quill. The free barbs at (a) are called the aftershaft.

INSECT FLIGHT

Most insects have wings in their adult stage (that is, when they are grown up), although they spend some time as caterpillars or larvae before they become winged insects, and they usually have a resting or pupal stage too. The dragon fly spends many months in water as a larva before becoming a winged insect.

All the insects shown on this page have four wings, but some insects, such as flies and bluebottles, have only two wings.

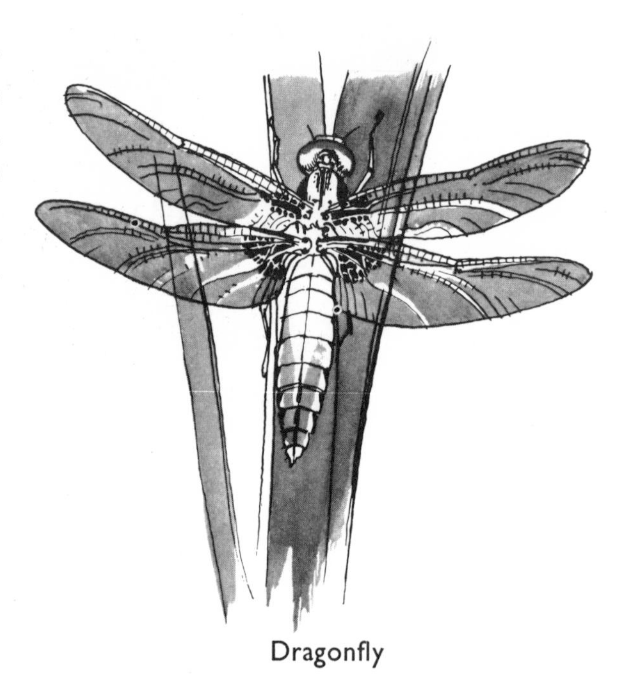

Dragonfly

Peacock butterfly

Hawk moth (Elephant)

Butterflies and moths have very beautiful wings with lovely colours and patterns by which you can identify them. Look at the " eye spots " on the wings of the peacock butterfly and look at the streamlined wings of the hawk moth.

Butterflies and moths have tiny scales on their wings and for that reason are called Lepidoptera, meaning " scale-winged " insects.

In *structure*, insect wings are quite unlike those of birds. Birds are the only animals in the world which have feathers. However, insects use their wings for the same *purpose*—to fly.

OTHER CREATURES WHICH CAN FLY

There are other creatures besides birds and insects which can fly. Here are some other flying creatures. The first drawing shows a flying fish. Large shoals of these fish will leap out of tropical seas and glide above the surface of the water for several hundred metres.

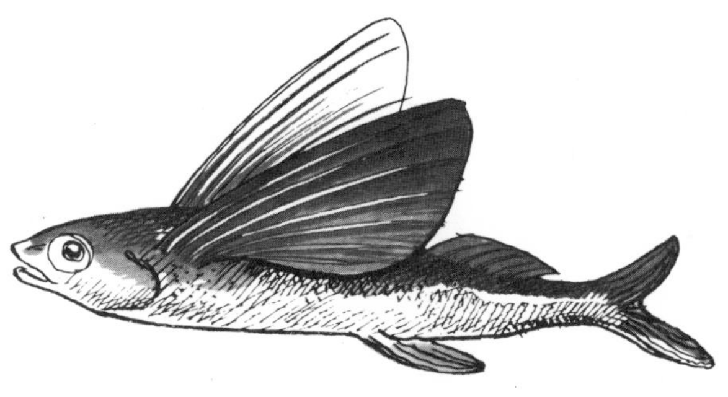

Flying fish

Below the flying fish, there is a drawing of a furry animal which can fly, the bat. Furry animals (mammals) do not usually fly, but the bat has wings of thin, leathery skin which stretch between the sides of its body and its arms and legs. You can see that these wings are not at all like the wings of a bird and there are no feathers on them.

Bat

Flying squirrel

Here is another furry animal, or mammal, which can fly a little. It is the flying squirrel, which is an American animal. You can see that it has flanges of skin and fur along each side of its body. It can glide slowly down from trees to the ground or from the branches of one tree to another.

You will see that the wing of a bird is a more perfect structure for flight than any of these. None of these other flying creatures can fly as well, or as far, as a normal flying bird. There are, however, a few birds which cannot fly. Can you name one ?

COLLECTOR'S PAGE

Wild columbine

Do you collect things? There are many objects to collect from Nature. Some people collect pressed flowers or grasses. You will see how to make a press for these on page 208.

Some people like to collect seaweeds. These are rather difficult to mount, but you will learn how to do this on page 232.

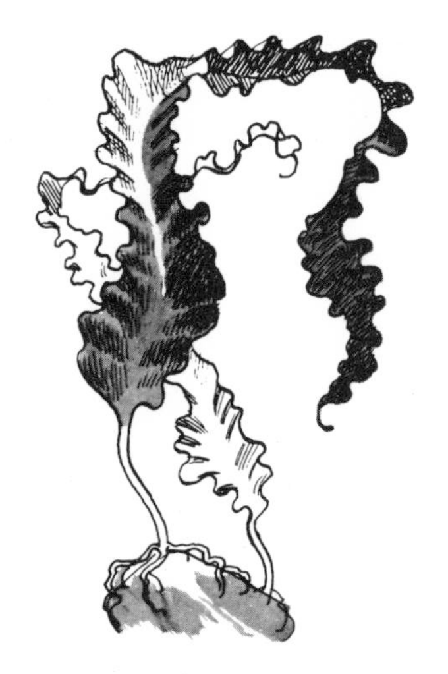
Laminaria

It is interesting to make a collection of animal footprints in the form of black diagrams like those shown on the right, or in the form of plaster casts from clay imprints, made in the same way as the casts of twigs described on page 159.

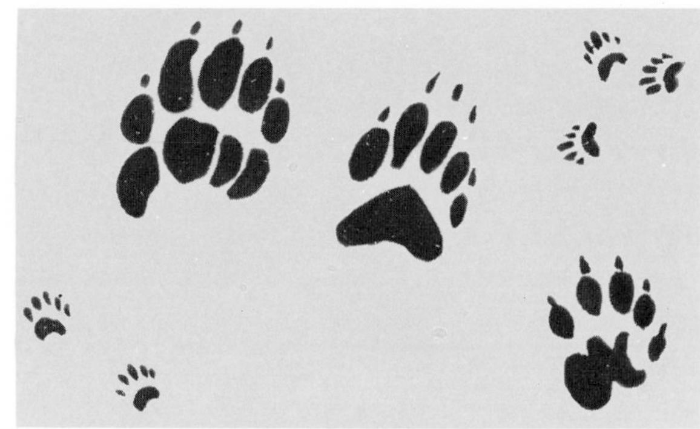

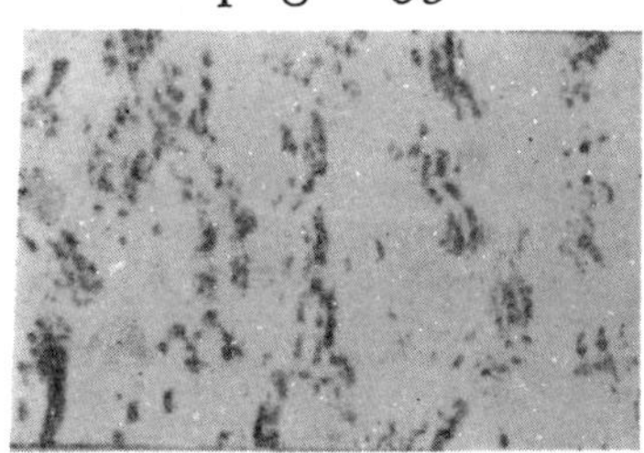

You could make an interesting collection of bark rubbings or leaf prints. (See pages 152, 153 and 156.)

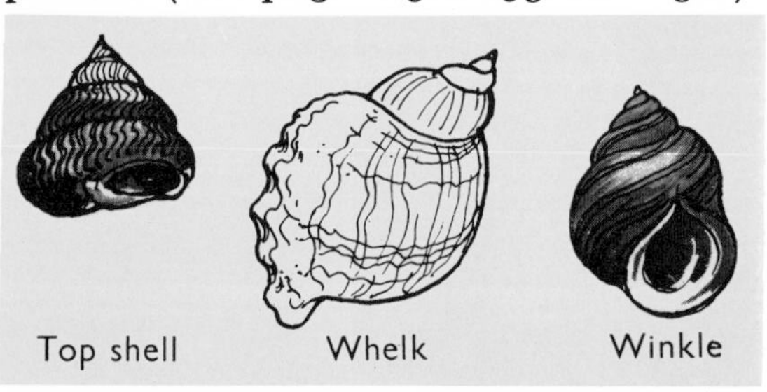
Top shell Whelk Winkle

Shells are interesting objects to collect. People who collect and study shells are called *conchologists*. Here are some types of shells you might collect. Perhaps you could make a cabinet for them.

HOW TO STORE OR DISPLAY YOUR SHELLS

Making a matchbox cabinet for your shells

If you stick together the covers of six or more matchboxes, you can make the framework of a little cabinet or chest of drawers. Round-topped paper fasteners make good handles for the drawers. You can keep different kinds of shells in each drawer. The drawers should each be labelled.

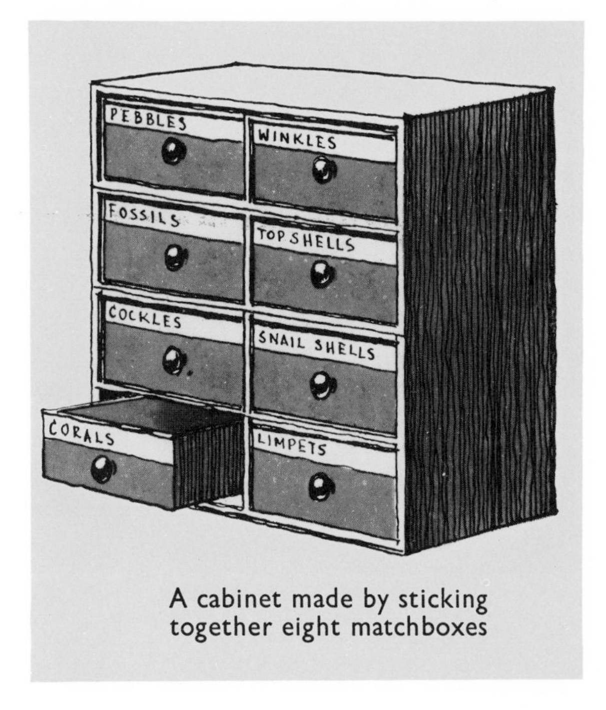

A cabinet made by sticking together eight matchboxes

You can put more matchboxes together to make a bigger cabinet or you can make several cabinets for different groups of shells. Perhaps you would keep only fossils in one, sea shells in another, and pebbles in another.

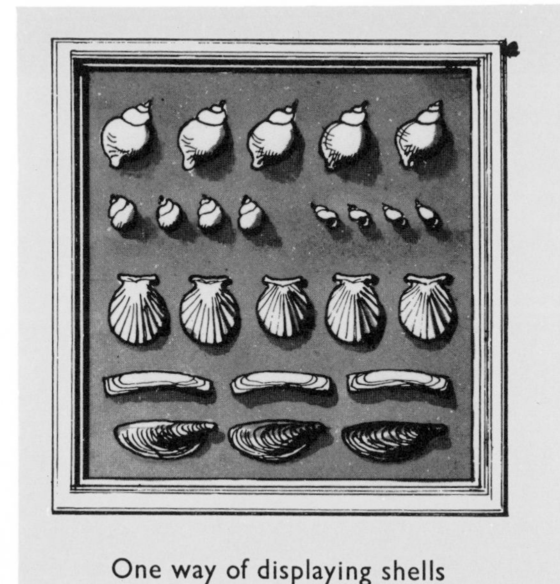
One way of displaying shells

Displaying your shells

There are various ways in which you can display your shells. The drawing shows you how to use an old picture frame, or simply a plywood board with picture moulding around it.

It is often more effective to have a dark background, since shells are usually light coloured, so stick black paper on to your board.

Stick the shells on to this background with a good cement adhesive and then paint each shell with clear varnish.

MOUNTING SEAWEEDS

Many seaweeds are delicate and are difficult to mount because they collapse and droop as soon as you take them out of water. Some are filamentous, such as coral weed (1) and some are membranous such as sea lettuce (3). Red and green seaweeds like these are usually more delicate than the tough brown bladder wrack types (2).

Seaweeds must be kept moist before mounting. The best plan is to place them in a shallow pan and cover them with a dilute solution of gum arabic. After soaking, arrange the seaweeds on a white card, using a camel-hair brush to spread out the moist fronds. As the card dries, the seaweeds will stick to it and will remain as you have arranged them.

Seaweeds are of three main varieties. They are brown, green or red. There is one of each type above.

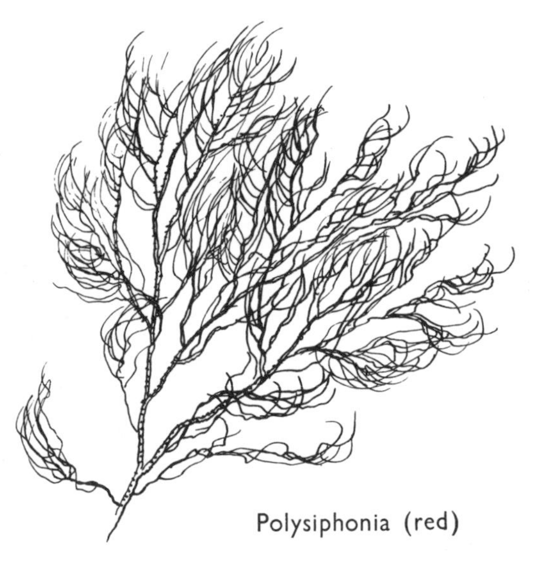

Polysiphonia (red)

On the left is a very delicate red seaweed which will look beautiful on the white card when you have arranged it with your brush.

KEEPING RECORDS

Below are some " sample " records from other children's notebooks.

Mallard in flight

EXPEDITION TO THE PARK LAKE IN FEBRUARY

Things seen :

Trees : willow, alder, ash.

Water birds : mallards (ducks and drakes), muscovy ducks, moorhens, swans.

Other birds : chaffinches, robins, sparrows, blackbirds, thrushes, water wagtail.

Notes and comments

The mallards have paired off. The female (duck) was a mottled brown but the drake was more colourful with green head, grey and rust body and white neck ring. Muscovy ducks are

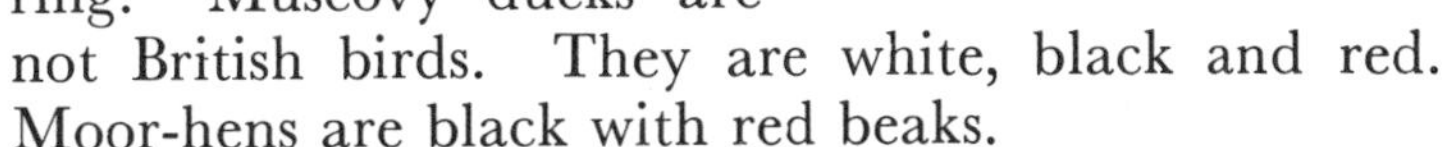
not British birds. They are white, black and red.

Moor-hens are black with red beaks.

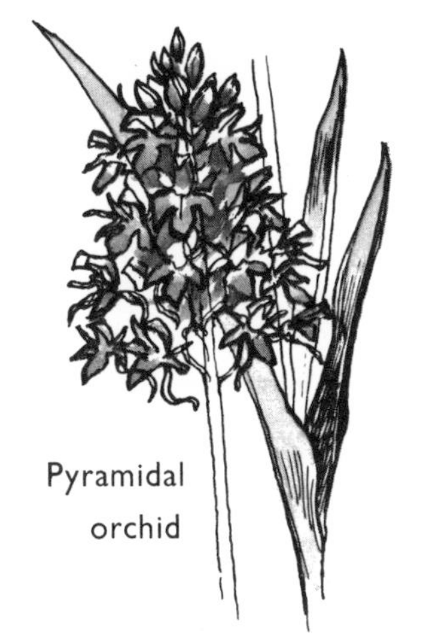
Pyramidal orchid

EXPEDITION TO DUNSTABLE DOWNS IN JUNE

Plant list :

Yellow rock rose
Pyramidal orchid
Butterfly orchid
Kidney vetch
Milkwort
Wild thyme
Wild mignonette
Quake grass
Knapweed (greater)

Notes and comments

We did not pick the orchids because they are rare.

These flowers are all called " chalk " flowers because they grow on chalk. The Dunstable downs are of chalk. The grasses on these downs are fine and wiry. We saw the chalk hill blue butterfly.

Make records of first appearances and their dates

The first coltsfoot.
The first aconite.
The first primrose.
The first violet etc.

The first hedgehog you see awake after its winter sleep.

The first bat, or the first time you hear the cuckoo, or the chiff chaff, or the call or song of any bird in spring.

Always give dates and localities of the things you find, like this :

Common name : Sea lavender
Date : July
Place : Salt marshes.

A label giving these details should stand beside each plant on your Nature table.

Coltsfoot

A PAGE OF CONTRASTS—ANIMALS

African elephant

Large and Small. There are animals which are very large, such as the elephant, and there are animals so small that they cannot be seen without the help of a powerful microscope.

The amoeba is one of our smallest animals. Its body consists of only one cell. The body of an elephant consists of millions of cells, many of them specialised to do particular work, such as digesting or carrying messages or protecting a surface.

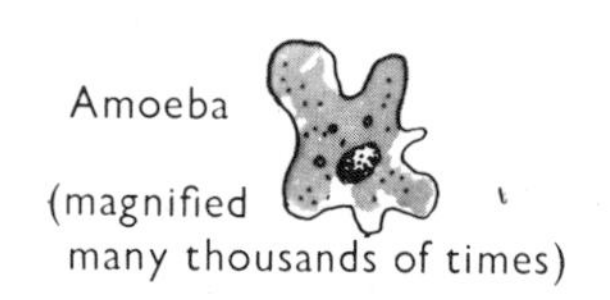
Amoeba (magnified many thousands of times)

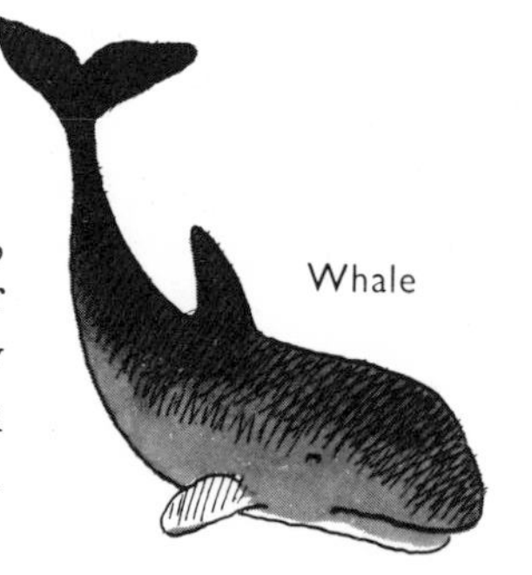
Whale

Land and sea. Creatures such as the elephant, tiger, giraffe, dog, cat, pig, deer, ox, rabbit, mouse, and all animals with either fur or hair on their bodies are called mammals. Although nearly all these mammals are land, or terrestial, animals, a few of them have become adapted to life in water and live rather like fish. The whale is one of these kinds of mammals.

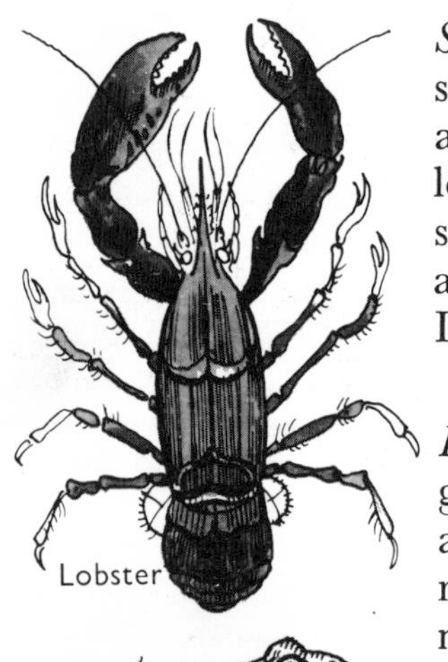
Lobster

Skeletons. Some animals have skeletons inside their bodies to support them. The elephant and all the mammals mentioned above have this kind of skeleton. Creatures such as the lobster, prawn, shrimp, crab, crayfish, etc., have supporting structures *outside* their body. These are jointed and so the animals are called Arthropods, which means " jointed limbs ". Insects and spiders are Arthropods too.

Fur, Hair and Scales. Animals show great variety in their body coverings, and groups of animal types are often recognised by these. For instance, mammals such as cats and dogs have fur or hair ; birds have feathers ; lizards and snakes (reptiles) have scales ; frogs, toads and newts have slimy skins and fish have oily scales.

Cat— fur

Bird— feathers

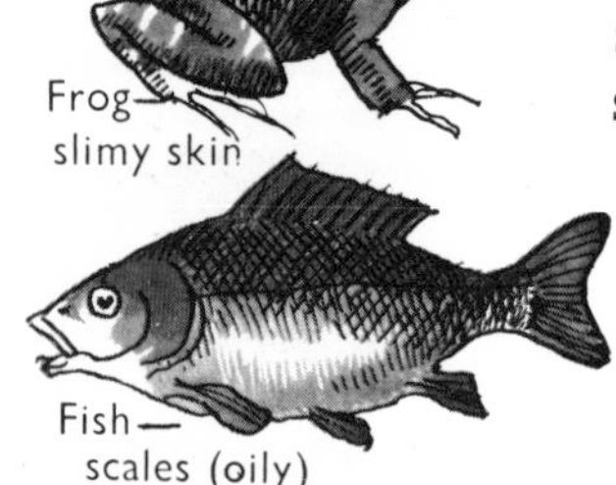
Frog— slimy skin

Fish— scales (oily)

Lizard—scales (dry)

A PAGE OF CONTRASTS—PLANT LIFE

Wild rose

en plant
ree)

Land plant
(bluebell)

There are plants which have flowers and which produce fruits and seeds, such as the wild rose, and there are plants which have no flowers and which produce spores instead of seeds, such as the fern plant.

There are plants which are leafy and green, and there are plants which have no green colouring matter and live on decaying food materials or are parasites on other living plants. Fungi (mushrooms, toadstools, moulds, etc.) are examples of non-green plants.

There are flowering plants which live on the land, and there are flowering plants which are adapted to live in water. Some of the water plants have floating leaves and flowers like the water-lily plant.

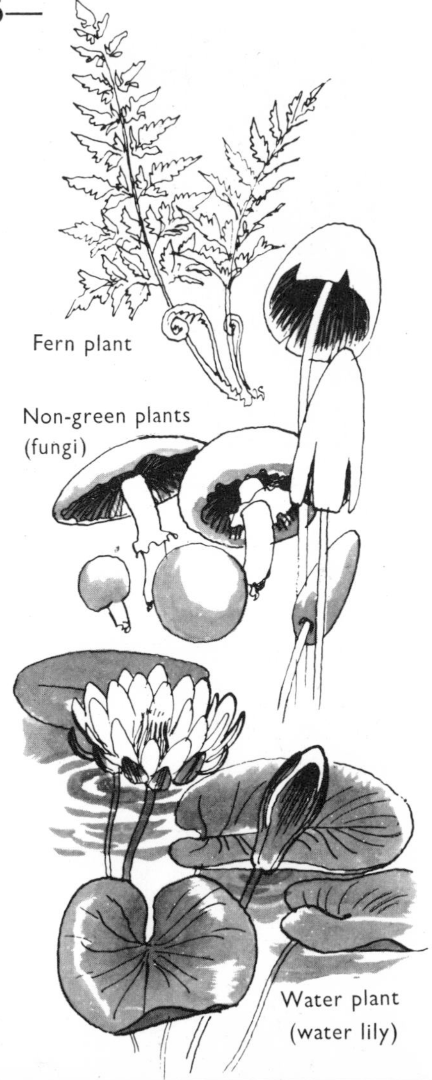

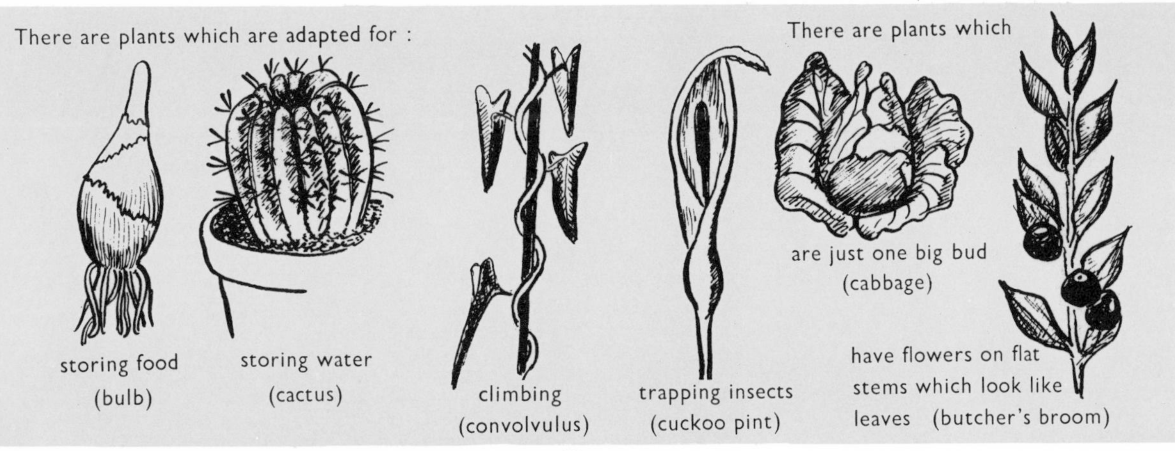

AMONG LIVING THINGS WE RECOGNISE

A PLANT KINGDOM

In the plant kingdom, there are :

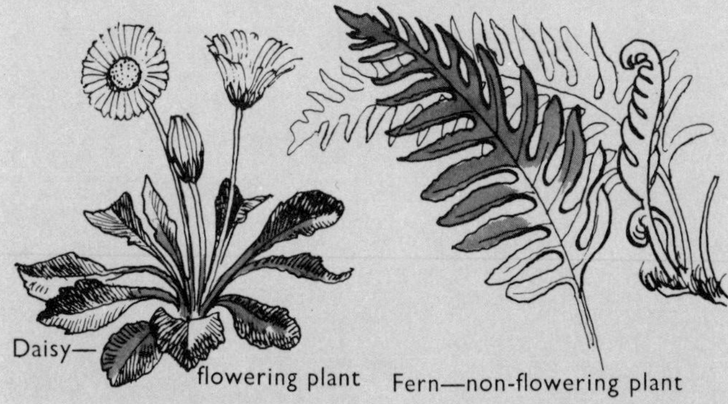

flowering plants and non-flowering plants

Among the flowering plants, there are :

plants which have net veined leaves and plants which have parallel veined leaves.

Among the non-flowering plants, there are :

Ferns, mosses, lichens, liverworts.
Algae (such as seaweeds) and fungi.

There are also :
very simple one-celled plants, such as protococcus.

AND AN ANIMAL KINGDOM

In the animal kingdom, there are :

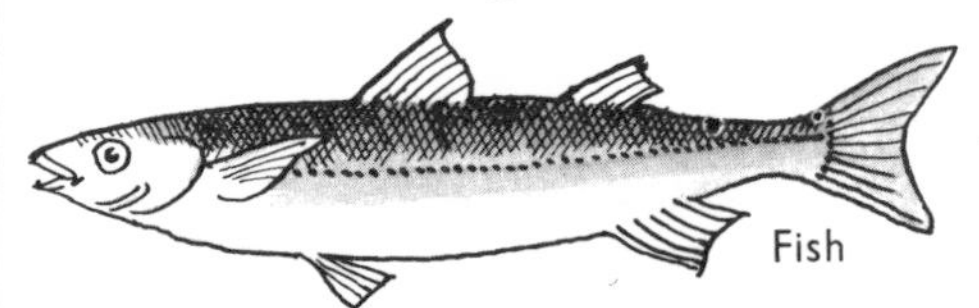

animals with backbones and

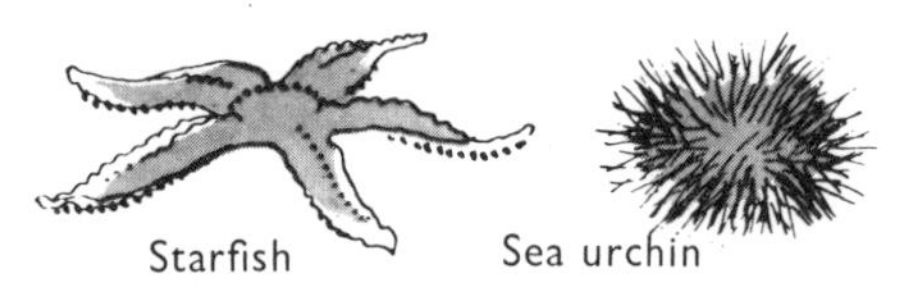

animals without backbones.

Among the animals with backbones there are :

animals with fur (mammals)
animals with feathers (birds)
animals with dry scales (reptiles)
animals with moist skin (amphibia)
animals with oily scales (fish)

Among animals without backbones, there are :

arthropods and crustacea
molluscs—soft-bodied animals with shells
worms
starfish, sea urchins, sponges
very simple one-celled animals, such as the amoeba

"WORKING WITH NATURE"

Score a point for each one of these tasks which you have carried out :

1. Made leaf prints
2. Taken bark rubbings
3. Made twig imprints or casts
4. Used a thermometer
5. Used a barometer
6. Made a graph
7. Tried out air-pressure experiments
8. Estimated wind speeds with the Beaufort scale
9. Seen an anemometer
10. Discovered weather signs
11. Tried soil and water permeability experiments
12. Tried an osmosis experiment
13. Made a wormery
14. Proved by experiment that green plants give off oxygen in sunlight
15. Tested oxygen gas with glowing splinter
16. Set up a well-aerated aquarium
17. Seen the gills of a fish
18. Grown some mould
19. Made spore prints
20. Tested foods for presence of starch
21. Tested foods for sugar
22. Tested foods for oil
23. Tested foods for protein
24. Tried out chemical experiments with a hen's egg
25. Looked at pollen grains with a lens or microscope
26. Found ovules in the ovary of a flower
27. Discovered how seeds disperse in several fruits
28. Tried to make an incubator
29. Learnt the names of the different parts of a bird's body
30. Made a bird hide and used it
31. Identified several birds
32. Kept a golden hamster
33. Made a hamster cage or rabbit hutch
34. Set up a bird table
35. Gone hunting for insects
36. Recognised at least seven insect types
37. Set up a caterpillar-rearing cylinder
38. Made a pond dipping net and used it
39. Made a press for flowers and grasses
40. Mounted flowers and grasses
41. Recognised several flower families
42. Made a plant distribution map or chart, or a plant profile
43. Kept a record of flowering plants
44. Studied ferns, mosses, liverworts and other non-flowering plants
45. Set up a moss garden
46. Germinated some seeds
47. Tried experiments with growing seedlings
48. Observed plant movements (tropisms)
49. Discovered how a worm moves
50. Discovered how a snail moves
51. Discovered how a fish swims
52. Looked at the structure of a bird's wing
53. Seen various kinds of five-fingered limbs
54. Discovered other creatures (besides birds) which can fly
55. Made a collection and displayed it
56. Tried mounting seaweeds
57. Kept a Nature diary or class record

NEW WORDS IN PART THREE

AERATE	to allow air to penetrate
ALGAE	very simple green plants such as seaweeds
AMOEBA	a very simple one-celled animal
ANEMOMETER	an instrument for measuring wind speeds
ANEROID BAROMETER	a barometer without mercury
AQUARIUM	a tank for water creatures
AQUATIC	living in water
ARTHROPOD	an animal with jointed limbs, e.g. an insect
BARBS AND BARBULES	tiny parts of a feather
BEAUFORT SCALE	a scale of wind speeds
BIRD HIDE	a tent for watching birds
BUNSEN BURNER	a gas burner used in laboratories
CARBOHYDRATES	starchy or sugary foods
CARBON DIOXIDE	a gas in the atmosphere
CARPEL	female part of flower—containing ovules or young seeds
CARTILAGINOUS	not bony
CELL	living unit from which plants or animals are built up
CHAETAE	bristles of an earthworm
CHALAZA	one of the twisted white cords holding up the yolk in an egg
CONCHOLOGIST	a person who collects shells
CRUSTACEA	animals such as crabs, lobsters, prawns, shrimps
CYCLOID SCALES	round scales found on bony fishes
EGG TOOTH	lump found on beak of embryo chick
ELYTRA	the hard outer wings of beetles
EMBRYO	very young plant or animal
FEHLING'S SOLUTIONS	two solutions used as a test for sugar
GERMINAL DISC	part of egg which could develop into a chick
HYDROGEN	a gas in the atmosphere
INCUBATING	keeping warm (eggs are incubated)
INORGANIC	not living
LICHEN	a grey-green plant often found on tree trunks
LOAM	a good soil mixture of sand and clay
MAMMALS	warm-blooded animals with fur

MEMBRANOUS	thin and delicate
MOULD	a simple fungus
MYOMERES	muscles which work the body of a fish
ORGANIC	living or once living
ORGANISM	a living thing
OSMOSIS	the passing of liquids through membranes or tissues
OVULE	young seed—not yet fertilised
OXYGEN	a gas in the atmosphere
POLLEN	yellow fertilising dust of flowers
PRECIPITATION	rain, hail, snow, etc.
PROTEIN	a food substance found in white of egg, meat, etc.
PUPA	the resting stage of an insect
SILHOUETTES	outlines showing shapes of objects, usually filled in with solid black
SPERMS	male cells
SPORES	the reproductive, seed-like bodies of non-flowering plants
STAMEN	male part of flower bearing pollen
STIGMA	top of carpel in flower
TELEOSTS	bony fishes
TEMPERATURE	the degree of heat or cold
THERMOMETER	an instrument which records temperature in degrees
TROPISMS	growth movements in plants

FLOWER IDENTIFICATION BOOKS

A Pocket Guide to Wild Flowers, D. McClintock and R. S. R. Fitter, Collins
Companion to Flowers, D. McClintock, Bell
Finding Wild Flowers, R. S. R. Fitter, Collins
Flowers of the Coast, Ian Hepburne, Collins
Garden Flowers, Vlastimil Vanek, Hamlyn
Pocket Encyclopaedia of Wild Flowers, M. S. Christianssen, Blandford
The Concise British Flora in Colour, W. Keble Martin, Sphere
The Observer's Book of Wild Flowers, W. J. Stokoe, Warne
The Oxford Book of Wild Flowers (Pocket edition), O.U.P.
Wild Flowers, John Gilmour and Max Walters, Fontana
Wild Flowers, Alois Kosch, Burke

PART 4

LOOKING AT LIFE

CONTENTS OF PART 4

ABOUT PART FOUR

During your school life, you are constantly learning about living things. You know, too, that living things, such as plants and animals, differ from non-living things such as stones, water, metals or gases.

Now is the time for you to discover what it is that makes living things differ from non-living things, and to think seriously about LIFE and all that life and LIVING mean.

You are in the state of being alive and so are the living plants and animals around you. How would you describe things that are alive ?

Here are some of the " properties " or " attributes " of living things.

They feed and grow and reproduce themselves (or " multiply "). They are also active and sensitive to outside conditions.

Non-living things do not have all these properties. What causes living things to behave in this way ?

Scientists tell us that it is because they possess a special, active substance called PROTOPLASM. This is found in almost every part of a living plant or animal. In fact, units of protoplasm, or " cells ", make up the bodies of animals and plants.

Protoplasm is a complicated substance, something like protein, which is a body-building substance found in meat and fish and white of egg.

Protoplasm is, in fact, a protein substance of a very special kind. In spite of knowing this, no scientist has yet been able to make living protoplasm.

ENERGY

What makes us certain that a plant or animal is really alive? You will probably say: "An animal can move if it is alive". Is this true of plants? Plants can, in fact, make certain movements but these are not always easy to detect. You can read about plant and animal movements on pages 258 to 266.

In growing, moving, playing (in young animals), competing for good living conditions, and so on, living things use up a tremendous amount of energy. When it is used up, they are tired or fatigued like the puppy in the picture. Unless it is ill, however, the puppy will not *remain* fatigued, because its energy will be renewed. How does this come about? Read on and you will find out.

Our beautiful plants and animals would die out if they could not multiply. You will read about this and see how, from small beginnings, the complicated structures of the cells, tissues and organs of living plants and animals develop.

Man himself destroys many living things, sometimes using them for food, sometimes crowding them out when he builds towns. Now, as you will read in this section of the book, man is making efforts to save some of this wild life.

FATIGUE

Living Things Need Food

Because living things are using up so much energy through growing or moving or struggling for existence, they need to restore that energy. A car will not go unless we give it petrol and a fire will not burn unless we provide it with fuel in the form of wood or coal. In a similar way, living things need their own special kind of fuel in the form of food.

With what kinds of " fuel " or food do you restore your energy ?

You eat starchy foods such as potatoes, bread or rice, and you eat fats such as butter, cream and fat meat.

Sugar is found in many kinds of food. It is very like starch but dissolves more easily and gets quickly into your blood stream.

Sugars and starches provide energy. Fats are also energy-giving.

As well as sugars, starches and fats for energy, you eat body-building proteins in the form of eggs, meat, fish, peas and beans. You have heard of protein before. It forms bone, muscle and blood, and becomes part of that wonderful living substance: protoplasm.

Fresh fruit and vegetables are good for you. They provide vitamins which help the organs of your body to act efficiently.

A car runs on energy provided by the consumption of petrol

The burning of a fire releases energy in the form of heat

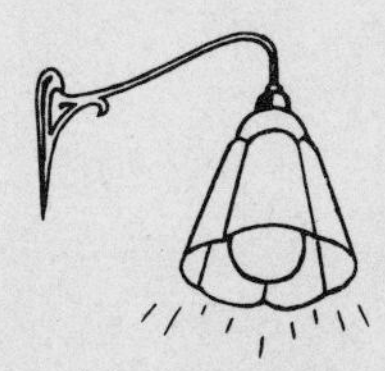

Electrical energy in an electric light bulb is shown in the form of light, but an electric radiator turns this energy into heat

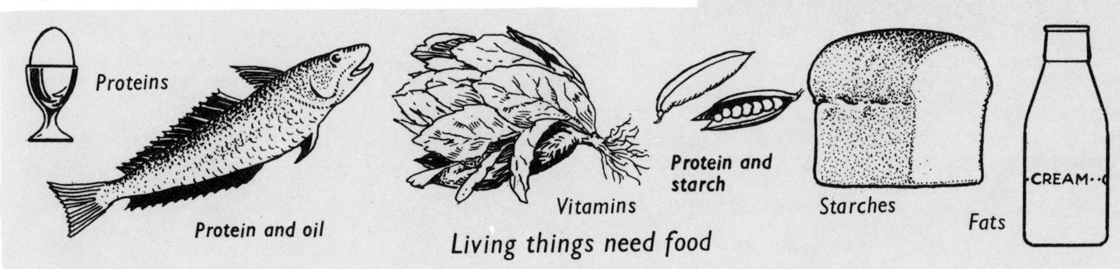

Living things need food

THE FOOD OF GREEN PLANTS

Green plants rely on the soil and the air for their food.

Animals cannot use these materials as food and must have *living* or " organic " food in the form of other animals or plants.

Plants make their food with the materials they get from soil and air. Very often they store some of the food they make. Diagram (a) shows a carrot root swollen with plant food. Compare this with an ordinary root as shown in diagram (b).

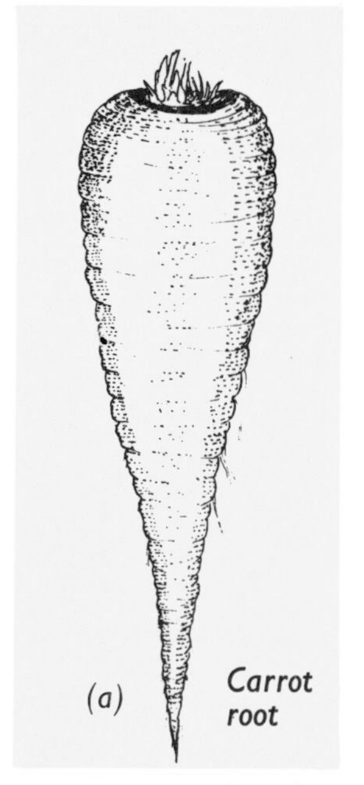

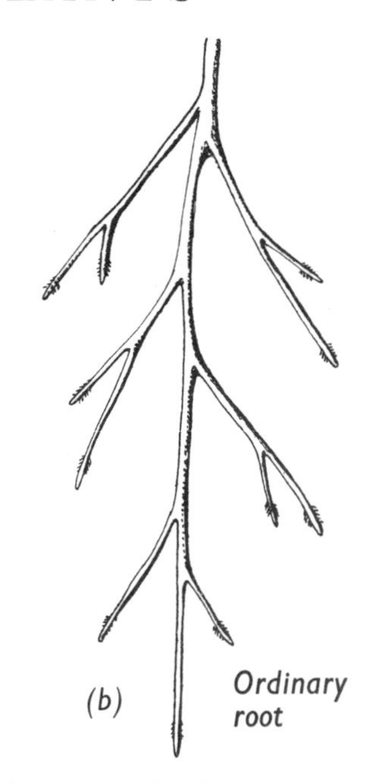

From the soil a plant gets water and mineral salts which are dissolved in the soil water.

From the air the plant gets a gas called carbon dioxide.

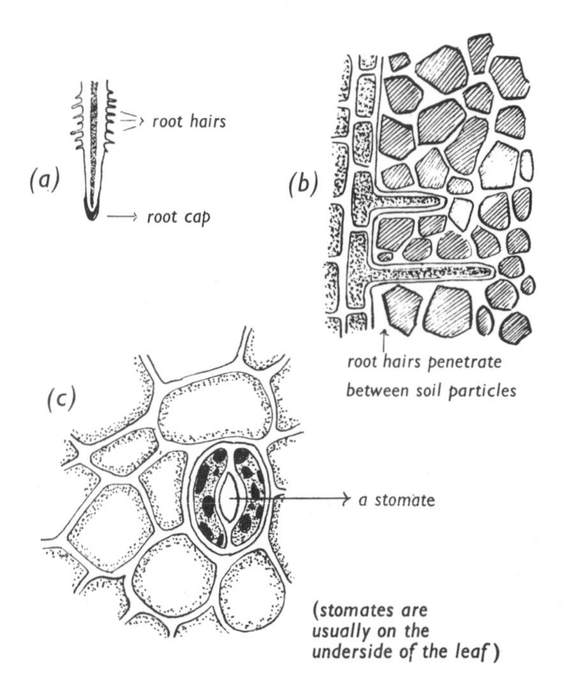

With these ingredients (water, salts and carbon dioxide) it makes its own special kind of food, some of which it uses at once and some of which it may store as you saw above.

On the left you will see, in diagram (a), an enlarged drawing of the tip of a root showing the root hairs. These absorb water and mineral salts from the soil. Diagram (b) shows two root hairs (much magnified) in between the soil particles.

Diagram (c) shows one of the pores through which gases such as carbon dioxide can enter the leaves of the plant. These pores are called *stomates*.

GREEN PLANTS AND THE SOIL

The Work of a Soil Analyst

If you were a farmer or a gardener, you would need to know a good deal about the kind of soil food or soil conditions which different plants require. If you were not satisfied with the standard of your crops, you would probably send samples of your soil to a soil analyst. After making certain tests, the soil analyst would be able to tell you whether your soil needed some kind of fertiliser to make it right for the particular crops you wished to grow.

Soil testing is an interesting process. We test soil mainly to discover how much acid there is in it and the diagrams on the right show how this is done.

1. This shows the soil being pulverised or ground to fine powder by a special machine.

2. Distilled water and barium sulphate are added to the powdered soil in a test tube, and shaken up. Barium sulphate, or barytes (sometimes known as "heavy spar") helps the soil to settle in the tube, leaving a rim of fairly clear soil water above.

3. Soil indicator—a bluish liquid—is then added and if we look at the "clear rim" mentioned above (c.r. in the diagram), we see that it takes on a certain colour—red, or yellow, or green, or blue, or some colour between these. By comparing this colour with a colour chart we can find out how acid the soil is. Red shows an acid soil. Blue shows an alkaline soil (the opposite of acid). Green shows that the soil is neutral.

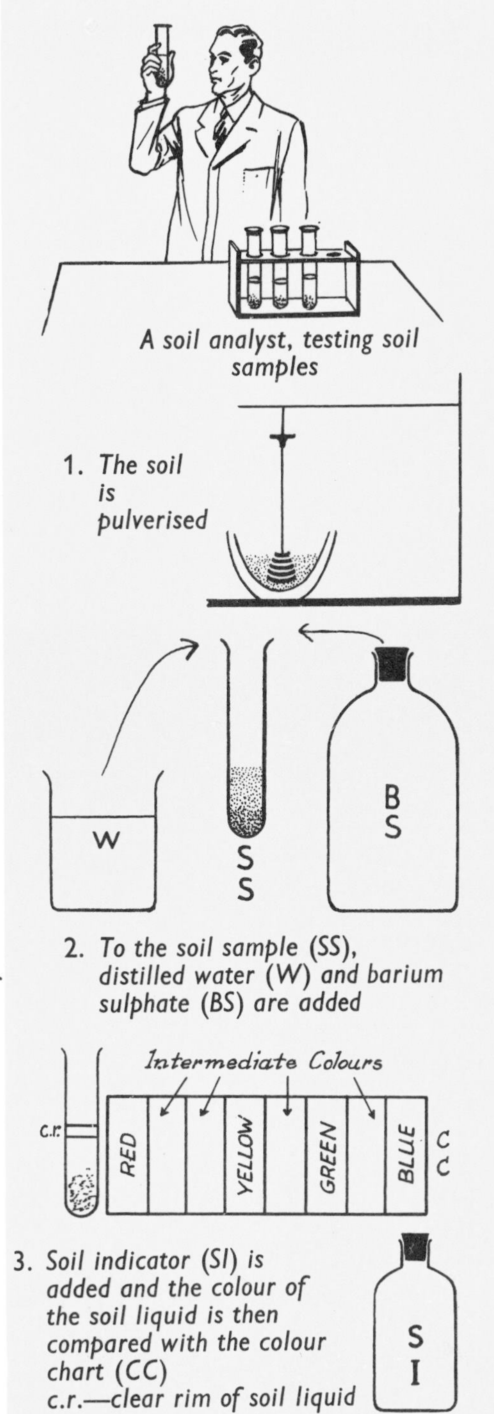

A soil analyst, testing soil samples

1. The soil is pulverised

2. To the soil sample (SS), distilled water (W) and barium sulphate (BS) are added

3. Soil indicator (SI) is added and the colour of the soil liquid is then compared with the colour chart (CC)
c.r.—clear rim of soil liquid

GREEN PLANTS—FOOD-MAKERS OF THE WORLD

The flesh-eating lion watches his chance to spring on a grass-eating buck. Very often it is the lioness who makes the kill

As you have read, green plants are able to make use of simple materials from soil and air to build up their own special kinds of food substances. These special food substances, such as starch, sugar, protein and oil, are " manufactured " by the green plant and the " factories " in which these foods are prepared are the leaves and the other green parts of the plant. You have also read how some of the surplus food of a plant is stored in its swollen roots or stems. In fact, plants are full of good food and the vegetable-feeding animals such as cows, sheep, deer and rabbits know this and make good use of their " food plants ".

All animals, however, are not vegetarians, or " herbivores ". Some are flesh eaters, or " carnivores ", and these regard many of the herbivores as their natural prey. Thus, the flesh-eating lion in Africa kills the gentle impala or some other grass-eating, deer-like animal. In the same way, foxes kill and eat rabbits or chickens, and stoats catch field mice.

The defenceless vegetable feeders, which form the food of the flesh eaters, could not exist without green plants, so we can truly say that green plants are the " food-makers of the world ".

THE GREENNESS OF GREEN PLANTS

Have you ever wondered why the commonest colour in Nature is green ? Can you imagine what it would be like if the grass and trees were some other colour—perhaps red ? There are, of course, some non-green plants such as mushrooms and toadstools. Apart from these, most of the common plants of the countryside and gardens have green stems and leaves, although their flowers and fruits are often of many other beautiful colours.

You will probably have noticed that only the aerial or " above ground " parts of plants are green. If you dig in the garden, you will find that roots and underground stems are usually of a white or cream colour, or sometimes brown or black. If you look at those parts of rhubarb or celery stems which are usually covered with soil, you will see that they are not green where they have been hidden from the light.

Is there then some connection between sunlight and greenness ? If you have guessed that there is, you are right. Plants cannot make their green colouring matter (called chlorophyll) without sunlight, and without chlorophyll they cannot make their food. This green colouring matter (or pigment) is usually carried in little bodies called *plastids* in the *cells* of the plant. Below you can see some plastids in a plant cell.

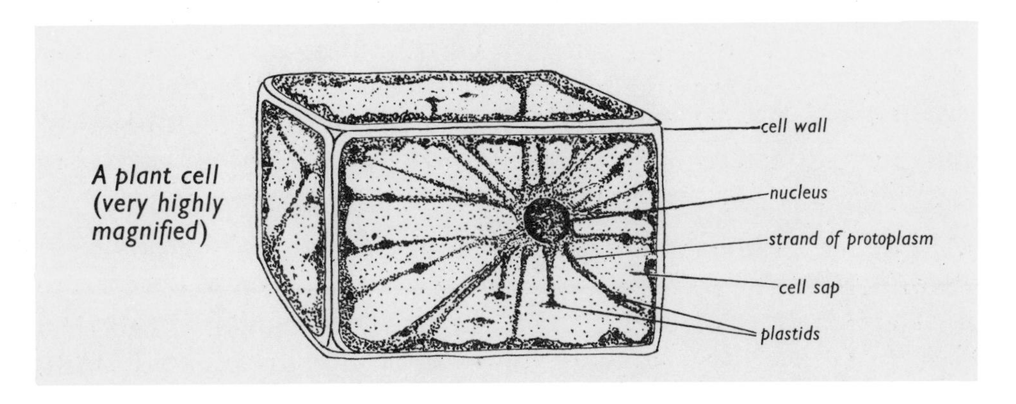

A plant cell (very highly magnified)

Look carefully. You are looking right inside this plant cell. It is a three-dimensional view. As well as the plastids, can you see the nucleus, which is the centre of control in the cell, and the strands of protoplasm ? The rest of the cell is filled with cell sap.

Spirogyra as it looks to the naked eye—a network of fine green threads

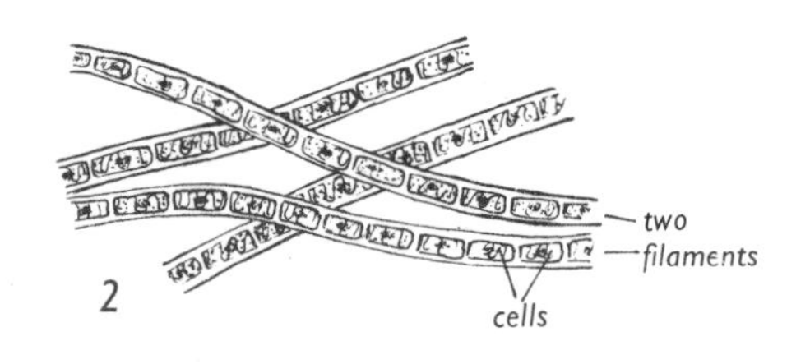

Some filaments of Spirogyra magnified to show the cells

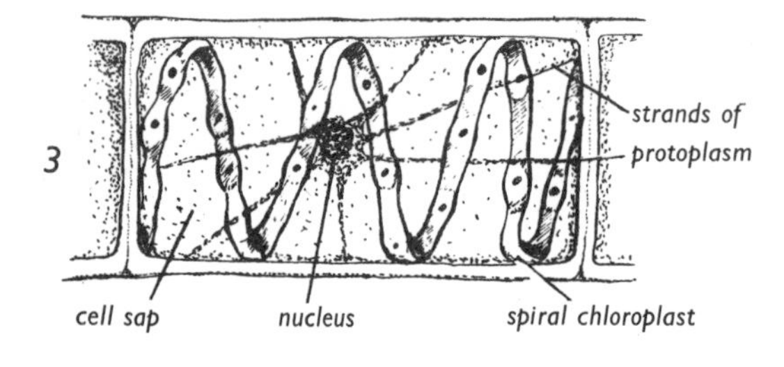

One cell from a filament of Spirogyra. This has been highly magnified to show the spiral chloroplast

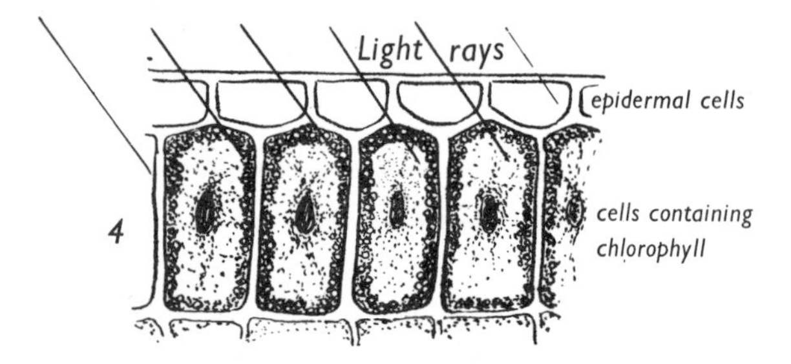

This diagram shows light rays penetrating through the cells of the epidermis of a leaf to reach the chlorophyll-containing cells

On page 251, you read that greenness (or chlorophyll) is usually carried about in small bodies called plastids in the plant cells. Some of these plastids are peculiar shapes. In diagram 3 below you can see a *spiral* plastid. "Spiral" means twisting round like a corkscrew. This type of plastid is found in a pond plant called *Spirogyra* which looks just like green scum on the top of stagnant pond water.

If you look closely at this scum you will see that it is made up of fine threads (diagram 1). Examine these fine threads, or filaments, with a microscope and you will see that they consist of strings of cells (diagram 2).

If you could magnify each one of these cells even more, you would see that each one contains a spiral green body (diagram 3). This is the *chloroplast.* Plastids are called chloroplasts or chloroplastids when they are *green*—that is, when they carry the green colouring matter, or chlorophyll, in a plant cell.

Light rays can pass through the cells of the upper epidermis (skin) of a leaf and so reach the green cells beneath, which contain chlorophyll. Light provides energy, and it is in these green cells that the food of the plant is made.

A CHEMICAL FACTORY IN A GREEN PLANT

Looking at the green leaf shown on the right, you would hardly believe that there was a chemical factory inside it, and yet this is true. At the bottom of this page you will see a diagram of a plant stem which has been cut across. The circle of dots which you can see at the top of the stem are the ends of the veins through which water and dissolved salts rise up to the leaves and down which the prepared food travels on its way to the various parts of the plant body.

A leaf of horse chestnut

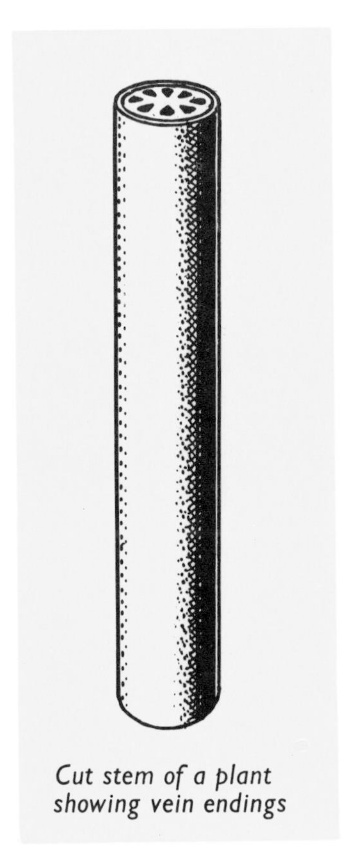

Cut stem of a plant showing vein endings

You know now that green plants take in raw materials from soil and air (water, mineral salts and carbon dioxide) and that, from these, they can manufacture more complex materials (starches, sugars, proteins and oils) in their leaf factories.

You know also that, in order to do this, green plants make use of energy from the sun.

You may know that in big chemical factories where products are manufactured, extra products, known as by-products, are often formed at the same time. This also happens in a " leaf factory ". In this case, the by-product is oxygen. While the plant is busy making its food, oxygen is being formed and given off into the atmosphere. The oxygen escapes through the pores or stomates of the leaf. This oxygen which plants *give out* replaces much of that *taken in* by animals and plants when they breathe.

A CHEMICAL FORMULA

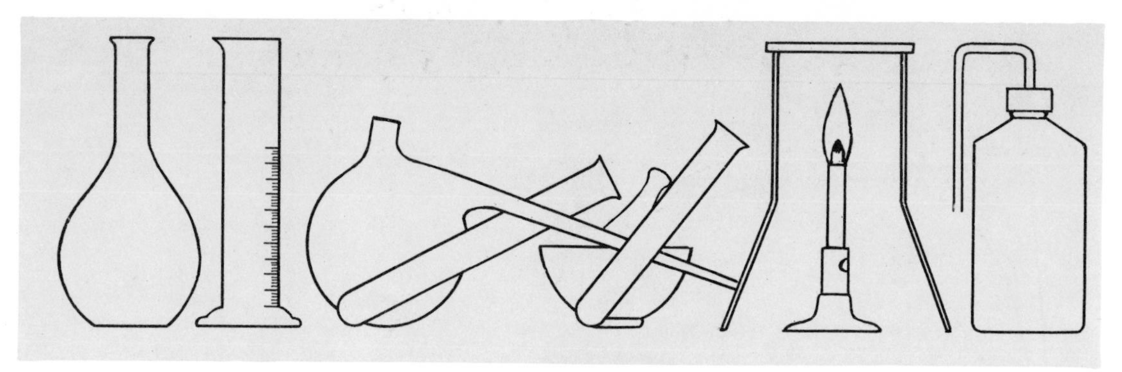

If you find this page difficult to understand, miss it out now and return to it later.

You have probably heard older children or adults talking about chemical formulae. These are short ways of describing chemical processes by using letters and numbers which are known as " symbols ".

For instance, the symbol for oxygen is " O ", for hydrogen it is " H ", nitrogen is " N ", and carbon is " C ".

Oxygen, hydrogen, nitrogen and carbon are single substances known as *elements,* and elements often come together in varying quantities to form *compounds,* or substances made of more than one element. The gas, carbon dioxide (CO_2) is composed of the elements of carbon and oxygen in the proportion of one part of carbon to two parts of oxygen, while water, the symbol of which is H_2O, is made up of the elements of hydrogen and oxygen in the proportion of one part of oxygen to two parts of hydrogen.

During chemical processes elements are often inter-changed and new substances are formed. The chemical formula below shows this :

$$CO_2 \text{ plus } H_2O \text{ gives } CH_2O \text{ plus } O_2$$

If you study this you will see that some of the elements in the carbon dioxide (CO_2) have combined with the elements in the water (H_2O), to make a new substance, formaldehyde (CH_2O), from which starch or sugar can be built. You will also see that oxygen (O_2) is given off in this process.

This chemical process takes place in the " factory " of the green leaf and the chemical formula above explains it.

GREEN PLANTS AND THE ATMOSPHERE

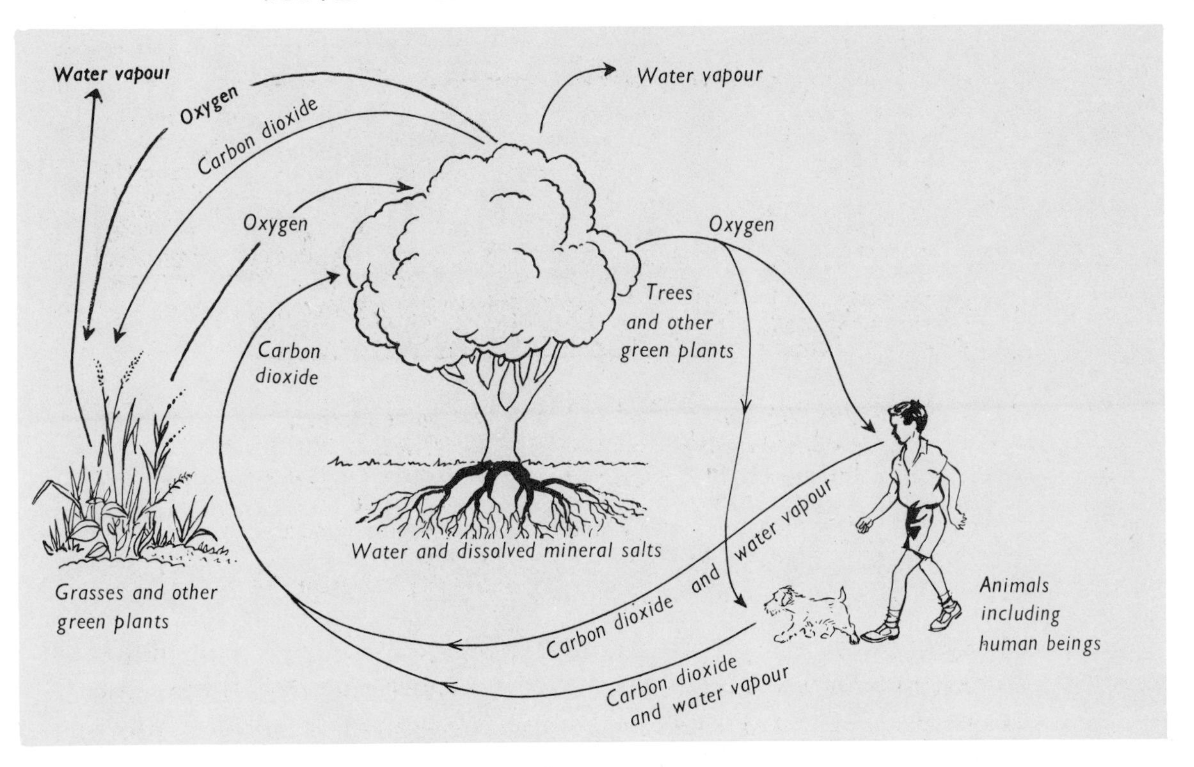

The picture above represents the interchange of gases between plants and animals (including human beings).

Trees and other green plants take in *water* and *mineral salts* from the soil and the gas *carbon dioxide* from the air. You have already read how a green plant uses these as ingredients in its food-making process. Green plants must also *breathe* and therefore must take in *oxygen* (study the direction of arrows in the picture).

People and animals also take in oxygen when they breathe. They do not need to *take in* carbon dioxide, but they *give it out* as part of their breathing process. You will see by the long arrows from the boy and the dog that the plant takes in this carbon dioxide, while other arrows show that the boy and dog breathe in the oxygen which the plant gives *out*. Notice that green plants, people and animals all give out carbon dioxide *and water vapour* during breathing.

A GARDEN IN A BOTTLE

Here is an example of a " balanced atmosphere " within a bottle.

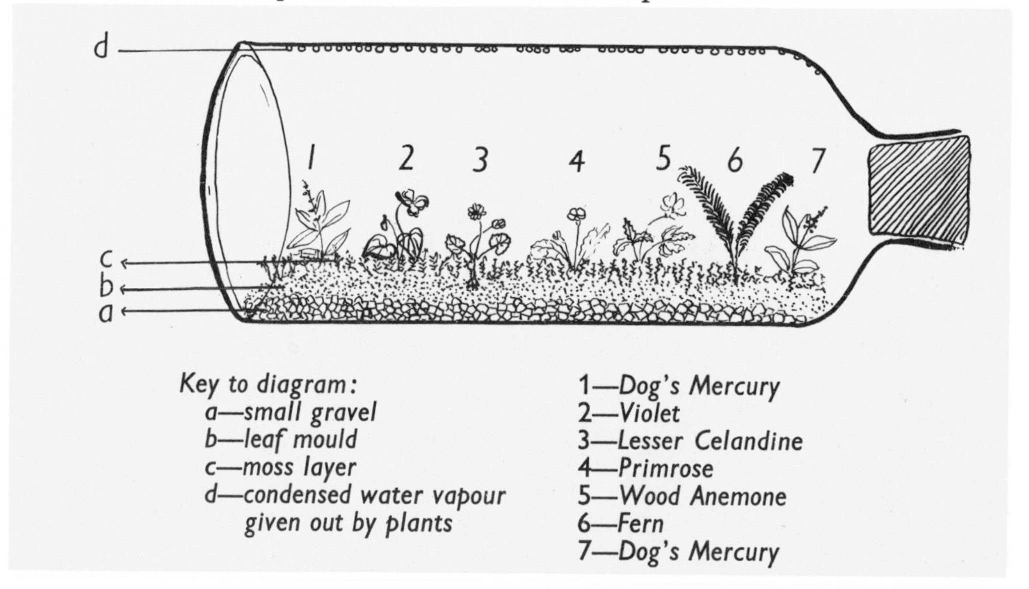

A little garden is set out inside a big sweet bottle. The plants are set in shallow soil or leaf mould with moss below and around them. In fact the plants are almost embedded in moss. Small woodland plants, such as violets and lesser celandines, are probably the best for this as they are used to moist and mossy surroundings.

The plants are watered once and then the bottle is sealed up. After this, *you will never have to water the plants again!* Does this surprise you? Furthermore, *you will never need to allow any more air into the bottle!*

If you have studied the previous pages, you will know why this is true.

Green plants can balance their own atmosphere. Although these plants in the bottle are using up oxygen when they breathe, they are also giving it out again in their food-making process. In the same way, while they are taking in carbon dioxide for their food-making, they are also giving it out in their breathing process. They also take in the water you gave them, but they give it out again (can you see the water droplets on the glass?).

COLOUR VARIATIONS IN PLANTS

COLOUR VARIETY IN NATURE

On pages 251 and 252, you read of the importance to plants of the green colouring matter : chlorophyll. Plants without this green substance are unable to make their own food and may have to take prepared food from some other living organism.

Some plants live entirely on another living plant, which is called the " host ". Such dependent plants are called parasites and an example of one, the *toothwort,* is shown on the opposite page (C). You will observe that it has no greenness but is coloured pink, ivory and brown. It is parasitic on hazel.

Look out for the bright and varied colourations of toadstools. On the opposite page you will see the *scarlet fly cap* (D), which lives on dead or decaying materials. Plants living like this are called saprophytes. They do not need a *live* host, but can feed on the dead or decaying remains of other plants, and sometimes of animals.

Most flowering plants have attractively coloured flowers, like the *primrose* (A). In many cases brightly coloured petals attract certain insects which help to pollinate the flower. Bees seem to prefer some colours, such as blues and yellows, to others. Colours in flowers are usually caused by colour pigments which are dissolved in the cell sap of the petals or are carried about in special colour plastids within the cells. You may have noticed that early spring flowers are often paler in colour, while the later summer flowers are, on the whole, much brighter. The strength of sunlight has much to do with the development of strong colour pigment.

There are some plants, such as *dog's mercury* and *herb paris,* both woodland plants, which have green flowers. You can see a green flower (herb paris) on the opposite page. The carpels are somewhat blue or purple as they ripen, but the flower appears mainly green.

The colour changes which we see in many leaves in autumn are due to leaf decay and to the breaking down of the green colour pigment. The yellow and russet colours are due to the decomposing of the green pigment.

GROWTH AND MOVEMENT IN PLANTS

The growth of a plant begins at germination, when the seed, stimulated by warmth and moisture, springs to life. First the radicle, or young root, grows downwards. If it is a root like that of the bean seedling in the diagram, other (secondary) roots will grow from it. Growth in length of roots is called "elongation" and this takes place mainly at a place just behind the root tip.

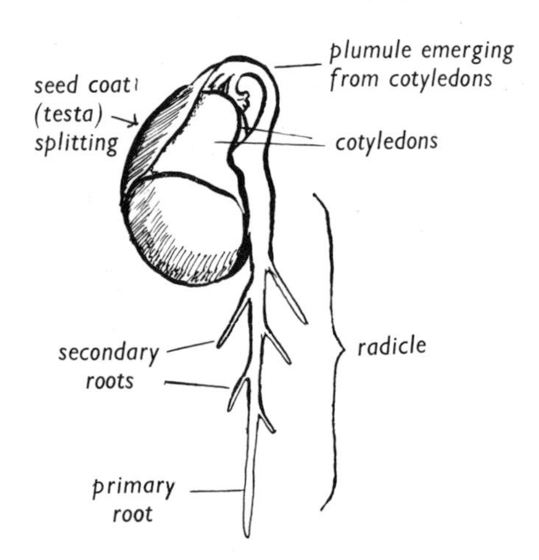

A bean seedling

The diagram below shows a seedling of oak grown from an acorn. You can see the acorn in the picture and the seedling is still drawing nourishment from it.

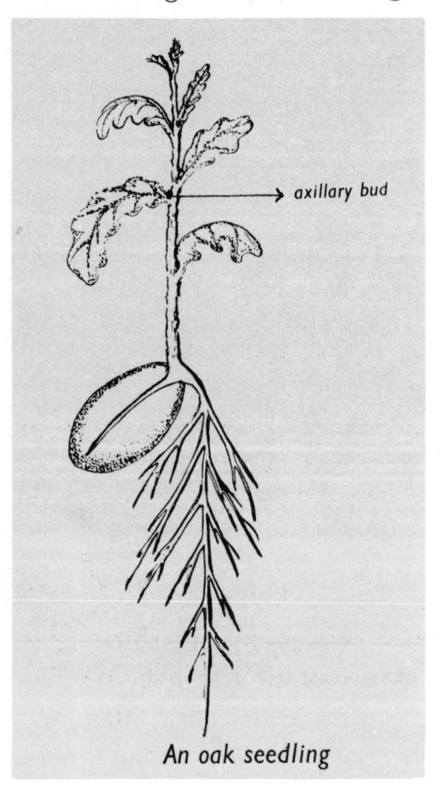

An oak seedling

Here you can see the downward growth of the root and the upward growth of the shoot. At the tip of the shoot there is a terminal bud and it is by the development of this bud that the shoot elongates, or grows longer. There are also buds in the axils (joining places) between leaves and stem. These are axillary buds from which side branches are formed.

We know that plants grow but do they move ?

Plants can move parts of themselves in various ways. The plant whose leaves are shown at the top of the next page, *Mimosa pudica*, is sometimes called the "telegraph plant" because of the peculiar movement it makes. Its leaflets are sensitive to touch. When touched they immediately close up against the leaf stalk.

Here you can see two views of the leaf of Mimosa pudica. *The first one shows the leaf as it usually is. The second shows what happens as soon as the leaf is touched*

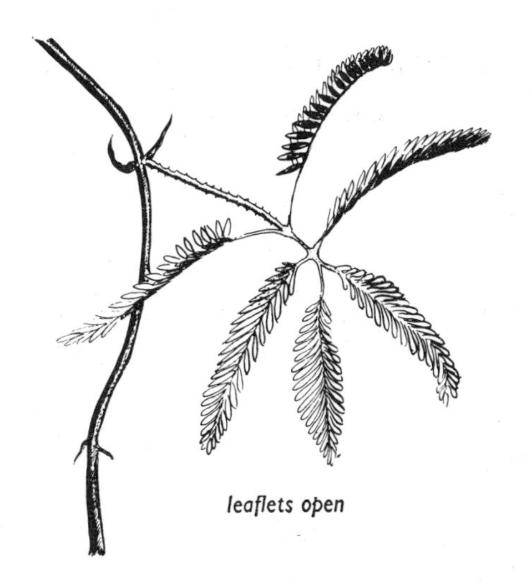

leaflets open

leaflets closed

Wood sorrel—leaves and flowers open

The pictures shown here of wood sorrel give another example of plant movement, this time a "sleep movement". The leaflets and flowers of this little plant close up at sunset or during very rainy weather.

Wood sorrel—leaves and flowers closed

You must have noticed sleep movements of this kind in various flowers. A good example of sleep movement is shown in the daisy. The white florets which are on the outside of this flower close over the inner yellow florets in the evening and we say that the daisy flowers have "gone to sleep".

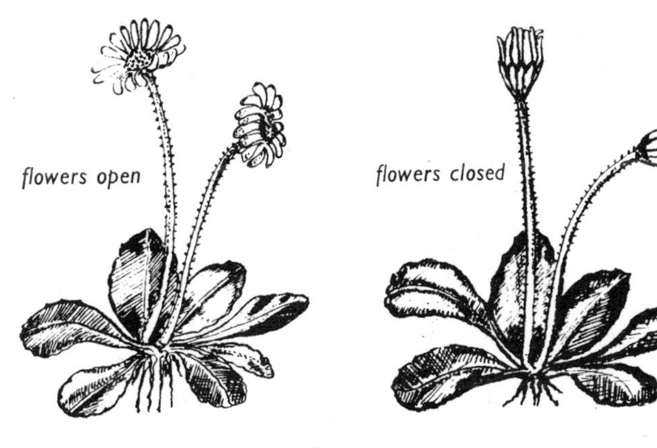

Daisy

Here is a record of an experiment carried out by a ten-year-old girl as she watched the opening of crocuses in springtime. You will find that Anne actually made crocus flowers close again after they had opened in the morning.

Experiment by Anne

"On Sunday, 13th February, I picked a bunch of yellow crocuses. I brought them indoors in the warmth and in no time they were wide open. I thought I would try and make them close up again. It was a cold snowy day and so I put them outside and soon all were closed. I tried to open the flowers again by bringing them into the warm room and they opened once more. Then I put them into a cold room and left them to close. I have drawn a crocus which I watched opening. The times at which I made these observations are given too."

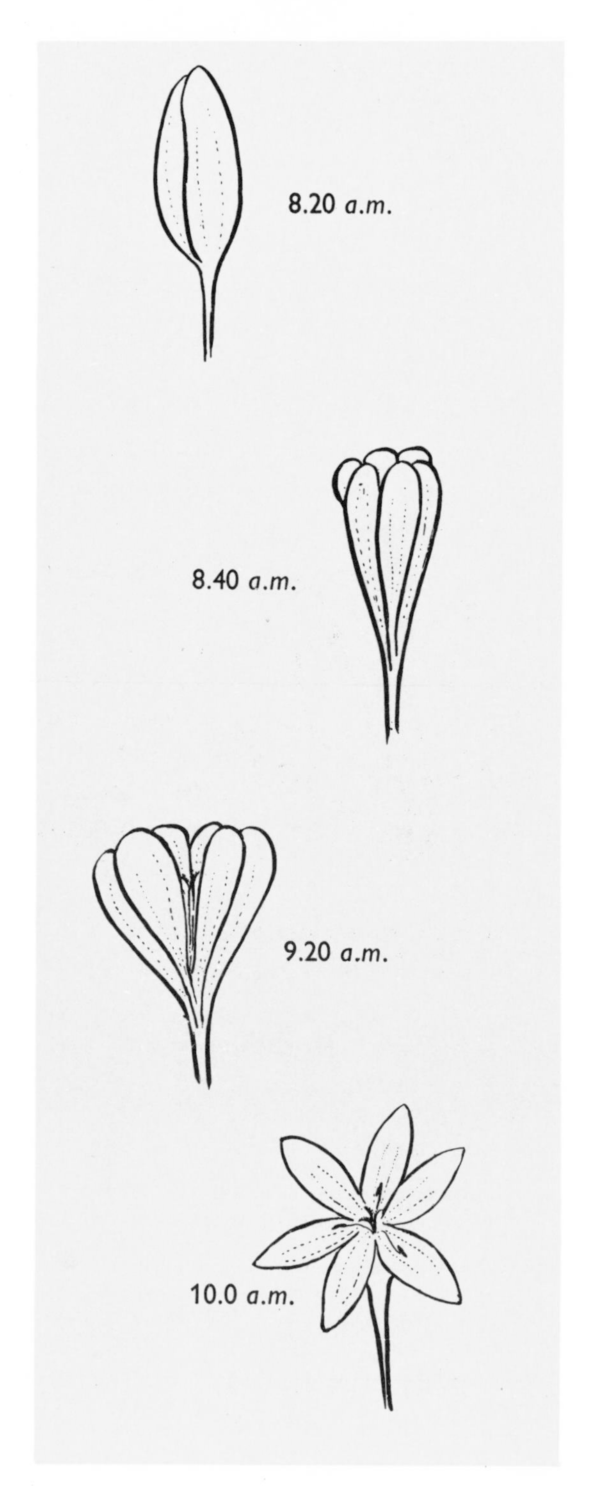

Perhaps you can try some experiments of this kind for yourself. Some plants which have "closing" flowers are daisies, marigolds, wood sorrel, crocuses and coltsfoot, but there are also many others.

Many plants of the trefoil (three-leaved) type such as shamrock, clover and wood sorrel have "closing" leaflets.

THE RESPONSE OF PLANTS AND ANIMALS TO STIMULI

When plants make growth movements, they are really responding to some sort of influence or stimulus (plural " stimul*i* ") from outside themselves. You have seen in Part 3 that plants respond to the stimuli of light, or water, or gravity (the " pull " of the earth). Some plants, such as *Mimosa pudica,* respond to the stimulus of contact, or touch.

Shoots grow and bend towards the light and are said to be " phototropic " (*photo* comes from a Greek word meaning " light "). Roots are *geo*tropic, that is, they grow downwards into the earth, responding to the influence of gravity. Roots are also *hydro*tropic (water-seeking) and sometimes go against the influence of gravity in order to grow towards water.

All living things respond in some way or other to stimuli. The simplest animals are those which consist of only one cell, like the microscopic amoeba or the *Paramoecium.* Even these simple forms of life can tell the difference between food particles and grains of sand. They will also move away from harmful fluids in the water, as you will see if you add strong salty or acid solution to water in which they are moving.

We might compare these movements of simple animals with the tropisms or growth movements of plants. In some cases the stimuli (water, food, light, contact, and so on) are similar, although different organisms may respond in different ways to them. Paramoecia, for instance, will move *away* from strong light. Most organisms respond to the nearness of food, but will repel, or move away from (or grow away from), anything which might be harmful.

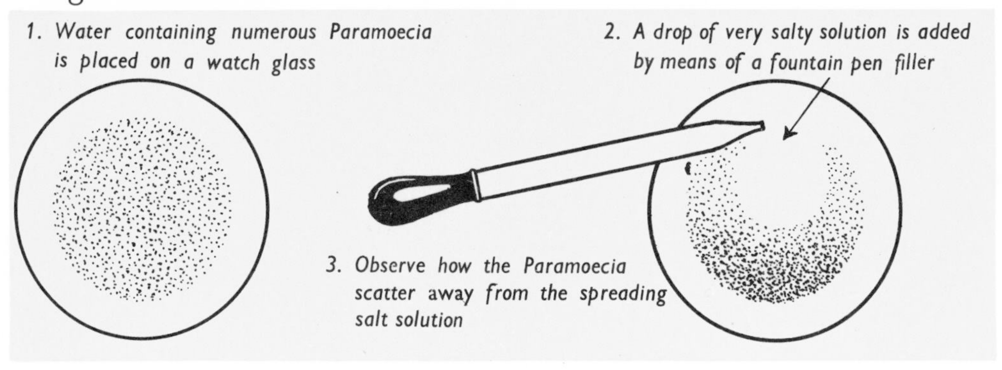

SENSE ORGANS

Higher animals have special organs to receive stimuli from outside. Our own eyes, our sensitive fingers, our noses, our ears and our tongues are all special organs of this kind. These are our *sense* organs and through them we receive impressions of light and colour, or sensations of touch, smell, hearing and taste. By means of this special equipment, we can test and appreciate the world of sights, sounds and smells around us. Animal equipment of this kind varies enormously and you could make an interesting study of animal sensory apparatus : such things as the antennae or " feelers " of moths and butterflies, the tentacles of snails, the whiskers of rabbits, and so on.

Sense organs, such as eyes, ears, fingertips and tongues, can *perceive,* that is, they can receive impressions from the outside world through their sensitive nerve endings. From these nerve endings, messages travel along nerves or neurons to the brain, or to the centre of the nervous system, and when this message is received, some *sensation* is experienced. This sensation may be of *taste*, *sight*, *sound*, or may be some form of *touch* sensation. If the touch sensation is caused through some sort of violent contact, or through contact with a sharp or hot surface, the sensation may be that of *pain.*

Plants and the more primitive forms of animal life have no nervous systems so that it is assumed that they cannot feel or experience sensations in the same way that higher animals do. Naked protoplasm is sensitive, however, as you will find if you prod the microscopic amoeba with a fine glass needle while you observe it on a microscope slide.

This diagram shows a small section of human skin (much magnified), with nerve endings and neurons. Try finding these nerve endings yourself by pressing the point of a needle gently on different areas of your own skin. Record what sensations you feel.

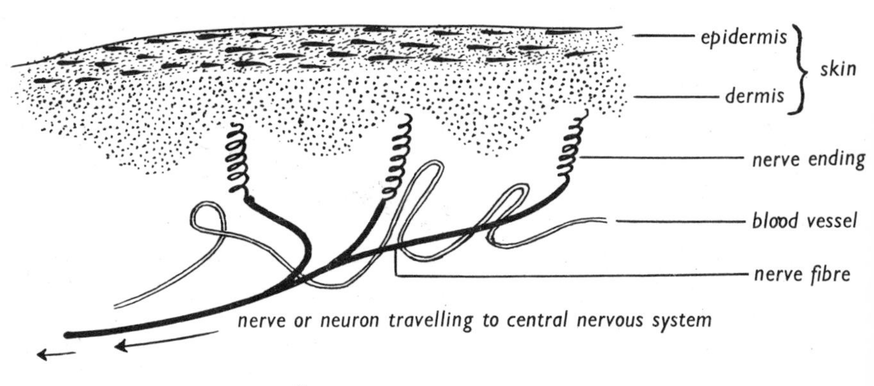

MOVEMENT IN ANIMALS

Animals are not usually fixed in one position and are therefore free to move about from place to place. This sort of movement is called " locomotion ".

There are, however, some primitive kinds of animals which do not move quite so freely as this and often remain in one place for periods of time. The pictures on this page show you three such animals : the hydra, the sea anemone and the limpet. These animals can move when the need arises. The hydra (which is very tiny and only just visible to the naked eye) can perform looping and somersaulting movements, but for long periods it remains attached to some water plant by means of its basal disc (or " base "), just waving its tentacles about.

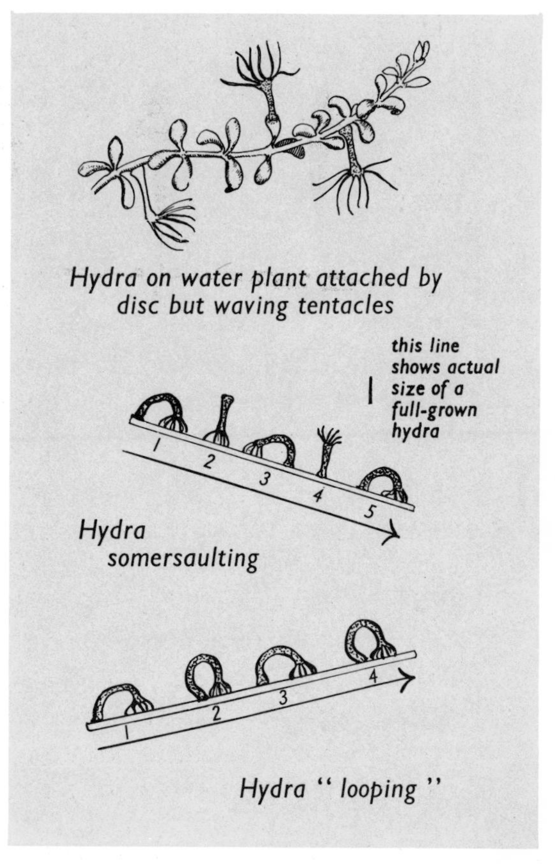

Hydra on water plant attached by disc but waving tentacles

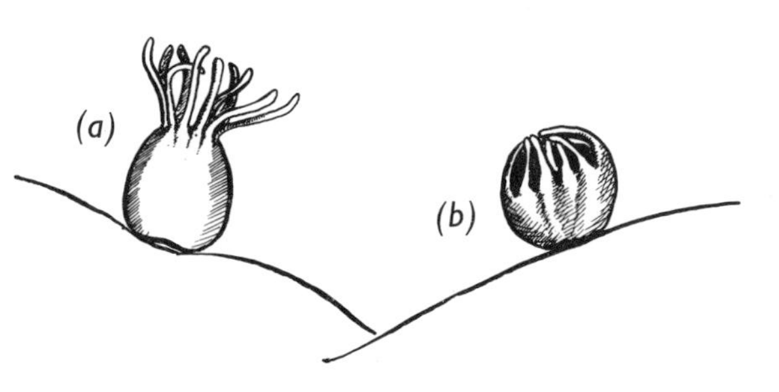

The sea anemone (a) open (b) closed
(diagrams slightly less than actual size)

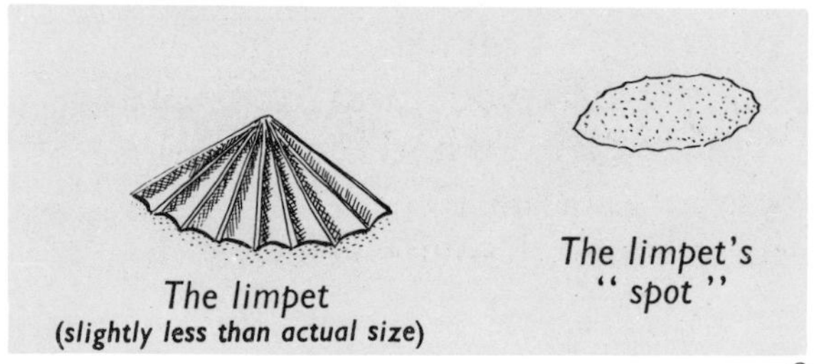

The limpet
(slightly less than actual size)

On the left you can see a sea anemone open (a) and closed (b). Like the tiny hydra, the larger sea anemone clings closely to rocks and waves its tentacles about in the water of its rock pool. It is trying to catch little water creatures on which it feeds, and when it has caught some, it closes its tentacles around them and proceeds to digest them.

Limpets remain for long periods on one spot. Even when they do move from it, they always return to the same spot.

MOVEMENT IN ANIMALS—Swimming, jumping, crawling

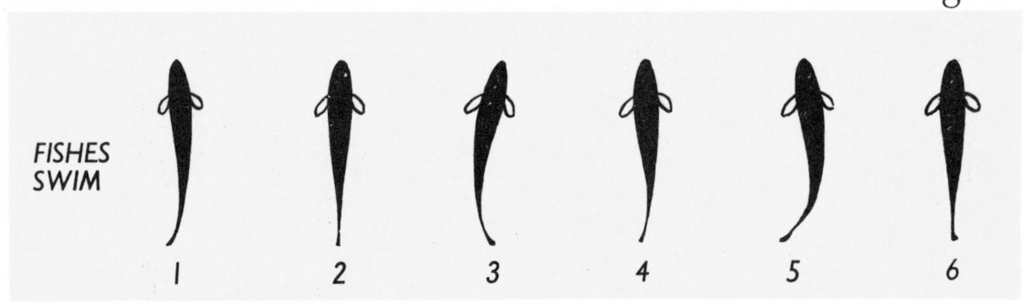

The fish has a muscular body with powerful muscles in the tail. In order to move along, the fish lashes its tail from side to side and this movement drives it through the water. (See also page 224.)

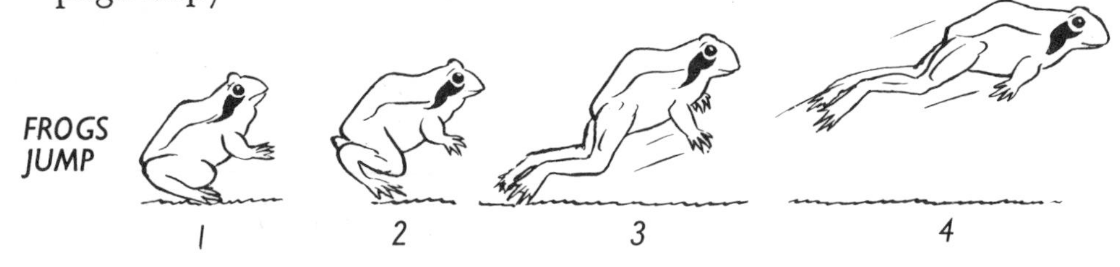

The frog is an excellent jumper. Here are four stages in its jump. The hind legs of a frog are very long and strong. You can see that it pushes off with these legs as it leaps. Its front legs are short and the animal uses them to break its fall when landing.

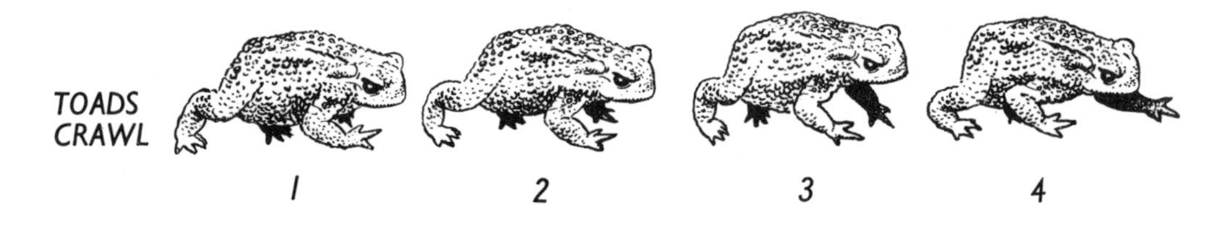

Toads walk along slowly and deliberately. You can see their leg movements in the diagrams above. Both toads and frogs are amphibious (living part of the time in water and part of the time on land) and both lay their eggs in water. The toad, however, is much more of a land animal than its relative the frog, while the frog is a better swimmer.

MOVEMENT IN ANIMALS—Running and leaping

The vegetable-feeding animals are usually very defenceless. They do not have sharp claws or tearing teeth like the flesh eaters. Since they are not good fighters they often have to run away from danger and they are usually swift runners. Members of the deer family are typical " running and leaping " animals.

Here are two African buck to show their springing and leaping movements.

Antelopes, horses and cheetahs are also swift runners and can keep up their speed over long distances. These animals use the strong muscles of their back, thighs and legs in running. Horses run on one undivided hoof on each foot, which is really an enlarged development of their middle toe nail (see page 289). Deer belong to the " cloven hoof " type of animal. You can see the split hoof in the drawing on the right.

The cheetah is a cat-like animal with the usual five claws, but its " instep " is well raised off the ground.

The athletic man can also become a good runner. By training, muscles of leg and thigh can become well developed. Speeds can be increased and speed records broken.

MOVEMENT IN ANIMALS—Flight

Several different types of animals are able to fly. Of these, however, birds are better equipped for flight than such animals as bats, flying foxes, flying squirrels or flying fish.

Gannet—gliding flight

These pictures of a gannet and a pigeon in flight show two kinds of flight movement in birds.

The gannet in the top picture has a gliding flight, making use of air currents and using its long, plane-like wings spread out rigidly.

Pigeon—flapping flight

The pigeon flies by *beating* the air with its strong wings. The pigeon can increase the speed of its wing beats and the " flight feathers " attached to its wing bones are remarkably well formed.

Bats have wings of skin and the wings of flying fish are extended fins.

(See also pages 228 and 229.)

Bat in flight

A flying fish

LIVING THINGS REPRODUCE THEIR KIND

If you pick up a stone you may be able to crack it in half and so get two stones from it, but it is unlikely that a stone would split of its own accord unless it received a violent blow or came under some sort of great pressure. Even if this happened, the two halves would not grow bigger or produce other small stones like themselves. Living things, on the other hand, can produce other living things like themselves and this process is called REPRODUCTION.

These two diagrams are of diamond crystals. B shows the natural diamond crystal and A shows one which has been cut. These diagrams are included to remind us that crystals are among the very rare non-living things which increase in size or *appear* to grow. This is because fresh layers of crystals are added to the original ones. The growth of living things is more complicated.

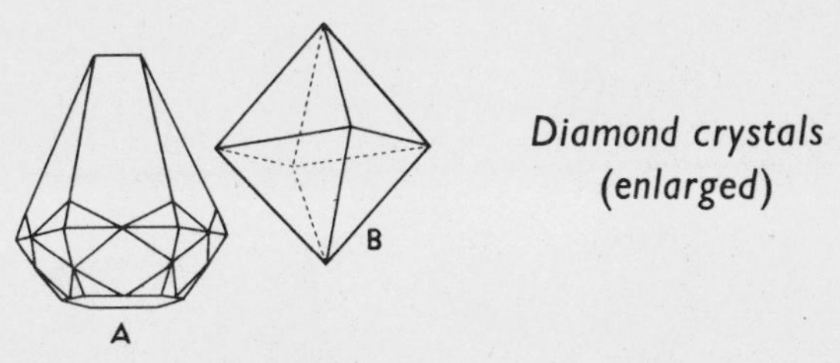

Diamond crystals (enlarged)

You can grow crystals for yourself from a solution of salt, ammonia and water poured over coal, or by evaporating slowly a warm solution of alum in water in a shallow dish.

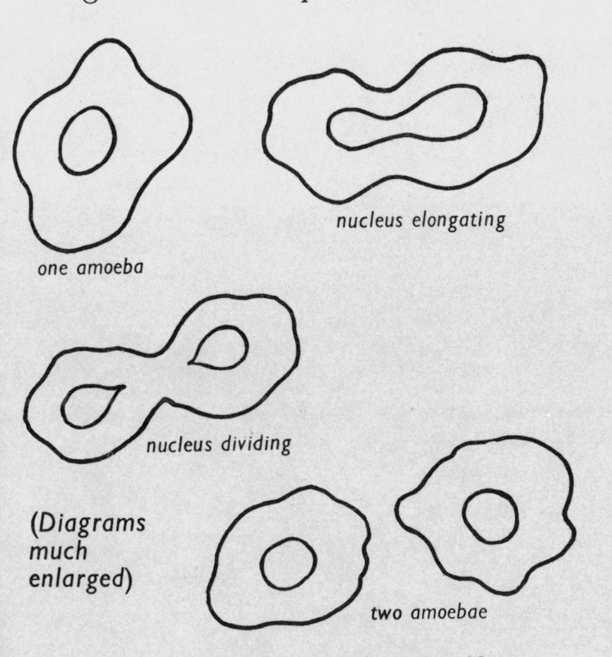

How an amoeba reproduces itself

Living things grow in size and reproduce themselves in a number of interesting ways. One way is simply by splitting into two. We call this *binary fission.* This kind of reproduction can be seen in one of the tiniest of living things—the amoeba—which is found in stagnant water.

The diagrams here show how an amoeba reproduces simply by dividing into two amoebae. Do you notice anything peculiar here? Do you notice, for instance, that the two new baby amoebae have no parent (or parents)?

REPRODUCTION IN OTHER ANIMALS

The simple kind of reproduction by dividing into two does not happen in higher types of animals. The young amoebae had no parents. In the higher animals there are two parents, a male or " father " parent and a female or " mother " parent. After the parents have mated, the female usually produces eggs which will develop into new individuals.

Eggs usually consist of only one small unit of protoplasm (one *cell*), although this living cell often has food or " yolk " with it. This single cell grows and divides into many cells until it has formed a little body called an embryo—or baby animal.

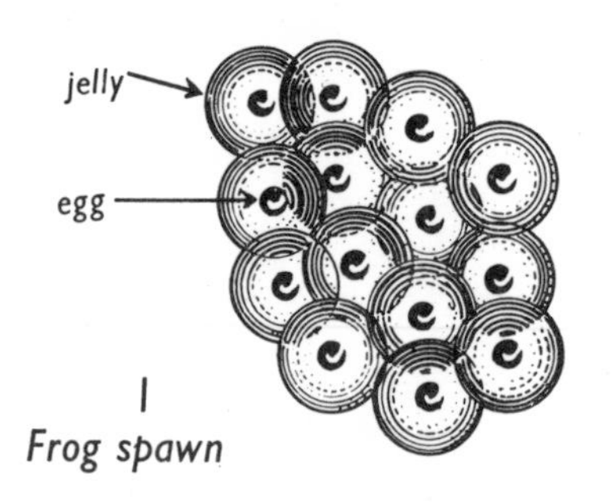

Frog spawn

The size, shape and appearance of eggs differ. Fishes, frogs, toads and newts lay their eggs in water. Frog and toad spawn (groups of eggs) is often seen floating in ponds, each egg surrounded by a ball of jelly.

Birds lay eggs which have shells and each baby chick can develop inside the protective shell in its own private " pond ". In furry or hairy animals the egg is even safer for it develops inside the body of its mother within a little sac or " pond " of its own.

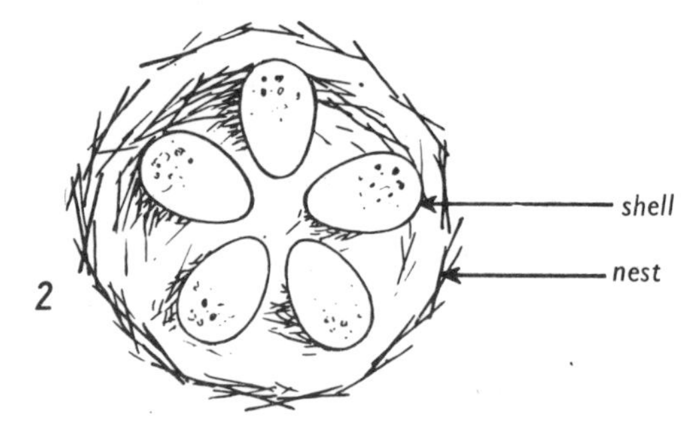

Birds' eggs in nest

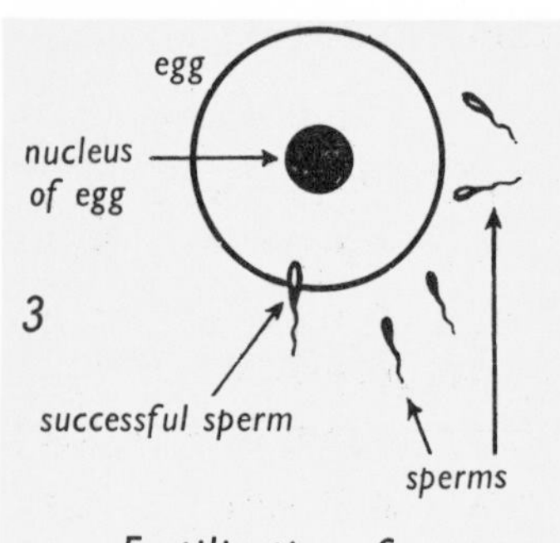

Fertilisation of an egg

Eggs will only develop when they have been fertilised by male cells or *sperms*. This is happening in diagram 3 and from this fertilised egg an embryo will develop (diagram 4).

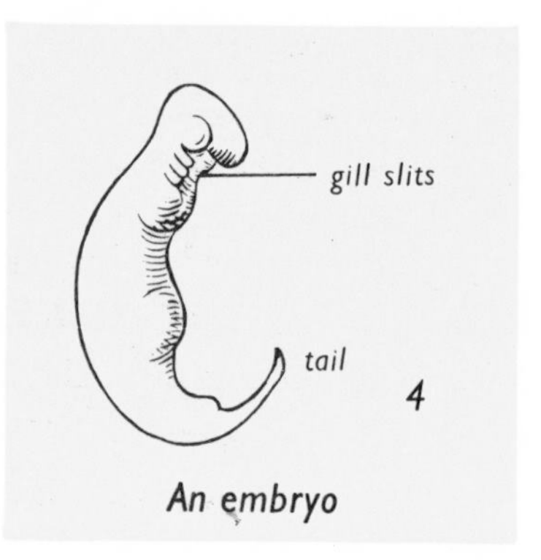

An embryo

WATCHING AN EGG DEVELOP

When an egg has been fertilised by a sperm, it is said to be " fertile ". Let us look at some of the earlier stages in its development into an embryo.

On page 267 you saw that the simple amoeba reproduces by dividing into two. The egg of a higher animal begins its development in a very similar way—by dividing into two cells. These two cells divide again so that now there is a little group of *four* cells. These go on dividing until they have formed a tiny body with a hollow in the middle (see diagram 4).

One side of this little body curves inwards, just as an india-rubber ball might be pushed in to make a two-layered, hollow body (see diagram 5). Complicated changes now take place during which a mouth and food tubes are formed, eyes, a brain, and all the organs necessary to a young animal. The baby animal is usually fully formed before it is born or hatched out of the egg.

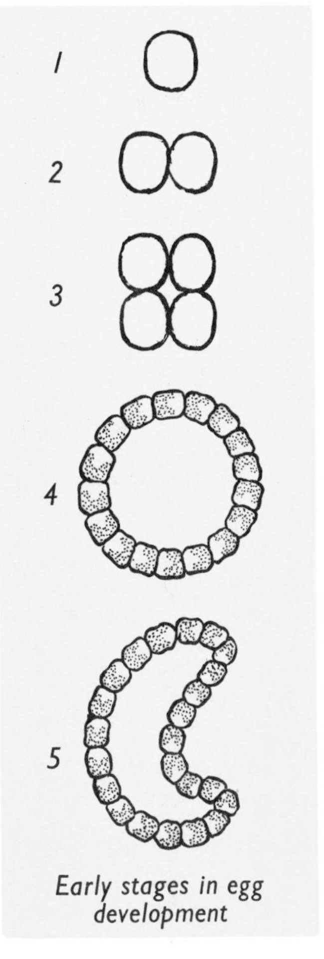

Early stages in egg development

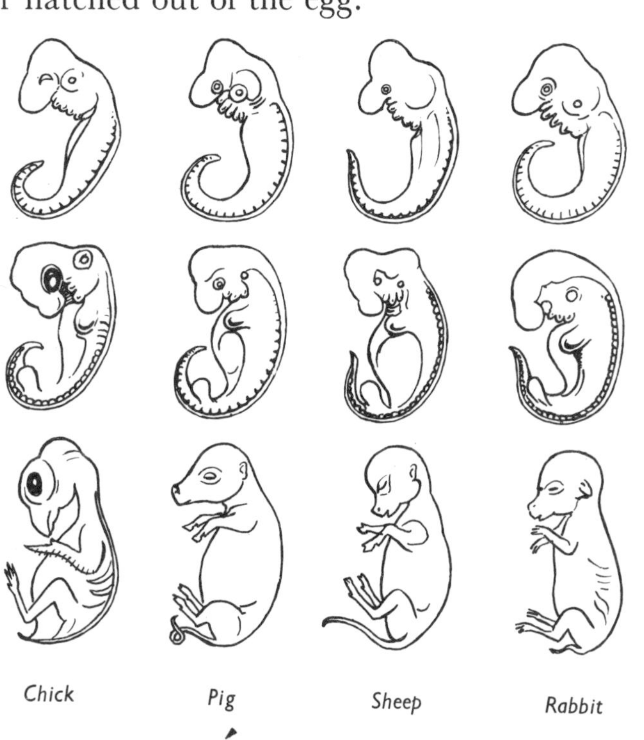

Animal embryos

As you can see from these diagrams, the embryos, or early baby stages, of many different animals are very much alike. Although the chick, pig, sheep and rabbit are easily recognised in their adult (grown up) state, you would find it difficult to distinguish their embryos from one another.

HOW SALMON GROW UP

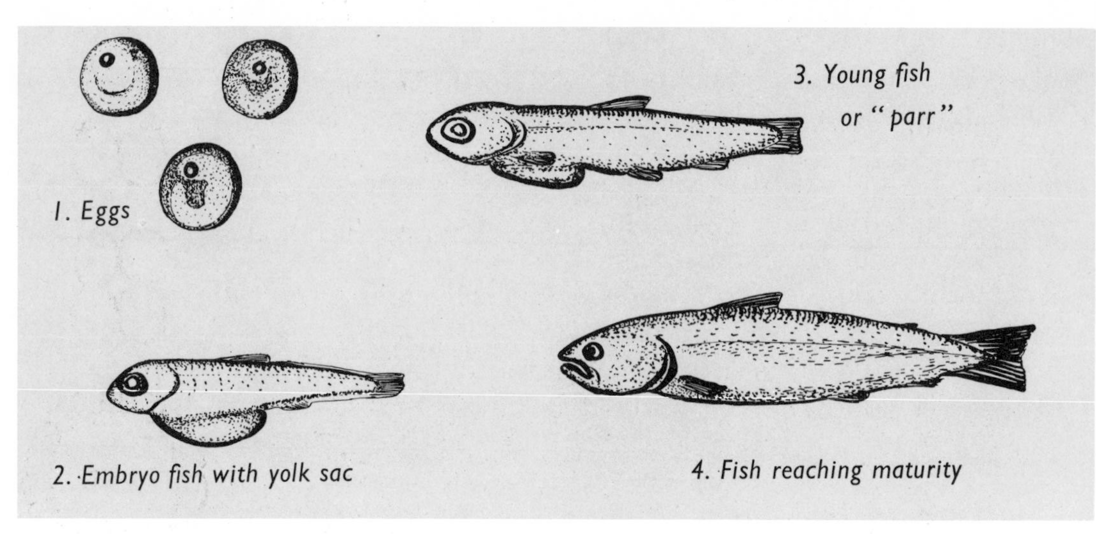

Above you see four stages in the development of the egg of a fish—the salmon.

Salmon spend part of their life in the sea, but in the autumn of each year they leave the sea and swim up rivers in order to find the shallow waters of inland streams where their eggs can be laid with safety. The mother and father salmon have a hard time, trying to reach their breeding grounds. Often they have to leap over rocky falls and many of them are pushed back by strong currents or are dashed to death on the rocks.

The nursery of the baby salmon is a shallow gravel trench in the bed of a stream. This nursery is called a " redd ". The eggs are laid by the female salmon and then the male salmon pours a fluid called " milt " over them. In this milt are sperms (see page 268) which fertilise the eggs. The mother salmon then covers the eggs with gravel and leaves them to develop, while she and the male salmon go downstream again and eventually find their way out to sea and to their old feeding grounds.

You will see that when the young fish develop, they have a little yolk sac attached to them. This provides food for them until they can find it for themselves. The yolk sac is gradually absorbed and disappears, and then the young fish, or parr (see diagram 3 above), feed mainly on shrimps which they find in fresh as well as in salt water as they travel seawards.

PARENTAL CARE—The Stickleback

As you have seen, salmon do not take much care over the rearing of their young. The eggs develop by themselves and the young fish which hatch out, never know their parents. Most fishes leave their eggs in this way and do not worry about their offspring, but the stickleback is an outstanding exception.

Three-spined stickleback with mate in nest

It is the father stickleback who is the most active partner. He builds a nest which is as well-constructed as any bird's nest and which is made from the stems of water-weeds, stuck together with a sticky substance which the fish secretes (or manufactures) in his kidneys. The nest is barrel-shaped with a hole at both ends and when it is ready, the male stickleback goes to look for a mate. He is very handsome during the mating season, and is a greenish blue with a bright red breast. He often has to fight a few rivals before he can take possession of his mate and then he persuades her to enter his nest and she lays eggs in it. When the nest is full of eggs, he swims round it and defends it. He watches it day and night and fans fresh water over it with his tail, until the young fish hatch out.

This is one of the best examples of "good fatherhood" in nature.

PARENTAL CARE—Toads, crocodiles and birds

The highest forms of good parenthood are found in birds and furry animals (mammals). Frogs, toads and newts, which are called *amphibians,* behave rather like fish in leaving their young to develop alone. There are exceptions to almost every rule, however, and a good exception here is the midwife toad seen in the picture on the right. Again (like the stickleback) the father is the good parent. After the female toad has laid the eggs, the father or male toad, carries these eggs around with him, attached to his back legs, until they hatch out.

The midwife toad

Baby crocodile bursting out of its egg

The various kinds of reptiles usually leave their eggs in some sheltered place to hatch out alone. Snake eggs are often laid on decaying vegetation, piles of leaves or " manure " which gives out heat. Crocodile eggs are buried about 60 cm deep in sand. The mother crocodile gives her eggs a little more attention than most reptile mothers do, because she often lies on top of the mound of eggs and guards them. Baby crocodiles cut their way out of their eggs with a special " egg tooth " and almost immediately make their way down to the water and swim.

As already stated, birds make good parents. They build nests in which to rear their babies and work hard defending these and bringing food to their young. Often, both parents take turns in sitting on the nest, in order to keep the eggs warm and to " incubate " them (that is, to help them to develop through warmth). The cuckoo, which lays eggs in other birds' nests and then abandons them, is an exception to this rule.

Brown trout 25-40 cm.

Rudd 20-30 cm.

Pike 40-100 cm.

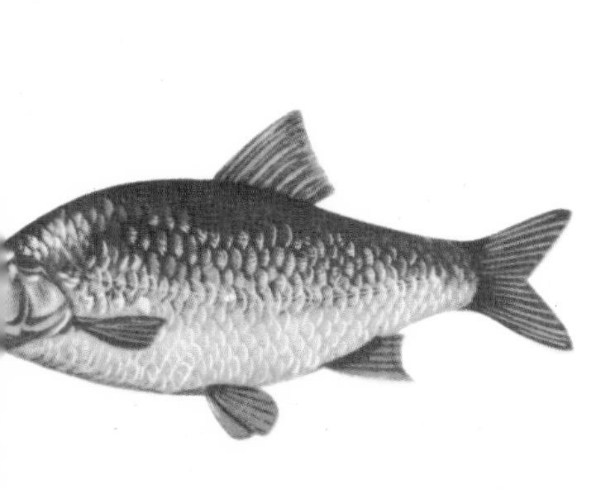

Roach 15-30 cm.

Bream 30-50 cm.

SOME RIVER FISH

With two exceptions, furry animals or mammals do not lay shelled eggs like those of birds. As stated on page 268, the young mammals develop inside the body of the parent.

The two furry animals which *do* lay eggs are the " duck mole " or duck-billed platypus of Australia and the echidna or spiny ant-eater of Australia and New Guinea.

Duck-billed platypus

The young of the duck mole emerge from their eggs in a very helpless state and feed on milk which exudes (comes out) from holes in the skin of the mother.

Kangaroo with baby in pouch

Another Australian mammal should be mentioned here—the kangaroo. Its babies (one at a time) are reared in a pouch in the skin of the mother's body until they are old enough to fend for themselves. You can see a well-developed baby kangaroo peeping out from its mother's pouch in the picture opposite.

Does good parenthood produce more intelligent children ? Furry animals or mammals are said to be more intelligent than other animals. This may have some connection with the care they receive from their parents in the early stages of their development, for their parents often stay with them long enough to give them some training. This long childhood is also true of human beings, for our parents often look after us until we are young men and women.

REPRODUCTION IN SIMPLE AND NON-FLOWERING PLANTS

nucleus
cell wall
chloroplast
ENLARGED VIEW OF PLEUROCOCCUS

1. *Pleurococcus—a very simple plant of one cell*

The simplest plants of all, like the simplest animals, are made of only one unit of living material known as a cell, which can divide into two in the manner described on page 267.

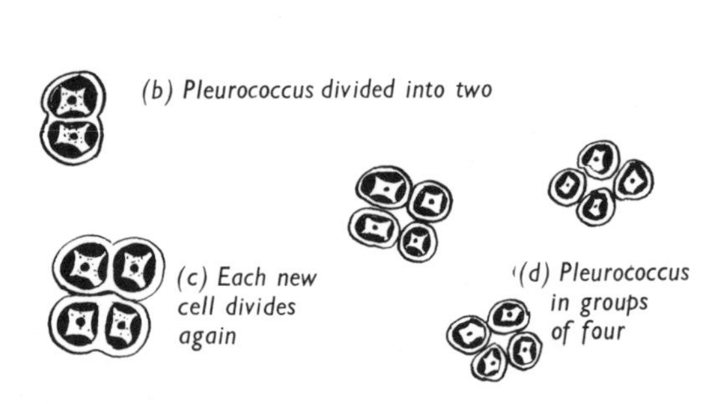

One very simple plant, often seen as a green, powdery covering on tree trunks and old palings, is the one-celled plant called pleurococcus (see above). The small diagrams show a single pleurococcus dividing into two and then into four cells. The groups of four cells stay together for a time.

gills
1
2

2. *Mushrooms and Toadstools*

These non-green plants produce spores. Spores are found in many non-flowering plants and act in the same way as seeds in that they can produce new plants. Diagram 1 on the left shows the gills underneath the head of a mushroom on which the brown spores develop. Diagram 2 shows a *spore-print* which you can make for yourself by placing the head of a mushroom, gills downwards, on white paper and leaving it for a few hours.

3. *Ferns and Mosses*

These are green non-flowering plants which reproduce by means of spores. In mosses, the spores are found in capsules, usually on the end of a long stalk (look back to page 75). In ferns, the spore groups called sori are found on the back of the fern frond at certain times of year. Sori are groups of tiny spore-bearing capsules, called sporangia, which can only be seen clearly with a microscope.

Part of a frond of Buckler fern showing the sori on the veins at the back of the frond

REPRODUCTION IN FLOWERING PLANTS

As you have learnt, flowering plants have *seeds*. If you study a flower (choose a fairly large one), you will see the various parts of which it is made up. A section of a typical flower, such as a buttercup, is shown in the diagram.

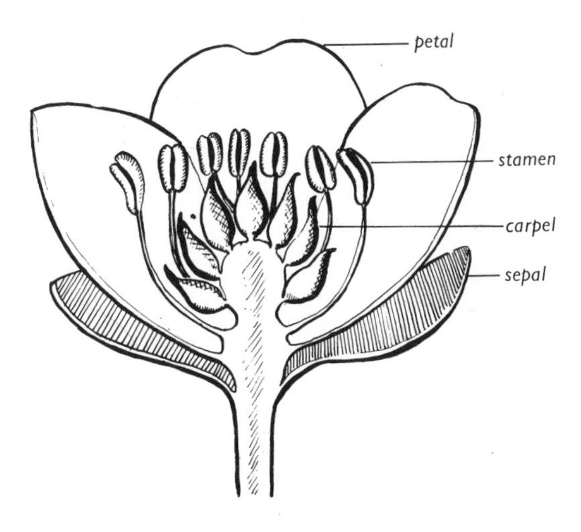

The arrangement of parts in a typical flower—Buttercup

There are four main flower parts :

1. petals
2. stamens
3. carpels
4. sepals

The petals and sepals are often brightly coloured and attractive. They also protect the more important parts of the flower—the male parts (the stamens), and the female parts (the carpels).

Diagrams 1 and 2 below represent these important organs of a flower. Diagram 1 shows a carpel with a seed or ovule inside it (some carpels have more than one seed), and diagram 2 shows the pollen sacs of the stamen. These contain a yellow dust, pollen, which scatters when the stamen is ripe and falls on the stigma of the carpel.

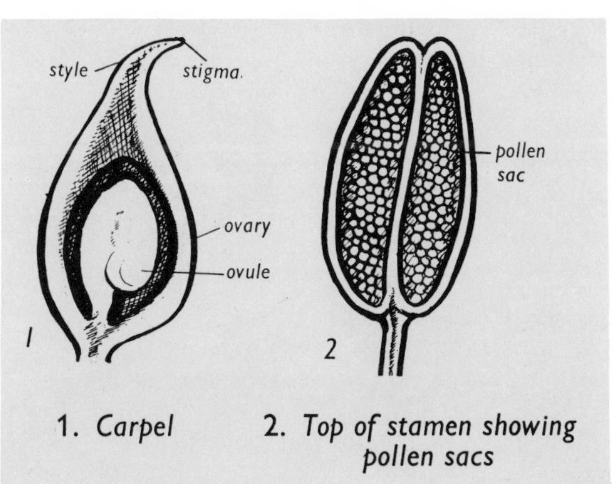

1. *Carpel* 2. *Top of stamen showing pollen sacs*

After pollen has fallen on a stigma, the ovules inside the ovary become fertilised and can then develop into true seeds. If the seeds are scattered on suitable soil and receive warmth and moisture, they will be able to develop into new plants. You can see that male and female organs are both necessary for seed production, just as they are in the beginning of new baby animals which you read about on pages 268 and 269.

Just as the fertilised ovule becomes a seed, the ripe carpel becomes a fruit. Some fruits have a pleasant taste and can be eaten. Often their seeds are thrown away and these may grow into new plants if they fall on suitable ground.

PLANTS HAVE OTHER WAYS OF REPRODUCING

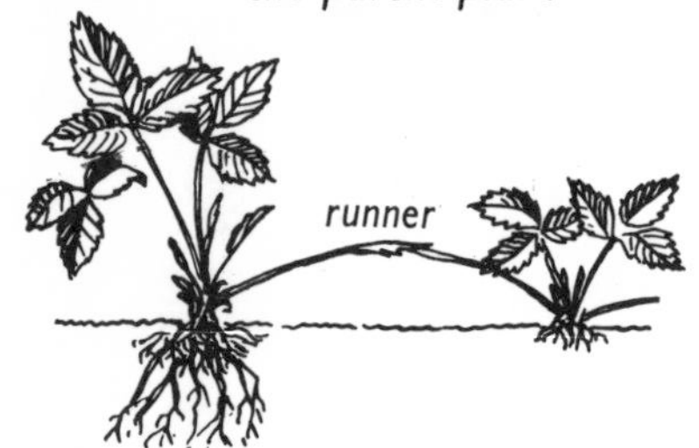

Below is a new young plant of strawberry propagated by means of a runner from the parent plant

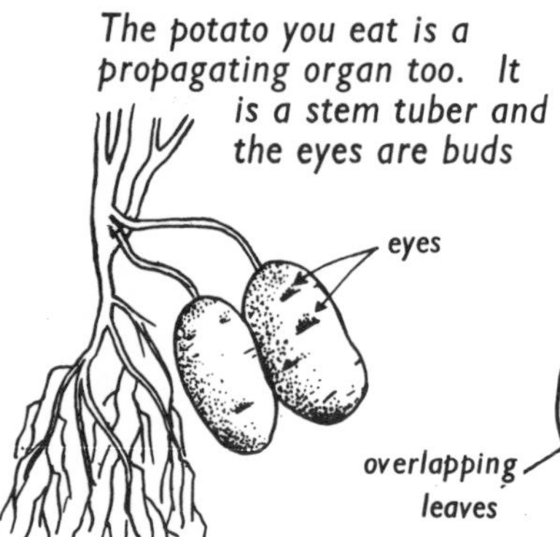

The potato you eat is a propagating organ too. It is a stem tuber and the eyes are buds

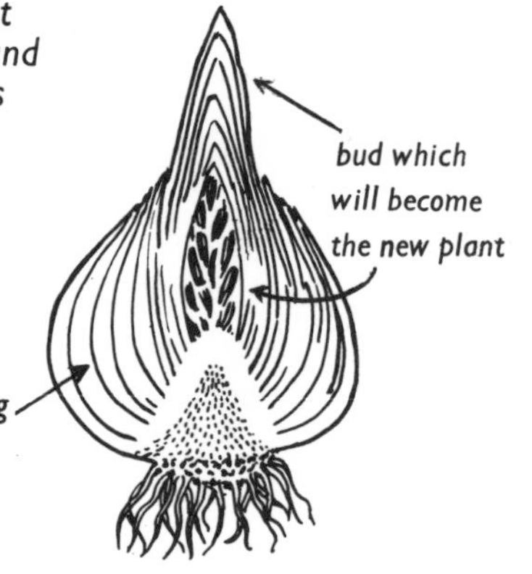

Bulbs are other types of propagating organs, with food stores in their overlapping leaves. This bulb is a hyacinth bulb

The plants which you can see above have flowers and seeds and they can reproduce by seed in the way already described, but they *also* have another way of reproducing themselves, that is by means of *runners* or *stem tubers* or *bulbs*. Below you see some more plants which have special organs of this kind.

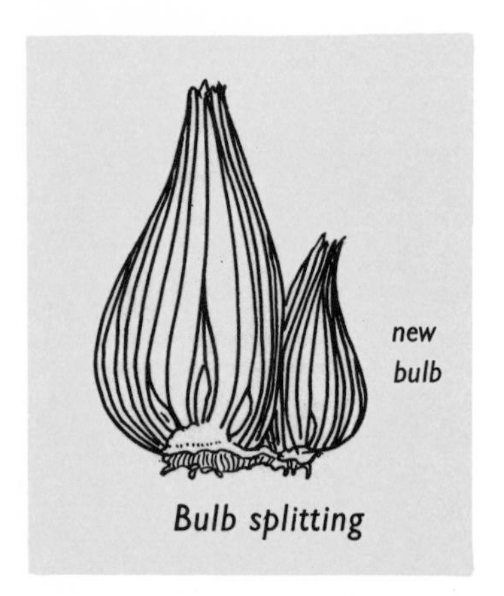

Bulb splitting

This kind of reproduction, which is really a kind of budding-off from the main plant, is called VEGETATIVE REPRODUCTION or VEGETATIVE PROPAGATION and the organs such as bulbs, corms or tubers which help in this are called VEGETATIVE ORGANS. You can see that some of these vegetative organs are quite fat. This is because they have food stored in them on which the new young plant feeds. The bud which gives rise to the new plant is situated very near to this food.

Sometimes new bulbs split off from the parent one. You can see this happening in the diagram above. In this way, a new little bulb is formed. The picture on the right shows two crocus corms. One corm has been cut in half. You can see that it is solid and hard inside, not made of folded, fleshy leaves like the bulb.

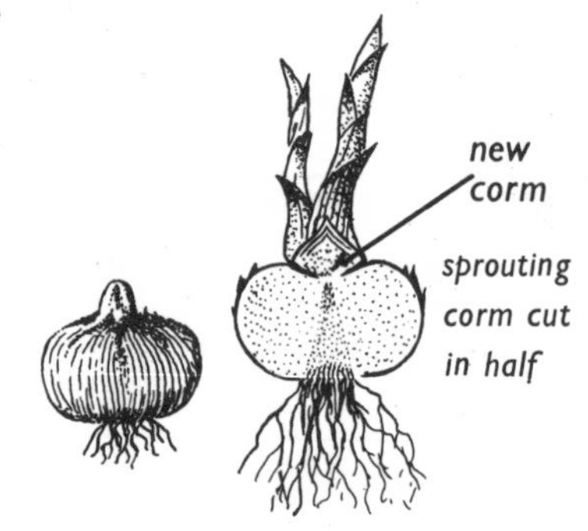

Crocus corms

CELLS AND TISSUES—The Hydra

The animal on the right is a hydra. You read about its way of moving on page 263. Its body is much magnified here and has been cut through the middle to show how it is made.

If you look at it closely you will see that the body is made up of two layers of *cells* (small units of protoplasm) with a layer of jelly-like substance between the two layers.

Layers of cells are called *tissues*. You can see in the diagrams that the outer tissue or " skin " is called the *ectoderm* and the inner layer which lines the inside of the body is called the *endoderm*. The cells of this inner layer or tissue are engaged most of the time in the work of digesting the food which the hydra has taken in through its mouth.

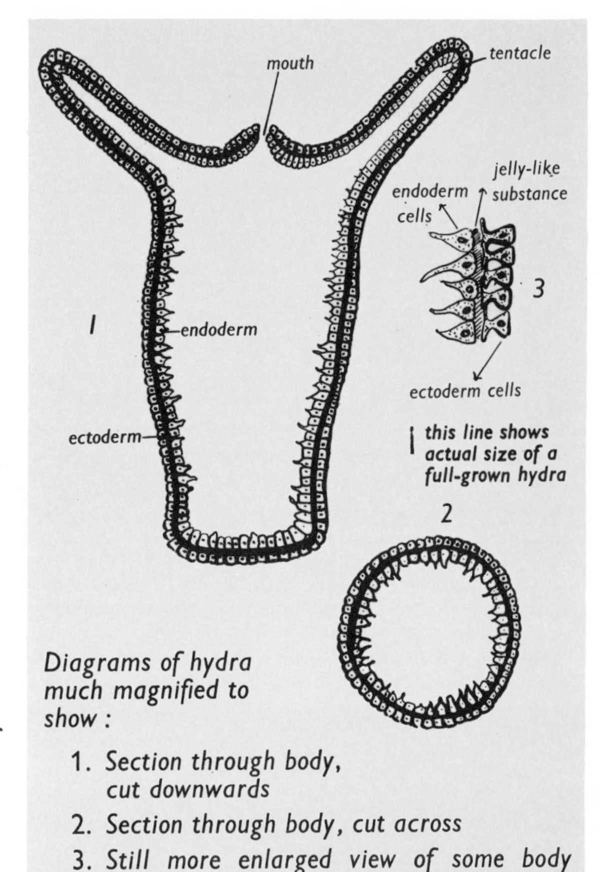

Diagrams of hydra much magnified to show :

1. *Section through body, cut downwards*
2. *Section through body, cut across*
3. *Still more enlarged view of some body cells*

The hydra feeds on tiny creatures such as water fleas. In the diagrams below you can see a hydra capturing a water flea and swallowing it.

Notice the bump in the last diagram after the flea has been swallowed. Notice too how the tentacles are used to trap the prey. You can see in the large diagram above that the tentacles are hollow like the rest of the body and also have two layers of cells.

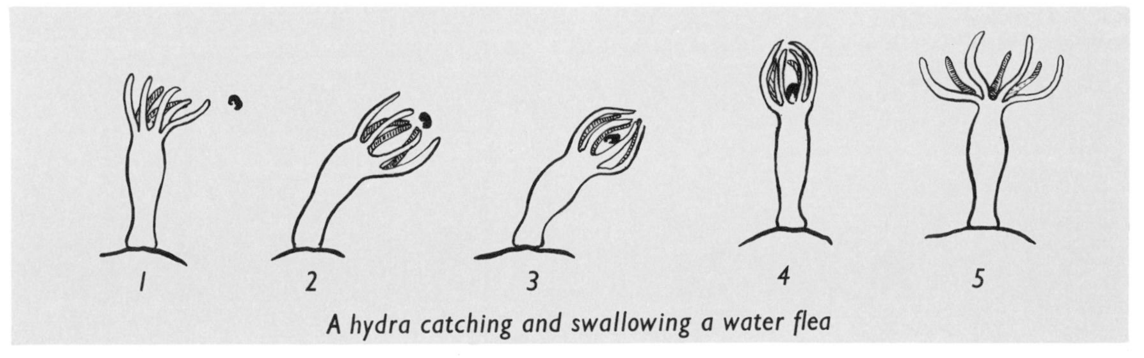

A hydra catching and swallowing a water flea

STUDYING TISSUES IN A LEAF

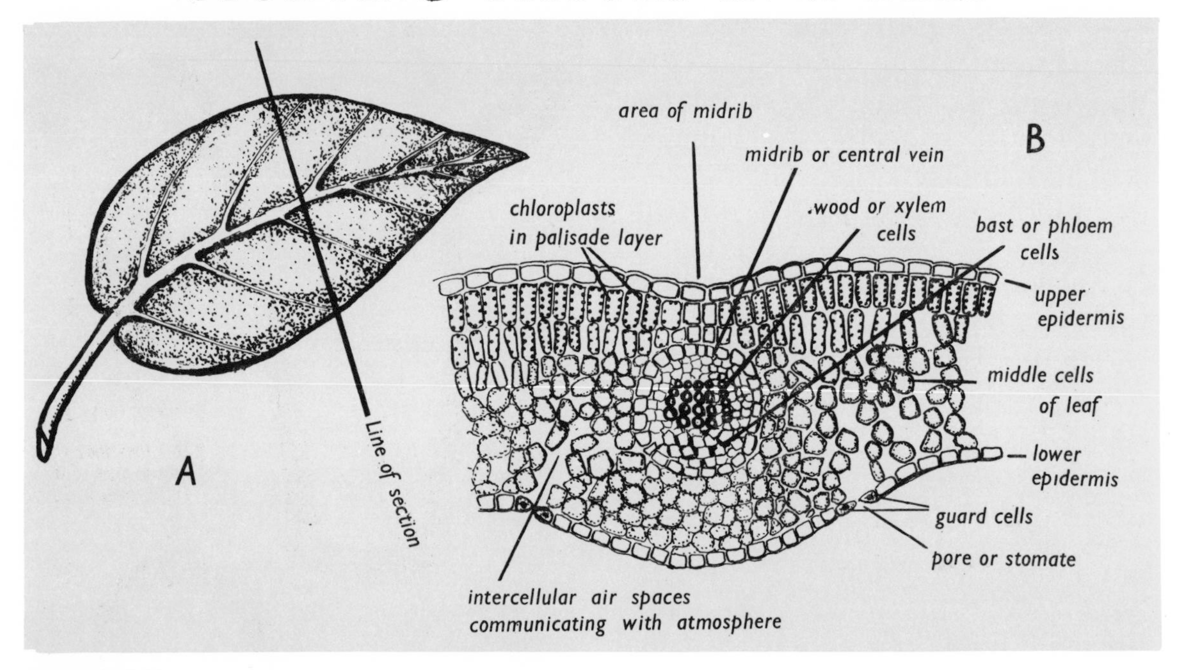

Above is a diagram (A) of a simple leaf. The dark line or " line of section " across the diagram indicates that the leaf has been cut across here with a sharp razor. Diagram B shows you what *part* of the cut edge of this leaf would look like under the microscope.

You can see that the line of section goes right across the midrib or central vein of the leaf. Find this " vein " in the middle of diagram B. You will see that it is made up of several different kinds of cells. Notice the ring of cells which goes round it and notice also the strong thick cells in the centre. These are the *wood* or *xylem* cells which form long tubes up the " vein " or *vascular strand* and carry water and dissolved mineral salts from the roots up to the leaves. In the lower part of this vascular strand (vein) try to find some thinner-walled cells with little " companion " cells beside them. These are the *bast* or *phloem* cells which carry the sap or manufactured food to the various parts of the plant.

You will see that the leaf has upper and lower skins—known as the upper epidermis and lower epidermis. Below the upper epidermis are some very regular cells forming the palisade layer. Find these and look for the small green chloroplasts (marked with little dots) inside them. These cells form the food-making apparatus of a green plant. Notice the air spaces between the cells in the lower part of the leaf above the stomates.

THE BODY WALL AND FOOD CANAL OF AN EARTHWORM

The diagrams below show three views of an earthworm. Diagram A shows the earthworm as we usually see it. Notice that the body is made up of a number of rings or segments. Look for the head end and the tail end and notice the patch called the *clitellum* (the " saddle "). This is a mass of glands which help to form the egg case of the earthworm. Counting from the front end of the worm's body, the clitellum is from the 32nd to the 37th segment. If you are careful, you can count the segments in the diagram, but miss out the first one—the prostomium—just in front of the worm's mouth.

Diagram B shows a section cut longitudinally (lengthwise) down the worm from head to tail. This shows the segments very well and it also shows the gut or alimentary canal (food channel) which stretches down the whole length of the body. Most of this canal is a long straight tube except at the forward end where there are stomach bulges. This tube opens at the front end of the worm as the *mouth* and at the hind end as the *anus* (or hind opening).

The small diagram (C) shows a section cut transversely across the worm. This shows that the worm has a body space between the food canal and the outer wall. This outer body wall is made up of skin and muscle tissues.

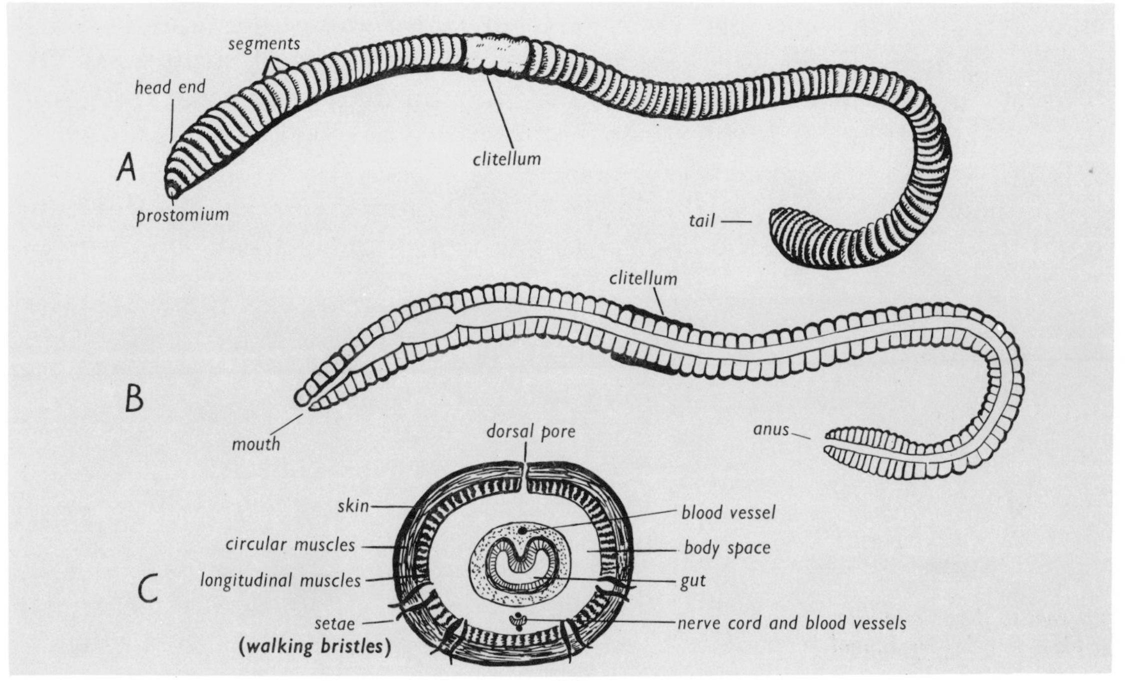

MORE ADVANCED STRUCTURES—Heart and brain

1. The Heart

As we work our way up through the plant and animal kingdoms, we find that more complicated arrangements of *cells, tissues* (layers of cells) and *organs* (structures made of tissues) are to be found. One of the most important structures in the body of a higher animal is the *heart.* The " heart " of an earthworm consists of five swellings of the main dorsal (back) blood vessel, one in each of five segments.

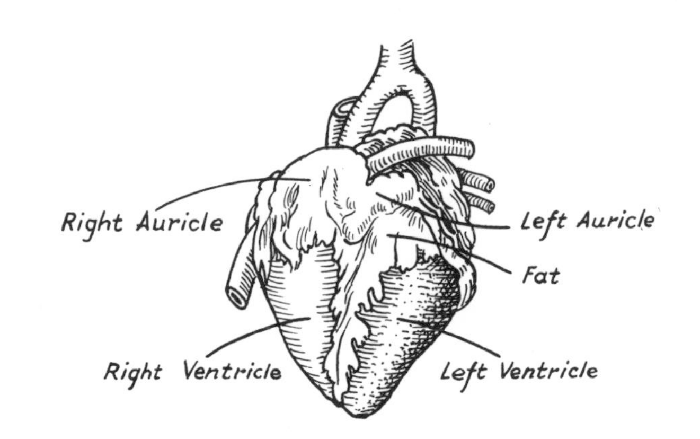

A sheep's heart viewed from the front

In the higher animals there is not so much of this repetition (as in the segments of the earthworm) and the organs are larger and more efficient like the parts of a machine. The heart of a higher animal, like that of the sheep in the picture above, is really a pumping organ. It is a large muscular structure, hollow inside, with blood vessels leading into it and going out of it.

The hearts of birds and mammals are divided into four compartments or " chambers ". These are the right and left *auricles* and the right and left *ventricles.* The walls of the ventricles are very thick and strong and the " beating " of the heart is largely due to the contraction of these thick and muscular walls. The " beating " of the heart drives blood from the heart to the lungs, where it receives fresh oxygen. The blood containing this fresh oxygen then comes back to the heart to be pumped out again—this time around the body through the arteries to give nourishment to all the tissues.

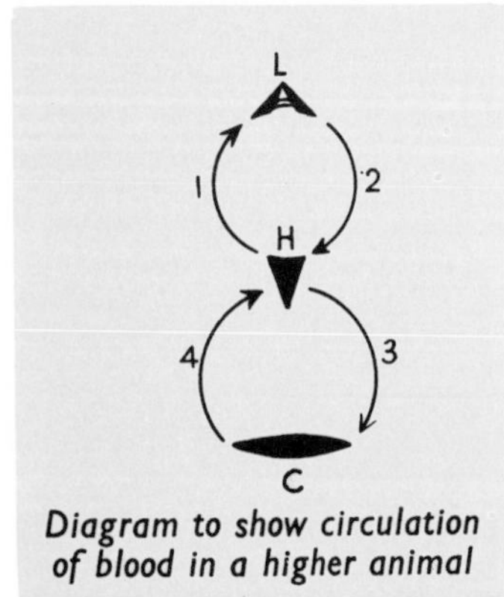

Diagram to show circulation of blood in a higher animal

These journeys of the blood to and from the heart are shown in the diagram on the left :

1. from heart (H) to lungs (L)
2. from lungs to heart
3. from heart to body—collected up by tiny blood vessels called capillaries (C)
4. back to the heart.

Note : all blood vessels going *to* the heart are *veins.* All going *from* the heart are *arteries.*

2. THE BRAIN

All animal organs show improvement as we pass from lower to higher forms of animal life and, as you have read, some of the most simple creatures do not have any complicated structures in their body.

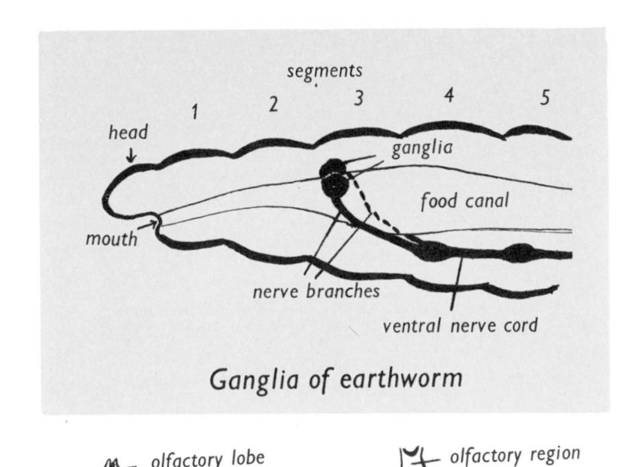

Ganglia of earthworm

Perhaps the most wonderful and complicated structure in Nature is the *brain* of a higher animal and the most advanced of these is the brain of *man* himself.

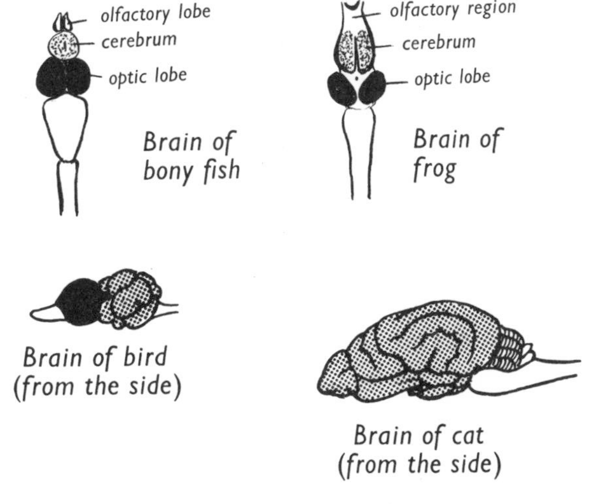

Brain of bony fish

Brain of frog

Brain of bird (from the side)

Brain of cat (from the side)

From the picture above you will see that even the earthworm has a kind of nerve tissue which serves as a simple brain. Two little white knobs or *ganglia* may be found in the third segment. From these ganglia, strong nerves encircle the gut or food tube, and then join together and travel towards the tail. This is the ventral (front) nerve cord.

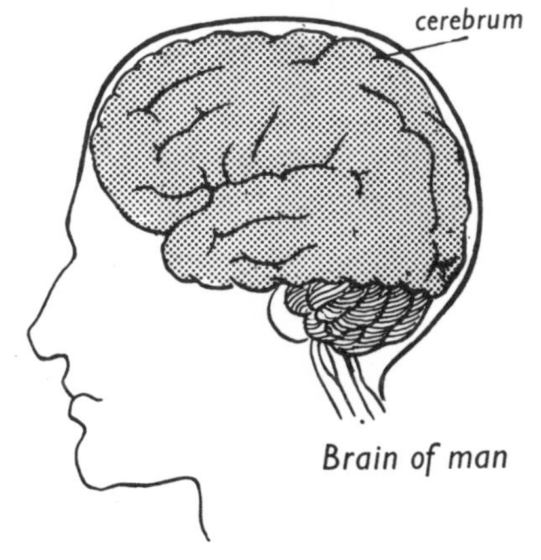

Brain of man

By comparing the five diagrams on the right, you will see that the shaded area, the cerebrum (loosely known as the "thinking area") becomes larger as we pass from the lower vertebrates, like fish, to the higher ones, such as cat and man. In higher animals than fish, the cerebrum becomes divided into two hemispheres. In man these cerebral hemispheres cover much of the rest of the brain tissue. Find also in three of the diagrams the optic lobes (coloured black) and the olfactory lobes. These parts of the brain are concerned with seeing and smelling respectively. Animals which see well usually have large optic lobes and those with a good sense of smell have large olfactory lobes.

THE INTELLIGENCE OF MAN

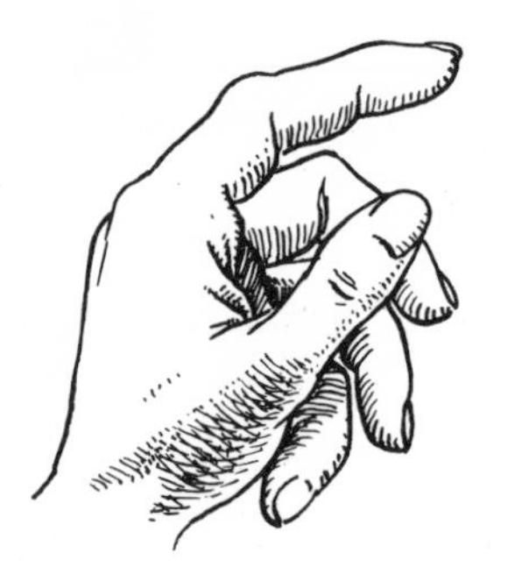
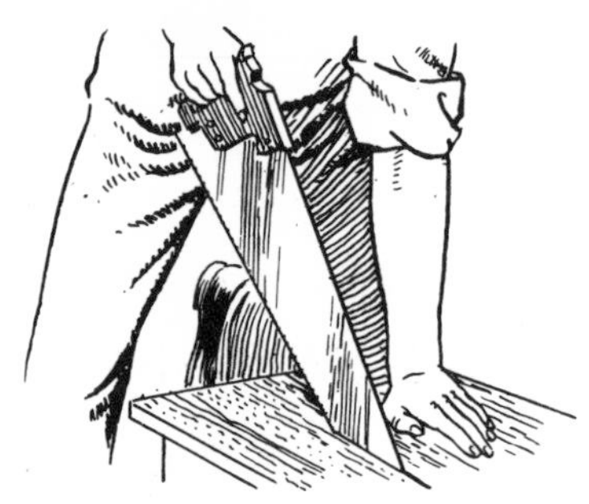

With superior hands and brain, man can construct many things

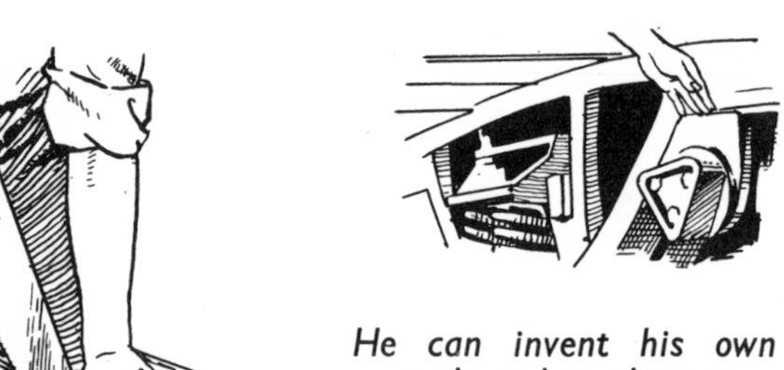

He can invent his own tools and machinery

Because of his good brain and high standard of intelligence and the way in which he can use his hands to carry out his ideas, man is easily recognised as superior to other forms of animal life.

Early man was not so advanced in ways of living and we learn about men of the Stone Age who lived in caves and hunted their food with primitive axes and spears.

Primitive man learned to use an axe which he made himself with flint stone

Man learned crafts such as the making of pots

Over thousands of years men learned to make pots, to work with metals, to write and to communicate with each other in more advanced ways. They also learned to make themselves good dwellings and to build bridges and roads, and they practised many other crafts and skills.

Physically, however, man is inferior to some animals. He has not got such great strength and is not nearly so well protected against enemies, either by defensive structures such as sharp claws or teeth, or by being able to run very speedily. Man is also quite helpless at birth and would die if not well taken care of by his parents. We have already seen that this good parental care is characteristic of the higher animals (and of man). Man learns from childhood to make use of the materials of his environment (the things around him) and to turn them to his advantage.

Dogs can be trained but no other mammal is as intelligent as man

THE EARLIEST LAND ANIMALS

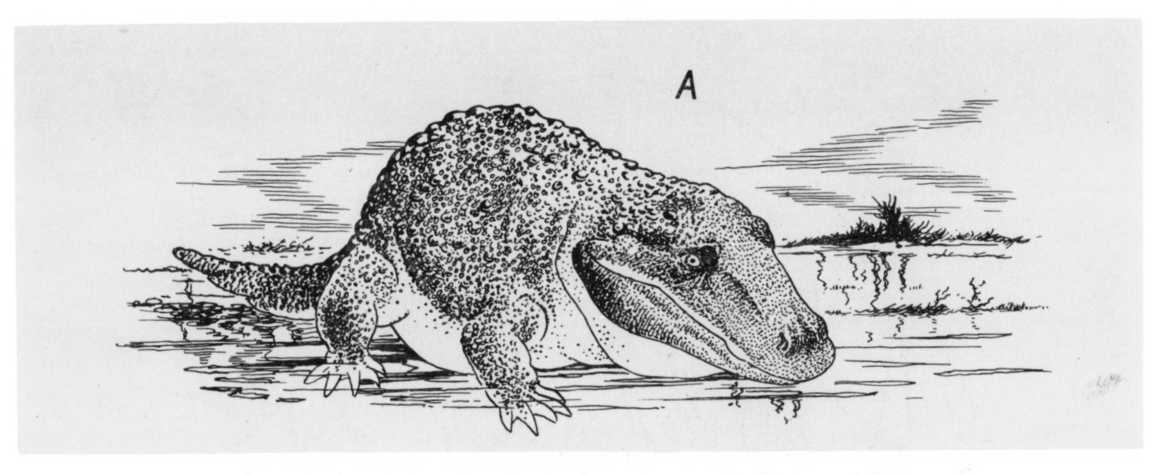

An artist's impression of a prehistoric, reptile-like amphibian

Scientists tell us that life began in the sea and as it is generally believed that at one time water covered nearly all the earth, we can believe this to be true. There must have been a time in the world's history, however, when some creatures came out of the water and began their life on land. What were these early land creatures like ?

Perhaps it will help us to imagine these first creatures if we remember that, even today, we have some animals and even some plants which are *amphibious,* that is they are able to live *in* water or *out* of water. Frogs, toads and newts are like this. That is why they are called " amphibians ". Some reptiles too, such as alligators and crocodiles, are equally at home *in* or *out* of water. Among plants, the water crowfoot is an amphibious plant. There is a picture of this on page 37.

It is quite possible then that the first land animals looked something like our amphibious creatures today, though much larger, and the reconstructed and fantastic creature shown in the picture above represents an artist's idea of one of these animals. There may have been fishes, too, which were able to spend short periods out of water. A lobed fish similar to the one below has been discovered and zoologists think that it may be an important link between water and land creatures.

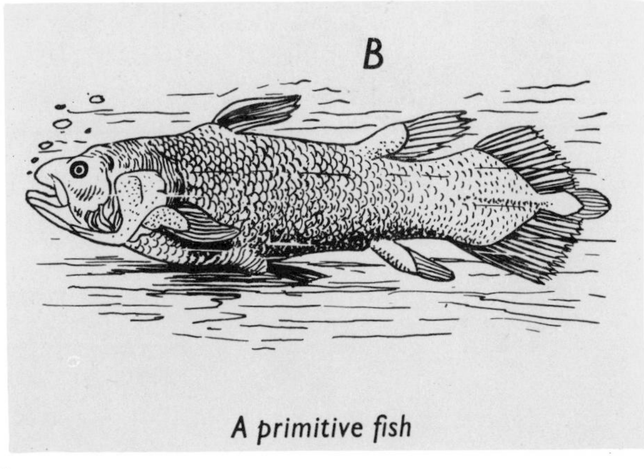

A primitive fish

Drawing of a mammoth by a Stone Age man

From the fossil bones which have been dug up, it appears that many of the animals which were on the earth in the life of prehistoric man were very large indeed. The drawings made by Stone Age men on the walls of their caves suggest also that this was so. The drawing above of a mammoth (a type of huge elephant) is a copy of one made by a Stone Age man.

The Archaeopterix

You read on the previous page that animal life emerged from the sea to the land. Eventually some animals must have taken to the *air* or how did our first birds appear ?

A fossil of the first bird-like creature has been found. It has been called the *Archaeopterix* and it looks like a mixture of reptile and bird, with long tail and wings. An artist's picture of an archaeopterix is shown on the left. This first " reptile-bird " had teeth which the birds of today do not have. There seems no doubt, however, that our earliest birds must have been flying reptiles.

These descriptions of early forms of life from past ages must have suggested to you that we learn much about primitive creatures and plants from the evidence of their fossils and from the imprints they have left on the soft clays and sands which later hardened into rocks.

The pictures on the next two pages explain this more clearly. The first primitive forms of life almost certainly arose in water. Over a long period of time, during which much of the water covering the earth gradually subsided, many living things developed the ability to make use of dry oxygen and to live on land. At first many of these living things must still have been semi-aquatic or amphibian, spending much of the earlier part of their life in water and returning to it for food.

As time went on, a great number of plants and animals became able to live completely on land and some creatures even became airborne, for, as you have read, a certain type of reptile became our first bird.

All these developments took millions of years to happen, during which time many forms of life became buried beneath layers of sedimentary rock. This rock was formed from sediment dropped by seas, lakes and rivers of bygone ages and piled up over the existing rocks, just as our rivers today bring down mud and silt and leave it on their banks as they near the sea.

These layers of sediment, well pressed down by the weight of fresh layers on top of them, hardened into rock. We are able to see these rock layers or strata today in places where they have become tilted or broken up by movements of the earth or by volcanic eruptions many years ago. We can also see them in places where rivers have channelled deep grooves into the rock and where man has uncovered surfaces and sections of ancient rock in making quarries. In some of these strata, as, for instance, in the Grand Canyon of Colorado (see page 288), we can actually read the story of the earth's history for we can see fossils and imprints of organisms (plants and animals) which were alive when their own stratum of rock was uppermost.

THE STORY OF THE ROCKS

Start reading this picture story of the rocks from the **bottom** of these two pages. As you do so, you will pass through millions of years and see the different forms of life which appeared on the earth through these ages. Notice how the plants and animals become more developed or " evolved " as you go upwards from the lowest layer to the top layer.

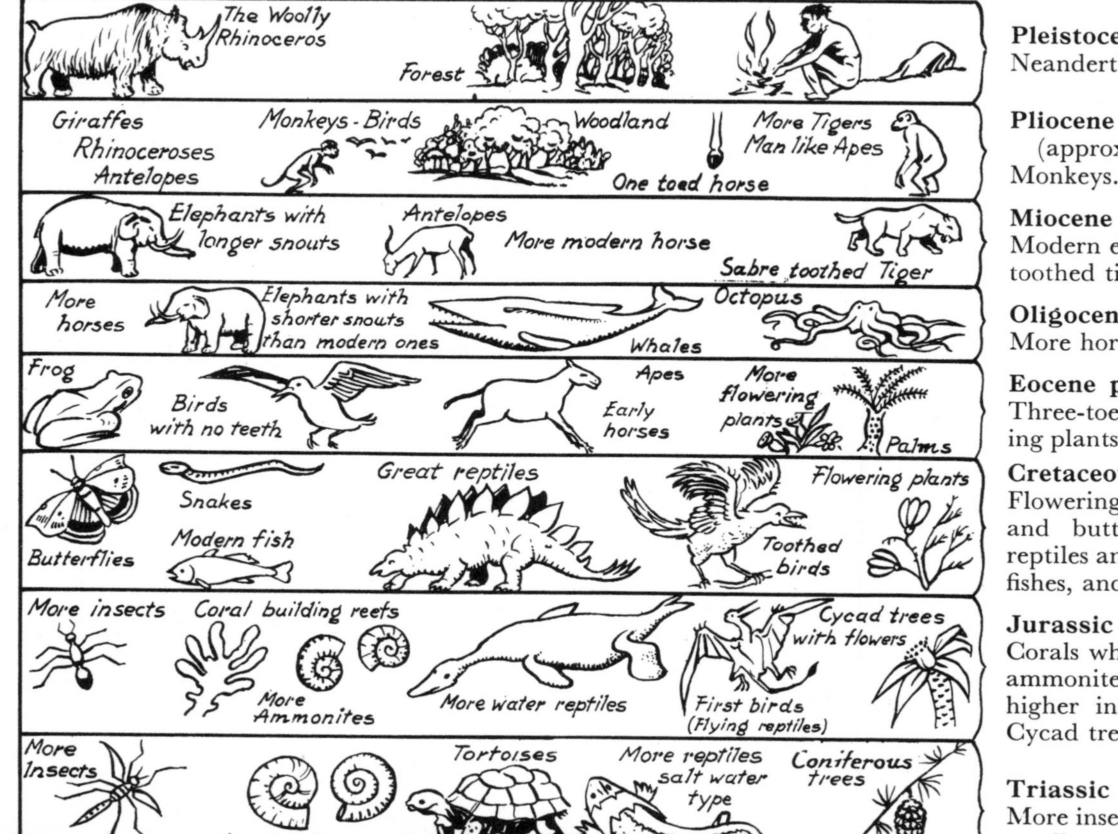

Pleistocene period. Duration 1 million years. Neanderthal man. Woolly rhinoceros.

Pliocene period. Duration 11 million years (approx.).
Monkeys. Man-like apes. One-toed horse.

Miocene period. 26–12 million years ago. Modern elephants. Antelopes. Horses. Sabre-toothed tigers.

Oligocene period. 34–26 million years ago. More horses. Whales. Elephants. Octopi.

Eocene period. 58–34 million years ago. Three-toed horses. More modern birds. Flowering plants and palms. The first ape type.

Cretaceous period. 125–58 million years ago. Flowering plants, and insects, such as moths and butterflies, visiting flowers. Old heavy reptiles and modern wriggling snakes. Modern fishes, and birds with teeth.

Jurassic period. 168–125 million years ago. Corals which built reefs in shallow seas. More ammonites and water reptiles. Development of higher insects such as ants. Flying reptiles. Cycad trees with flowers.

Triassic period. 200–168 million years ago. More insects. Ammonites. Tortoises and more

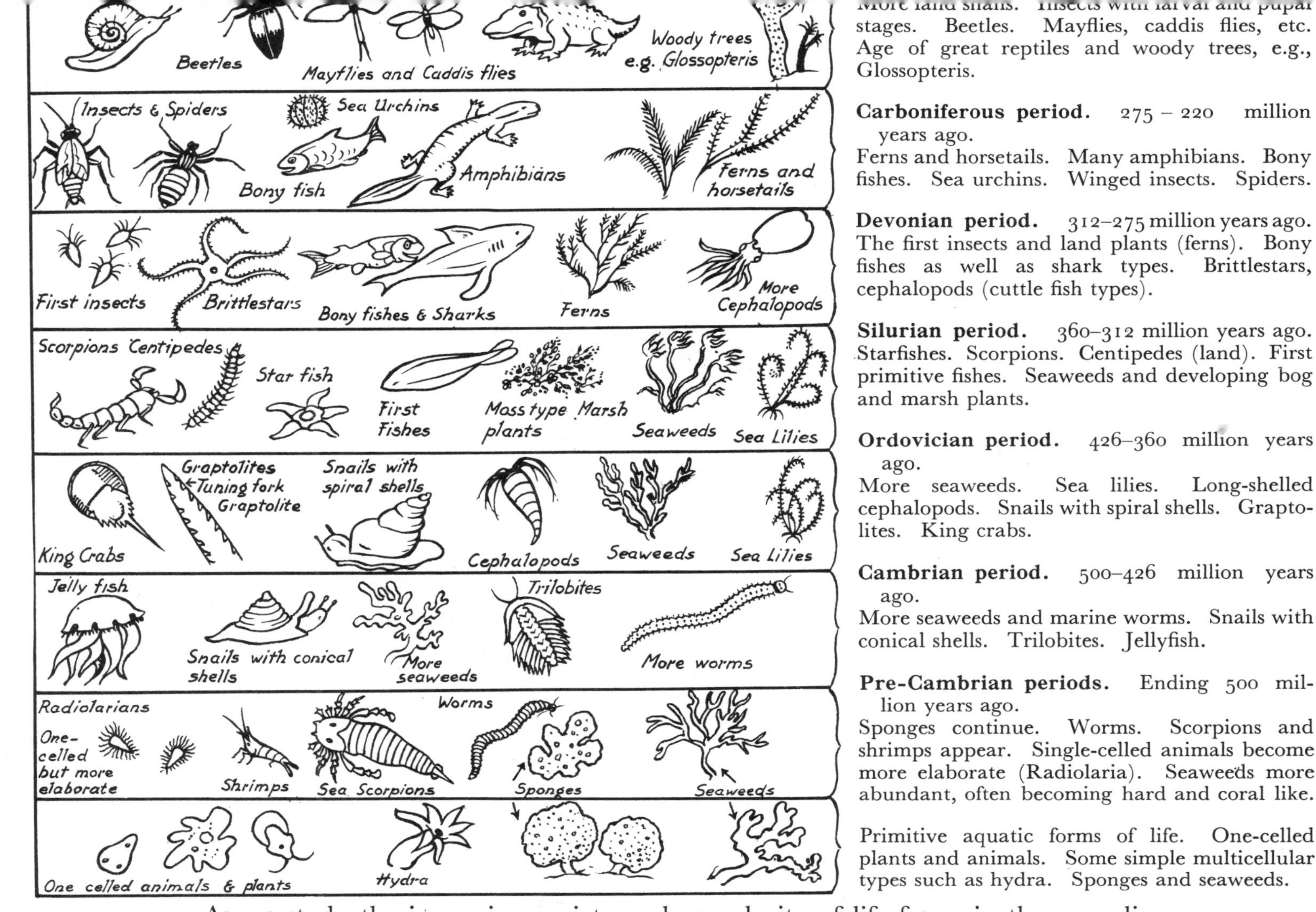

More land snails. Insects with larval and pupal stages. Beetles. Mayflies, caddis flies, etc. Age of great reptiles and woody trees, e.g., Glossopteris.

Carboniferous period. 275 – 220 million years ago.
Ferns and horsetails. Many amphibians. Bony fishes. Sea urchins. Winged insects. Spiders.

Devonian period. 312–275 million years ago. The first insects and land plants (ferns). Bony fishes as well as shark types. Brittlestars, cephalopods (cuttle fish types).

Silurian period. 360–312 million years ago. Starfishes. Scorpions. Centipedes (land). First primitive fishes. Seaweeds and developing bog and marsh plants.

Ordovician period. 426–360 million years ago.
More seaweeds. Sea lilies. Long-shelled cephalopods. Snails with spiral shells. Graptolites. King crabs.

Cambrian period. 500–426 million years ago.
More seaweeds and marine worms. Snails with conical shells. Trilobites. Jellyfish.

Pre-Cambrian periods. Ending 500 million years ago.
Sponges continue. Worms. Scorpions and shrimps appear. Single-celled animals become more elaborate (Radiolaria). Seaweeds more abundant, often becoming hard and coral like.

Primitive aquatic forms of life. One-celled plants and animals. Some simple multicellular types such as hydra. Sponges and seaweeds.

As we study the increasing variety and complexity of life forms in the ascending strata, we are reviewing the evolution of life on this planet.

THE GRAND CANYON OF COLORADO

There is permanent evidence of the " story of the rocks " in the wonderful rock strata of the Grand Canyon of Colorado in the U.S.A. An artist's drawing of this is shown on the right. The canyon is a great gash in the earth's crust. It is over $1\frac{1}{2}$ kilometres deep. It varies from 6 to 29 kilometres across and it is 350 kilometres long.

When you look into the canyon you are really looking into part of the interior of the earth which has split open to show fossils and imprints of plants and animals which lived on the earth millions of years ago. Each layer, or stratum, which you see was once on the surface before being covered by the mud of the next layer. These layers, or strata, are of various colours, some grey, some brown, some red.

A section through part of the canyon is shown below. The Colorado river (C.R.) and gorge is shown at the bottom and the rim of the canyon at the top.

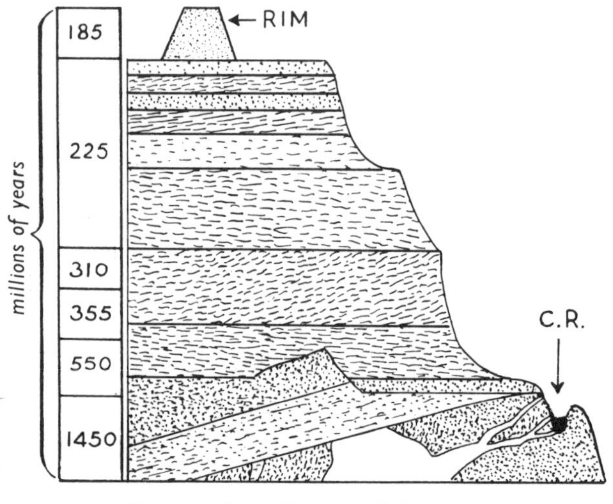

Section through part of the canyon

Tracks of some primitive land animals are found in the upper layers ; below these are fossil imprints of early insects and below these are fossils of shelled creatures, seaweeds and so on.

HOW ANIMALS CHANGE—
The story of the horse

The previous pages have taught you that we learn much about animals and plants from fossils and imprints which we find in the rocks. We learn other things too. One important fact, which we can discover from fossils and fossilised parts of prehistoric animals, is that ever since living creatures of any sort appeared on the earth, they have been constantly undergoing changes so that the descendants are often very different from their first ancestors. This is clearly shown in the story of the horse.

When horses first appeared about forty million years ago, they did not look like the animal we call " horse " today. We have found fossil bones of the first horse and we know it by the name EOHIPPUS. It was much smaller than the horse today, more the size of a large dog, and it had four good toes and one rather short one on each foot. Its toes were " splayed " or spread out and this would help it to get about better in the marshy lands of those days.

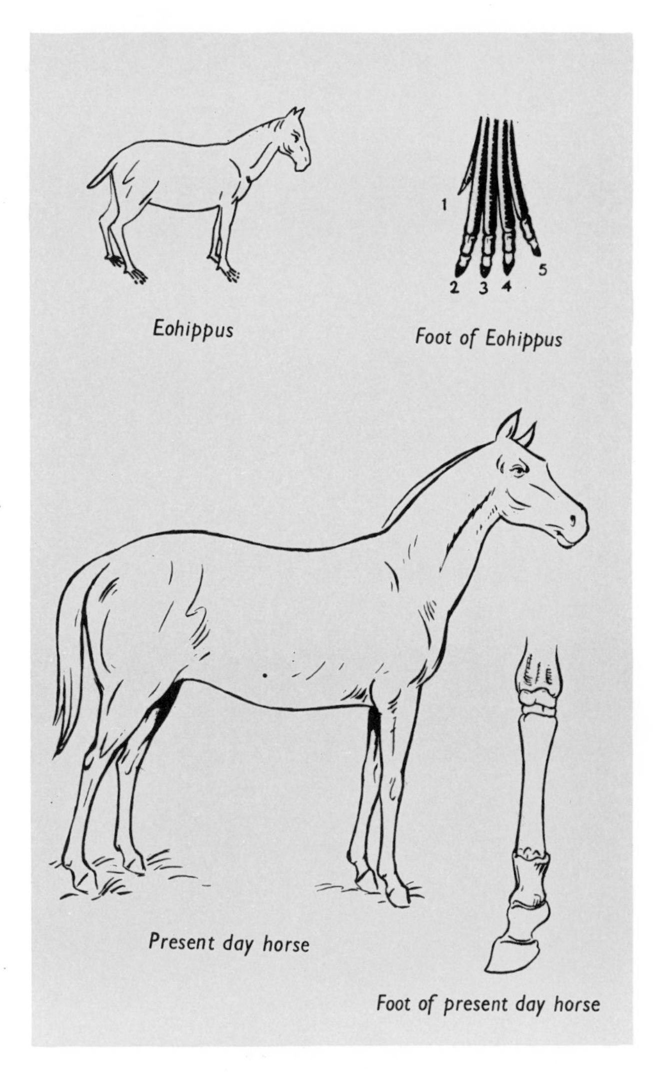

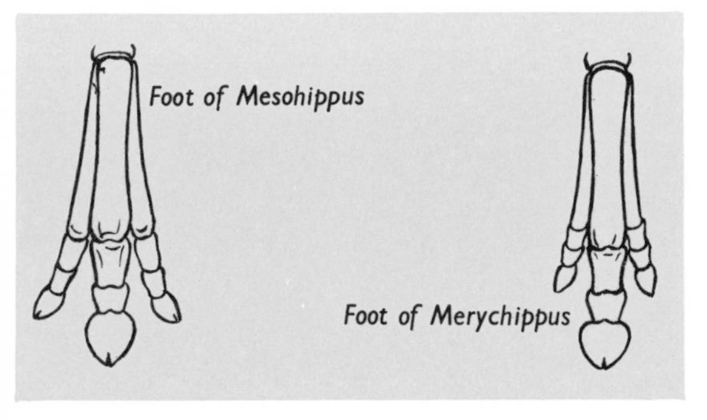

Fossil bones of later horse types (Mesohippus and Merychippus) show that the animal gradually got larger. They show also that the number of toes gradually became less (study the diagrams) until the evolution of the present day horse which walks on the much enlarged toe nail (hoof) of its middle toe.

HOW ANIMAL STRUCTURES VARY

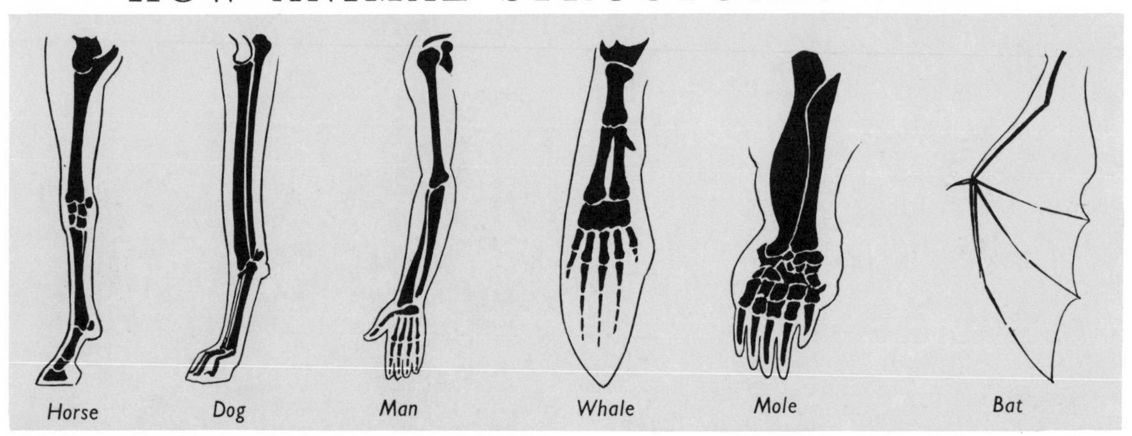

Look at the animal structures shown in the diagrams above. They show the arrangement of bones in the limbs of six different vertebrates (back-boned animals). The animals named here are also mammals (hairy or furry animals).

Backboned or vertebrate animals have two pairs of limbs and in most cases their limbs are of the five-fingered type. The limbs of different types of animals show many modifications and variations of pattern and arrangement, however, although the basic structure is often similar. Study the pictures above. Man has what we might describe as a " normal " pattern of limb with strong bones of the upper and lower arm (or leg), small wrist and hand bones and five digits (fingers or toes) ; the horse, as you have learnt, walks on one toe ; the dog has toes but has a " lifted " instep like the horse. The whale, which is not a fish but a marine mammal, has its whole limb covered with a flipper. The mole has a shortened and very powerful fore-limb which it uses in digging, although its hind limb is more normal. The bat has skin stretched between the " fingers " of its fore-limbs and uses these limbs as wings.

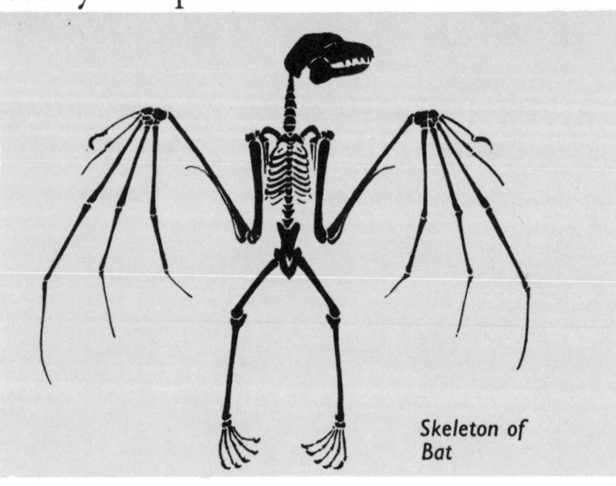

Skeleton of Bat

The lower picture shows the skeleton of a bat. Here you can see the elongated finger bones, between which the skin is stretched. This thin skin, which forms a membranous " wing ", is also attached to the side of the body and to the hind legs. The bat is a flying mammal ; there are no feathers on a bat and it is certainly *not* a bird.

A PAGE OF FOSSILS

If you have been interested in the information about fossils which you read on pages 285 to 288, you may like to know a little more about fossil shells and imprints, some of which you may be able to find.

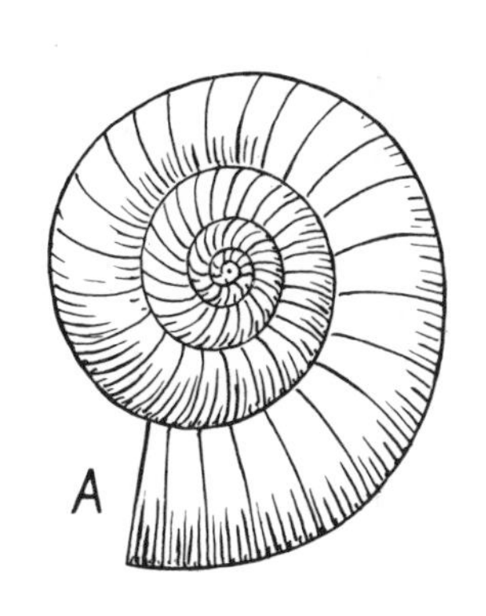

Probably the best known fossil to collectors is the type known as an *ammonite* (see A). There were different types of ammonites and these animals lived in a spirally coiled shell. Ammonites may be found in certain areas of chalk rock. Monster ammonites have been found near Aylesbury.

The *gryphoea* or " Devil's toe nail " (see B) is the fossil of an oyster-like shell which appeared at only one period in the world's history. Look in shale or limestone rock for these.

Fig. C shows a picture of a *belemnite,* once thought to be a small thunderbolt. It is now known to be the inner shell of a soft-bodied creature like a small squid.

Fig. D shows a fossil of a *fern* from the coal measures. You may find a leaf print on a piece of coal in your own coal cellar.

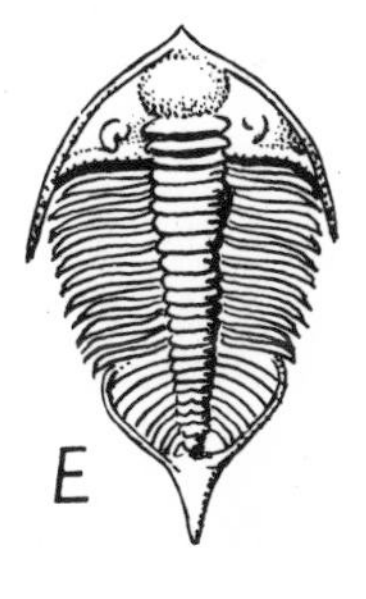

The *trilobite* (Fig. E) was a type of small crab which swam in the seas about 550 million years ago. It is one of the earliest fossils found. Trilobites are found in rocks in the Lake District, in Wales and in southern Scotland.

EVOLUTION AND CHARLES DARWIN

The last few pages must have helped you to realise that ever since the world began, things have been changing and that all life on the earth has been passing through a series of changes which have been going on for millions of years.

These changes in the characters of living things are known generally by one word, EVOLUTION, and evolution is still going on. A naturalist named Charles Darwin, who was born in 1809, wrote a book called the *Origin of Species* in which he tried to show something of the story of these gradual changes in plant and animal life. The story of the horse, for instance, which you read about on page 289, was one of the examples used by Darwin to explain some of the changes which we call evolution.

Many people were a little disturbed by this book of Darwin's because they had probably imagined that animals and plants had always been the same since they first appeared on the earth. Darwin, however, provided different kinds of evidence to prove all his statements, much in the same way that you might give evidence to prove your case in a court of law. The evidence or the facts which he gave to prove *his* case came from fossils (fossil evidence), from comparing structures of living things (evidence from comparative anatomy), from the world distribution of plant and animal types (geographical evidence), and so on. Darwin gained much of his geographical evidence on a voyage round the world which he made on a ship called the *Beagle* during the years 1831 to 1836.

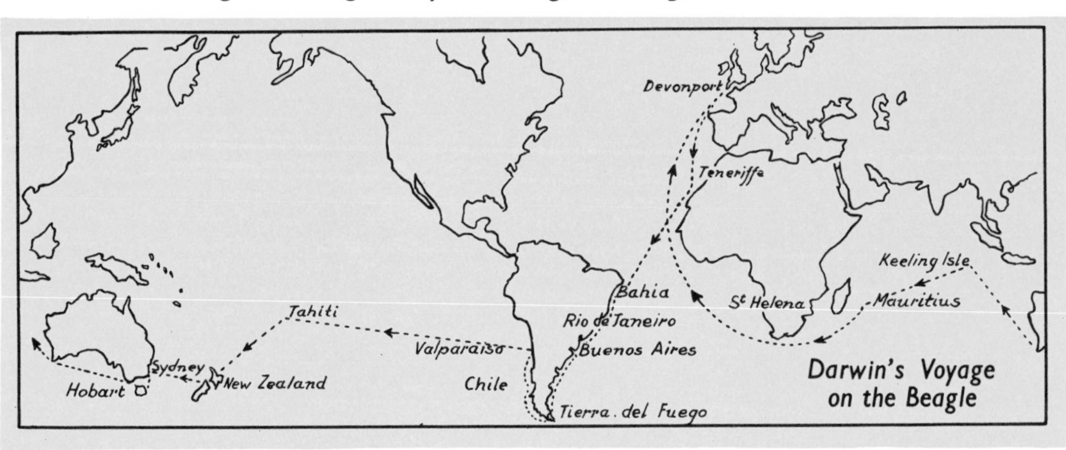

HEREDITY AND GREGOR MENDEL

Charles Darwin's evidences for the evolution of plants and animals are accepted by the majority of people today although they may not all interpret them in the same way. Great advances have also been made in understanding the ways in which characters of parents are passed on to their young. This science is known as *heredity* and the man whose work helped us most in those early days of scientific discovery was Gregor Mendel, an abbot who spent most of his life in a monastery in Brunn in Austria. Mendel was born in 1822 and was living at the same time as Darwin, but Mendel's work did not become widely known until after his death in 1884.

Like Darwin, Mendel was interested in the *varieties* of plants and animals and spent many hours studying the growth and development of plants in the garden of the monastery.

Mendel carried out experiments in crossing different varieties of sweet pea plants (fertilising the seed of one type of plant with pollen from another). The results of his many experiments were carefully recorded and were so mathematically correct that they have been used by plant and animal breeders ever since. The laws of inheritance which Mendel explained from his experiments are known as *Mendel's laws.*

Here is a picture to show you how Mendel's laws apply to the breeding of rats. The mother and father rats, one brown and one white, each come from a long line of pure-bred brown and white rats respectively. Notice that their two babies are both brown. This is because brown colour is dominant to white. These hybrids (cross breeds) may still be able to produce a white rat, however, and you can see that when two of these hybrids are mated, they produce a white rat in the next generation. The proportion of brown to white is three to one.

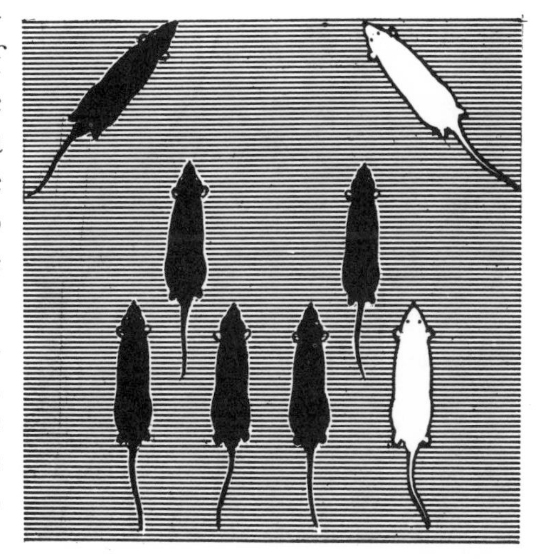

INVISIBLE CREATURES AND LOUIS PASTEUR

For many years people did not know that in air, in water and in soil, and even in our own bodies, there are millions of creatures so small that they are invisible to the naked eye. The man who revealed these " invisible armies " of bacteria to us was Louis Pasteur who in 1854 was head of the science department at Lille University in France.

Louis Pasteur (1822–1895) *at work in his laboratory*

One of Pasteur's first discoveries was concerned with *fermentation.* He discovered that there were little spores of yeast fungus in the air ready to settle on any sugary liquid which was exposed.

Pasteur also isolated the germs which cause the disease of *anthrax* in sheep and of *rabies* which causes madness in dogs.

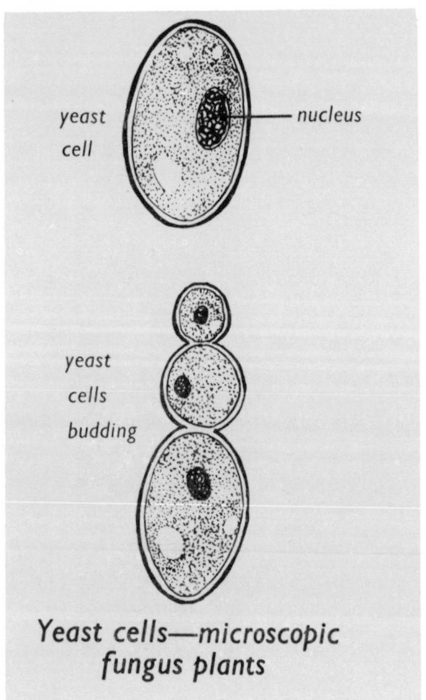

Yeast cells—microscopic fungus plants

With his new knowledge about bacteria, Pasteur was able to find remedies for some of the diseases caused by them. He introduced inoculation which could give a mild form of a disease but at the same time build up the body's resistance against it.

Pasteur's studies of fermentation and of yeast cells (see diagram) helped the manufacturers of cheese, vinegar and beer to understand the fermenting processes by which their products were made.

Because of Pasteur's work we now have foods in sealed cans and glass jars which are airtight to keep out bacteria, or we store foods in refrigerators at temperatures which are too low for the bacteria to develop.

ANTISEPTICS AND JOSEPH LISTER

Pasteur's discoveries about these invisible microbes which are always present in the air were studied with great interest by the surgeon whose photograph you see here, Joseph Lister, later to become Lord Lister.

Joseph Lister (1827–1912)

In those days many people died after surgical operations because their wounds became infected. Lister realised that, in order to prevent this, harmful bacteria must be killed. He studied chemicals which might prove to be " bacteria-cidal " or " bacteria killing ". Among these bacteria-killing chemicals he found carbolic acid, permanganate of potash, thymol and others.

Studying Pasteur's discoveries and seeking constantly for new methods to prevent the infection of wounds, Lister revolutionised surgery. His insistence on scrupulous cleanliness and the use of the new chemicals in performing operations resulted in a great saving of life.

These chemicals were the first of our *antiseptics* and eventually they were used in all our hospitals, not only in surgery but in general nursing as well. Florence Nightingale, the first really great nurse in our country, used the antiseptics discovered by Lister during the Crimean War and the lives of many soldiers were saved. The death rate in hospitals gradually declined after the use of these antiseptics became universal.

Lister much admired the work of Pasteur and admitted that the new knowledge of bacteria had helped him to study the treatment and prevention of disease.

Lister and Pasteur would not, however, have found it easy to study their " invisible armies " had it not been for the invention of the microscope, which you can read about on the next page.

THE MICROSCOPE

In order to see the tiny living things you have just read about, you would need to use a microscope. The first simple microscope was made by a Dutchman named Anthony van Leeuwenhoek in the 17th century, but the first *efficient* compound microscope, like the one on the right, was made by Robert Hooke, an Englishman, also in the 17th century.

You can see that the modern microscope shown below is built on the same plan but has more complicated fitments.

It was because he was able to use a microscope that Pasteur discovered so much about the tiny organisms we call bacteria and was able to study their growth and behaviour.

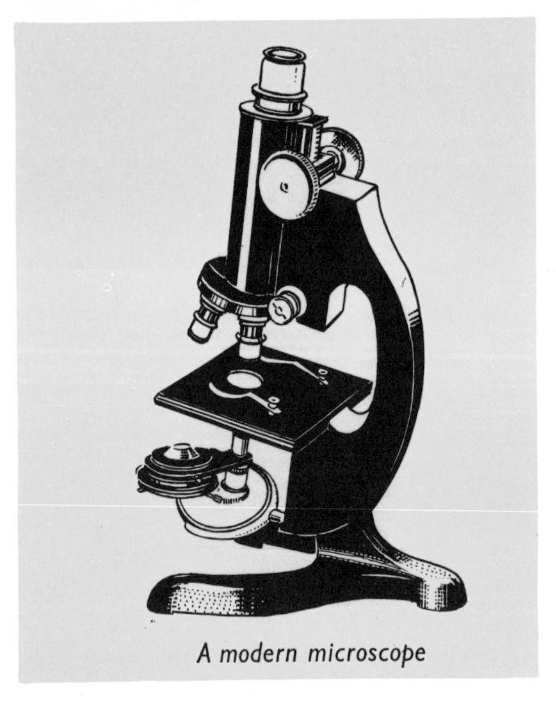

A modern microscope

An early microscope

Stinging hairs on a nettle leaf (much enlarged)

Some of the first things looked at by Robert Hooke under a microscope were the stinging hairs on the leaf of a nettle. If you have a good hand lens, you can see these hairs for yourself.

A triple hand lens

ALEXANDER FLEMING AND PENICILLIN

A scientist of our own day who has studied bacteria is Sir Alexander Fleming, the discoverer of the drug, penicillin. Penicillin can destroy several types of germs, including those which cause bronchial pneumonia, and many lives have been saved through its action.

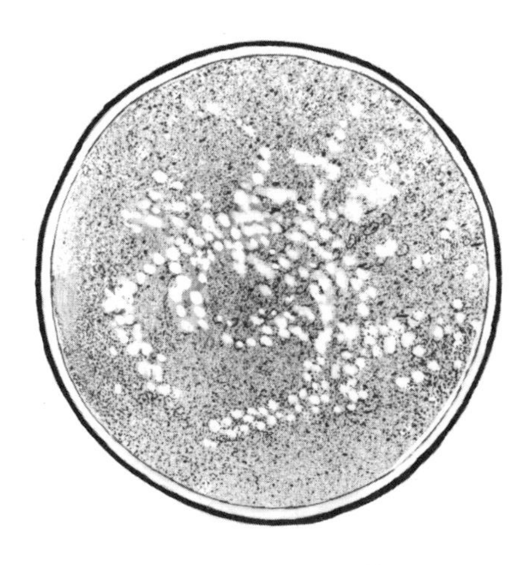

Bacteria growing in jelly on a glass dish

The way in which Fleming discovered penicillin is interesting. Scientists often develop cultures of bacteria in round glass dishes called petrie dishes like the one shown in the picture on the right. The bacteria cultures are reared on a nourishing jelly upon which they feed and grow. If the glass dishes are left exposed, it is possible that other germs or spores from the air might fall on to the jelly and grow alongside the original culture.

Something very like this happened in Alexander Fleming's laboratory. A " foreign " spore from the air must have started to grow on a germ culture and, more important still, it was seen to be killing the original germs or bacteria which had been growing there before. The new invader was a greenish mould called *penicillium* which you can grow for yourself on stale fruit. An extract from this green mould was made and was named penicillin. This wonderful drug was tested and was found to kill certain bacteria known as *cocci*. Two well-known types of *cocci* which penicillin attacks are streptococci and staphylococci.

Streptococci

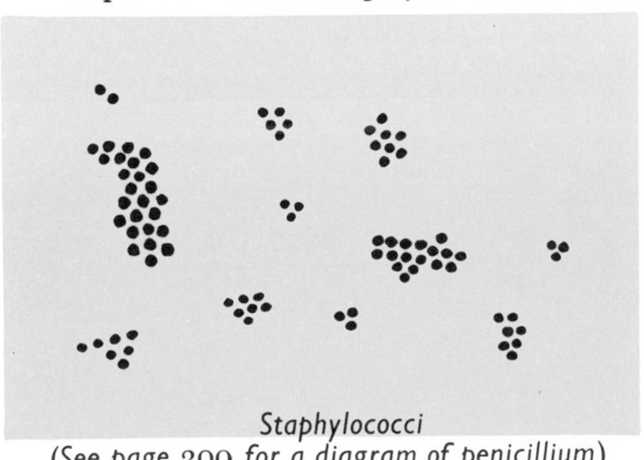

Staphylococci

(See page 300 for a diagram of penicillium)

MORE ABOUT BACTERIA AND DISEASE

Bacteria are often carried about on the hairy bodies of flies. The enlarged diagram of the house fly on the right shows you the hairiness of its body.

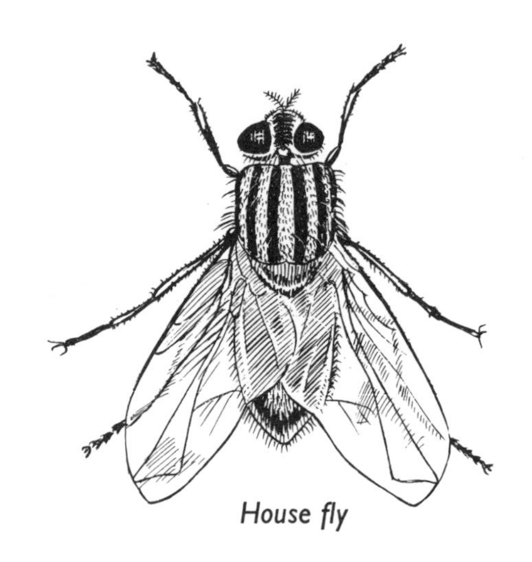
House fly

Apart from the *cocci* type of bacteria which you read about on the previous page, there are many other kinds. The typhoid bacillus, shown here highly magnified, is like a thick rod with little hair-like appendages called *flagellae.*

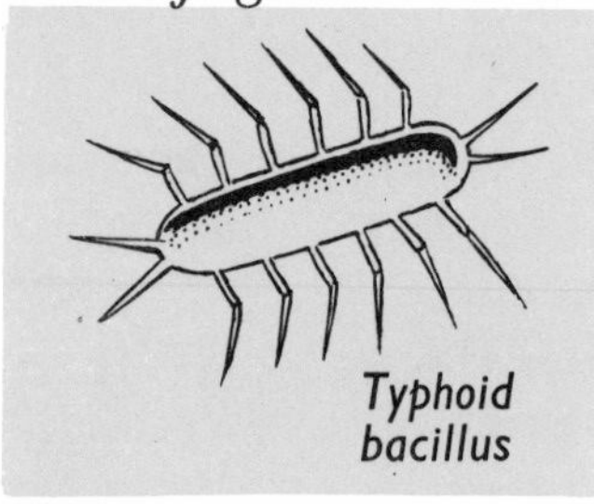
Typhoid bacillus

People have caught typhoid fever after drinking milk or water contaminated by these bacilli. We hear, too, of " typhoid carriers ". These are people who may not appear to have the disease but can pass it to others. The handling of food by such persons is a source of danger.

When disease germs enter our body, our blood has its own way of dealing with them. Certain cells in our blood known as *leucocytes,* or white blood corpuscles, deal with the invading germs and devour them.

On the right, you can see a picture, much enlarged, of a white blood corpuscle devouring disease germs.

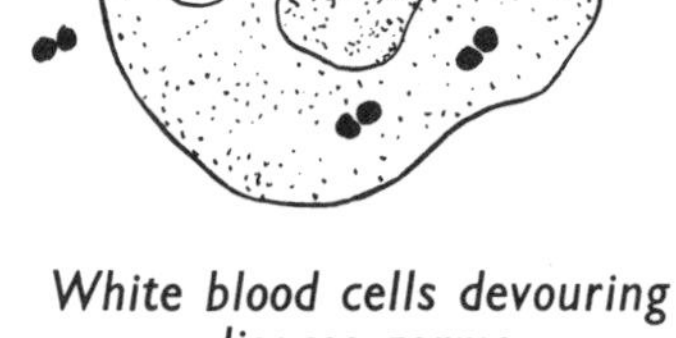
White blood cells devouring disease germs

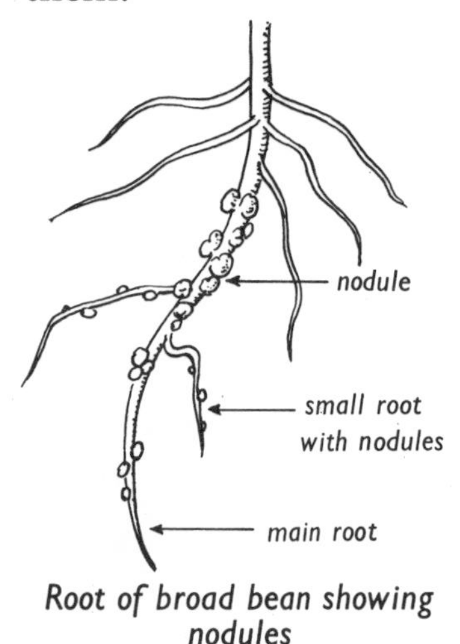

Root of broad bean showing nodules

Although as you have seen, many bacteria are harmful and can cause disease, many others are helpful. The little swellings on the broad bean root in the diagram on the left are called *nodules* and each nodule contains many bacteria which help the plant to obtain its essential supply of nitrogenous food from the soil. Nitrates are important plant foods.

FIGHTING DISEASE—Malaria

You have read that flies often carry disease. One fly, the mosquito, (not the one found commonly in this country), carries a dreaded and painful disease called *malaria*. Malaria occurs in tropical countries, especially in swampy districts where the mosquitoes breed.

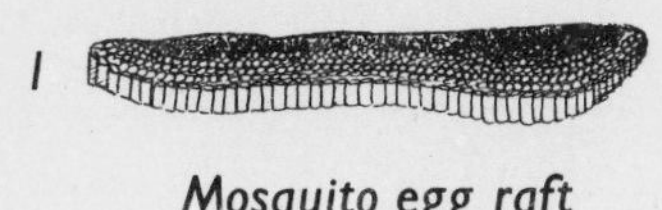

Mosquito egg raft

Two English doctors, Ronald Ross and Patrick Manson, helped by American doctors, studied the habits of the mosquito which carries malaria and helped to stamp out the disease in certain areas, although wherever these mosquitoes are allowed to breed in hot wet countries there is still much risk of contracting the disease. Perhaps it would be wise to follow the work of these doctors and first study the life and habits of the mosquito.

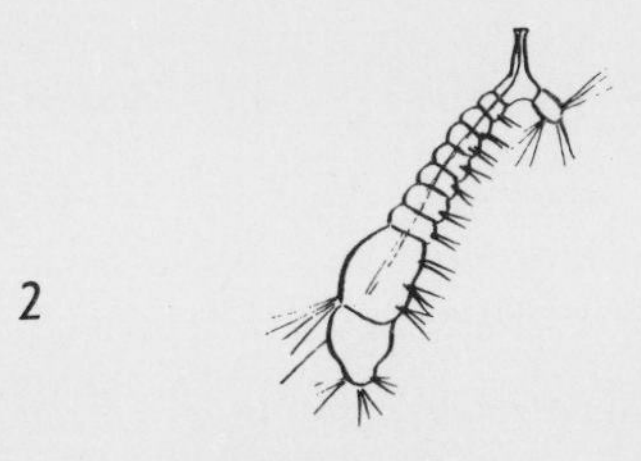

Mosquito larva

Adult mosquito

The *eggs* of the mosquito are laid in the form of a floating raft (Fig. 1). The tiny *larvae* (the young mosquitoes) are aquatic and swim about until they change into an individual called a *pupa*. The mosquito pupa has a big head. The pupa eventually comes up to the surface and splits, and out of its big head steps a dainty little mosquito (Fig. 3) with wings, ready to take to the air.

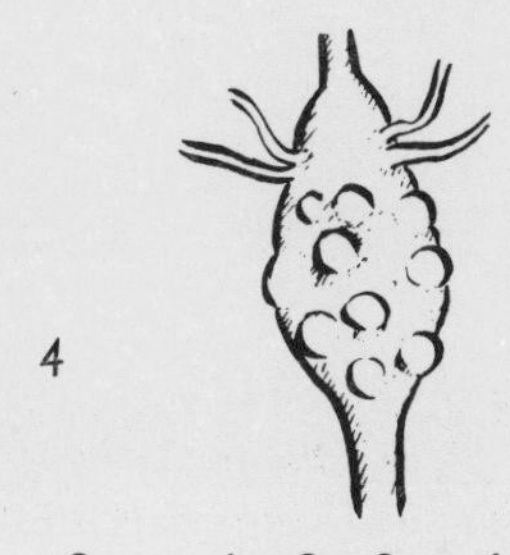

Stomach of infected mosquito showing swellings

Female mosquitoes suck the blood of their victims. They can spread malaria by sucking the blood of a person with the disease and then passing it on by biting another person. Mosquitoes also get malaria themselves and Fig. 4 shows the stomach of a mosquito swollen out with the disease. When an infected mosquito bites, it can inject the disease germs into a human being.

Section through salivary gland of mosquito showing parasites in its cells

6 Blood cell of a human being with malarial parasites in it

HELPFUL AND HARMFUL MOULDS

Here is a picture (much magnified) of the mould *penicillium*, which you read about on page 297—the one which produces the wonderful healing drug penicillin.

Many moulds are known to scientists. We can develop some of them for ourselves by the simple experiment of placing some damp bread in a saucer of water with a tumbler over it and leaving it for a few days. During this time mould will form on the bread.

Most of this mould will be a white one called *mucor*—the " pin head mould " you read about on page 186 which will, in time, look as if it is covered with black pin heads which are the sporing bodies.

Penicillium has a different kind of sporing head as you can see in the diagram.

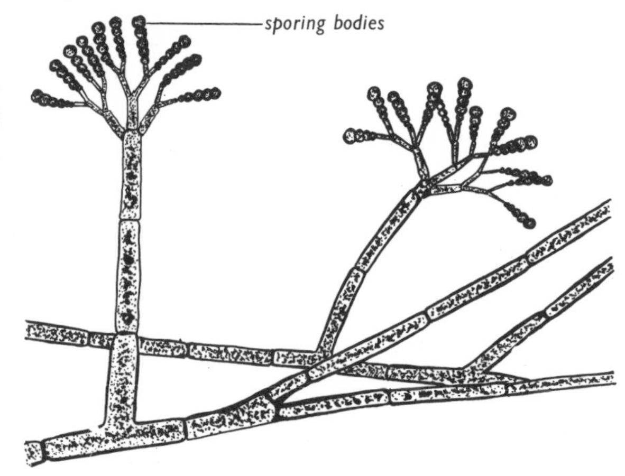

Penicillium (much magnified)

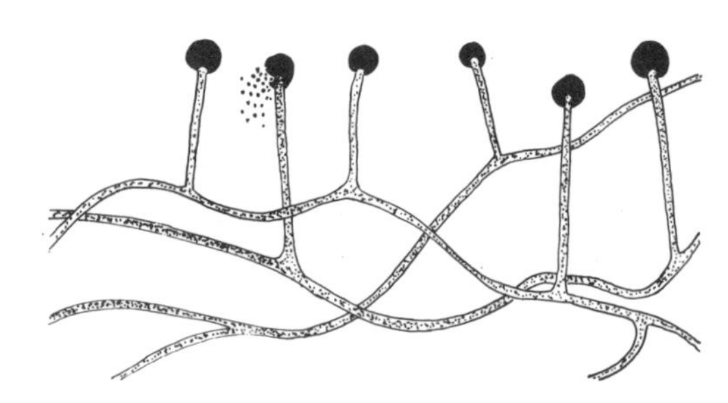

Enlarged view of the mucor " pinheads " growing from the white tissues of the mould

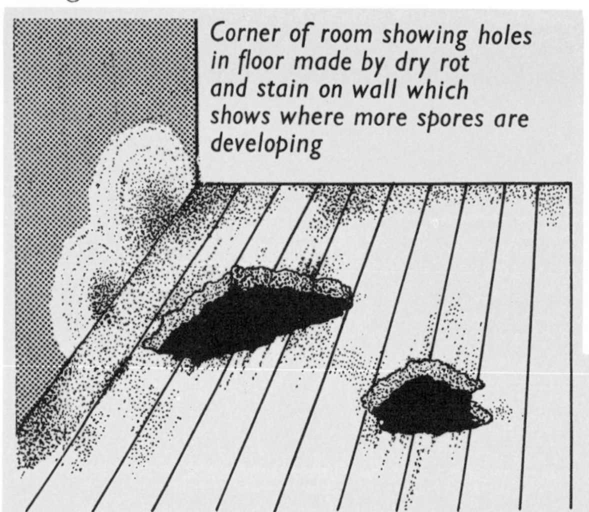

Corner of room showing holes in floor made by dry rot and stain on wall which shows where more spores are developing

The picture on the left shows a very unpleasant form of mould fungus known as " dry rot ". This affects walls and floors in old buildings which have become damp. This fungus produces spores which develop and grow into the wood, causing decay in the manner shown in the picture.

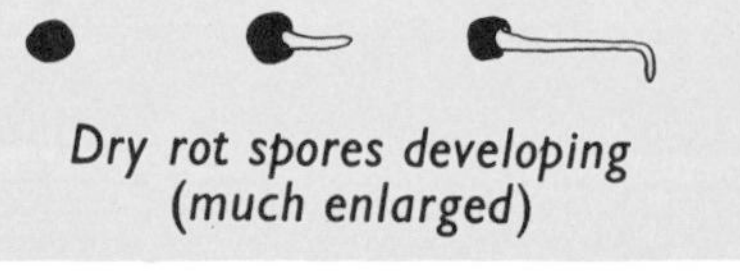

Dry rot spores developing (much enlarged)

HIGHER TYPES OF FUNGI

The moulds about which you have been reading are " lower " kinds of fungi. This means that they are very simple. They consist only of white fluffy threads and of the tiny fruiting bodies which they produce from time to time. The " higher " fungi such as mushrooms and toadstools, however, have a more complicated structure and the fruiting bodies which they produce are larger.

Here are two fungi which you can eat.

A is the well-known mushroom. It has a smooth round cap called the pileus (p) and pink gills underneath (g) which turn brown later. When in bud, the mushroom looks like a small round white button because its top part is joined in one with its stalk, or stipe (st). As it grows, the cap breaks away from the stalk, leaving a little frill round the edge of the cap, called the velum (v) and a ring or annulus (a) round the stalk or stipe.

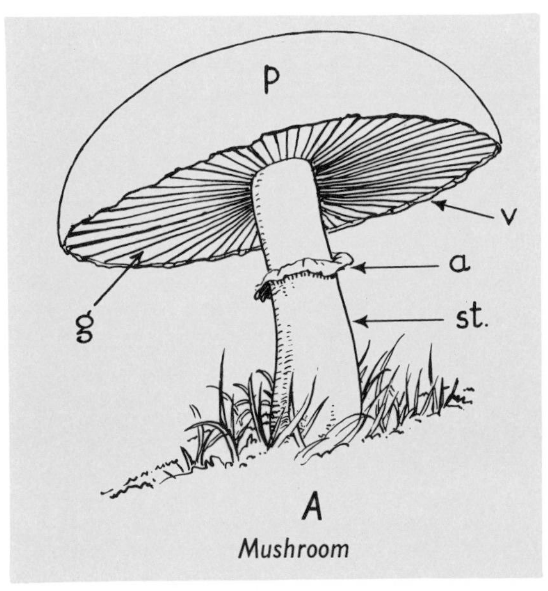

Mushroom

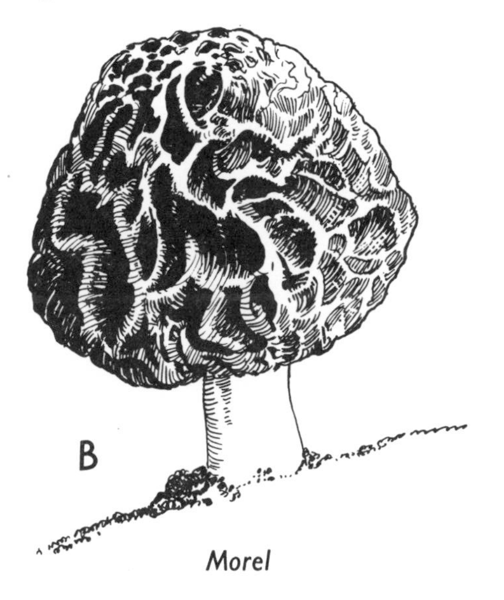

Morel

The other picture (B) shows a fungus of a different shape—the morel. This consists of a stem and a cap. The cap is egg-shaped and is yellow ochre in colour (look for this colour in your paint box).

The morel has been used as food from earliest times and is still a great delicacy. It is often dried and used as a flavouring, or cooked with meat. Morels grow in the spring in shaded places.

TWO POISONOUS TOADSTOOLS

Among the " higher " fungi of the toadstool type there are many which are poisonous. Some of these are very attractive to look at and may be beautifully coloured.

Scarlet fly cap

The scarlet fly cap which you can see on the right has a bright red pileus (cap) with white flaky spots all over it. The gills are white. You can see it in colour opposite page 257. Its name comes from the fact that it was once commonly used to trap and kill flies, and in parts of Europe it is still put on kitchen window-sills to discourage flies from entering the kitchen through the open window.

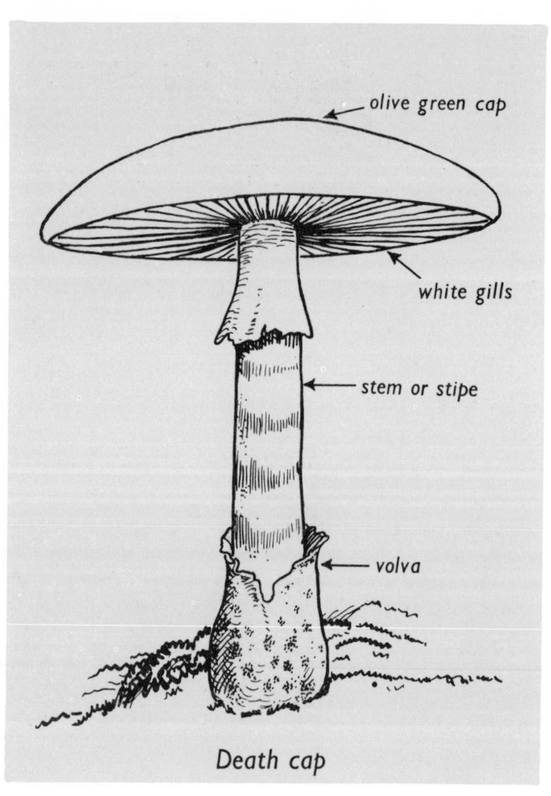

Death cap

Probably the most dangerous poisonous toadstool of all is the deadly white amanita, the Death Cap, shown in the lower picture. When seen growing in a dark wood it has a sinister greenish-white radiance about it.

The cap is olive green but the rest of the toadstool is greenish-white. The stipe (stem) grows to a height of about 12 centimetres and at its base is a sort of bulbous case, or *volva*, out of which the fruiting structure grows.

This toadstool is found in damp woods in late summer and autumn and if you see one *do not touch it at all !*

Here are some more pictures of higher fungi or toadstools of different types. You should recognise the first three which you have already read about. They are A, the mushroom, B, the morel and C, the scarlet fly cap.

D and E are different types. D is known as the " earth star " or *geaster* and looks rather like a round egg sitting in the middle of a starfish (except that a starfish has only five rays). It is a fungus rather like the " puff ball " type and its powdery brown spores are inside the central ball until this splits open and they escape.

E is the " bird's nest " fungus and you can see that the little containers with their spore capsules inside certainly give the appearance of small birds' nests.

How do fungi differ from other plants ?

In the first case they are not green. You will remember that green plants make their own food from the raw materials which they find in earth, air and water. Non-green plants cannot do this (refer back to pages 248 to 253). Because of this fact, moulds, as well as the higher fungi (mushrooms and toadstools), feed on decaying material which has once been alive. In other words they are *saprophytes*. A saprophyte is rather like a parasite except that while a parasite feeds from a *living* organism, a saprophyte takes food from a *dead* or decaying one.

Secondly, the bodies of these plants are not made up of cells in the same manner that the bodies of green plants are. Fungi are built up of a mass of fine threads or fibres called *hyphae*. If you pull or tear out the stem of a mushroom, you will see that it is made up of these tiny white threads.

A REVIEW OF THE ANIMAL WORLD

VERTEBRATES

If you study the pictures above, you may realise that each one represents a *group* of animals. The lion represents the class of warm-blooded, fur or hair-covered animals called *mammals*. The jackdaw represents the bird group, *Aves*. The snake (it could have been a lizard or a crocodile or an alligator) represents the group of reptiles or *Reptilia*. The toad, like its relatives, the frog and the newt, is one of those animals who live partly in water and partly on land, the *Amphibia*, while the little minnow represents the group of fishes or *Pisces*. You will read about some of these animals elsewhere in this book but now think about them all together. They really all belong to one family—the family of animals which have a backbone. In other words, they are *vertebrates*.

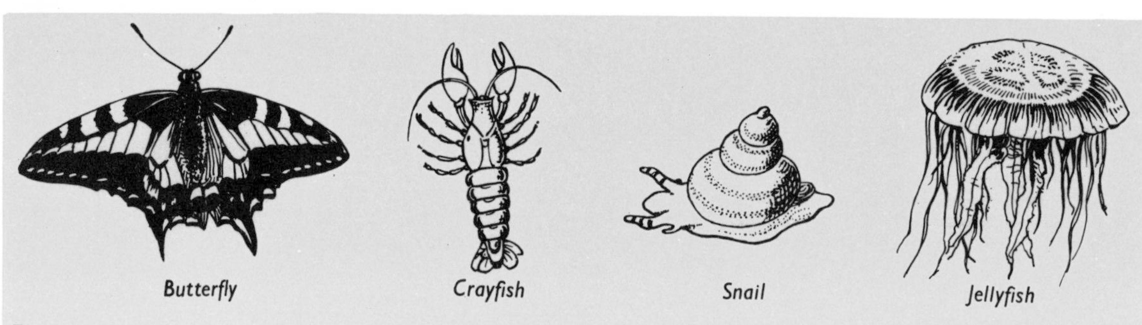

INVERTEBRATES

Now study the animals above. The butterfly represents the *Insect* group and the crayfish with its hard body covering is a *Crustacean*. These two groups are related because both contain animals with jointed limbs and they are known jointly as *Arthropods* (jointed limbs). The snail, a soft-bodied animal with a shell, belongs to the group of *Molluscs* (moll—soft). The jellyfish and its relatives sea anemones and corals are all very simple animals made up of a body wall with a space inside and are grouped as *Cnidarians*. You can also see another invertebrate on page 279—the earthworm, which belongs to the group *Annelida*.

A REVIEW OF THE PLANT WORLD

You have seen that animals can be classified into two main groups according to their possession or non-possession of a backbone. Invertebrates (non-backboned animals) are more primitive in many ways than vertebrates, although we know that some insects have highly specialised qualities. (One-celled animals belong to the group *Protozoa*).

Bulbous buttercup
an herbaceous plant

In a similar fashion, we can discover two distinct groups in the plant world. Again it is the possession or non-possession of a certain feature which helps with our classification. Some plants have flowers, some do not. All those plants which have flowers are called " flowering plants" or angiosperms. There are several groups of non-flowering plants. Some of these you have seen or read about. They are the seaweeds, the mosses, the ferns and liverworts, and the various types of fungi. All are green plants *except* the fungi, for although seaweeds are various colours they *do* possess chlorophyll or green colour somewhere in their plant body.

Sycamore
a woody plant

Most of the smaller flowering plants are herbaceous or non-woody but some larger ones like trees and shrubs have a woody supporting framework. They have flowers although (except in "blossom trees") their flowers are often greenish like those of sycamore. Conifers are not perfect flowering plants for their seeds are not completely sealed up in a fruit, but are exposed on the bracts of a cone. They may be considered as halfway between flowering and non-flowering plants.

CLASSIFYING A PLANT

When we cannot immediately recognise a plant, it often helps us to know what family it belongs to. This is a good step towards the *identification* of the plant. (To identify means to recognise each individual plant by its own special name.)

Each individual plant and animal is known as a *species* and its own special name is its *specific name*.

Plants have common or local names and they also have Latin names. The Latin name enables us to recognise the same plant or animal in different countries, for Latin is a universal language.

Let us now try to *classify* a specimen, or at least to find the group to which it belongs.

Here is one of our commonest plants, the daisy (Fig. 1), which you will recognise at once. If, however, you did not know its name, you could begin by putting it in some big group in the plant world and working from there. You could discover this big group by looking at some of the outstanding features or characteristics of the plant. For instance : Is this plant a flowering or a non-flowering type ? Obviously it has a flower (unlike mosses, ferns and other non-flowering types) and so we can classify it as an *angiosperm* or a flowering plant.

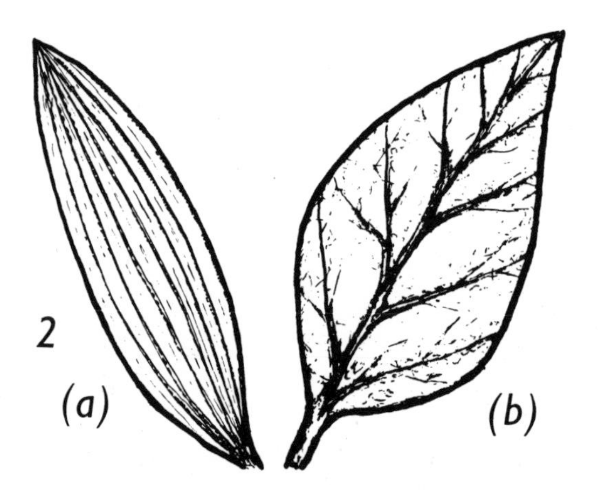

Angiosperms or flowering plants are themselves divided into two big plant groups. These groups are called *monocotyledons* and *dicotyledons* and Fig. 2 shows you the leaf of a monocotyledon (a) and the leaf of a dicotyledon (b). The monocotyledon has parallel veins and the dicotyledon has a network of veins branching out from a central vein. Look now at the leaf of the daisy plant. To which group does it belong ?

On the previous page you were asked to look at the daisy and to discover whether the plant could be grouped as a monocotyledon or a dicotyledon. This may require careful observation because the leaf of the daisy is thick and fat and the veins are somewhat embedded and therefore do not stand out at all clearly. However if you look carefully, you will see that the vein arrangement is like that shown in the diagram opposite and you will have no hesitation in classifying the daisy plant as a dicotyledon.

The flower head of the daisy gives us the next clue to its classification because this floral arrangement or " inflorescence " of the daisy shows it to belong to a large and successful group of plants known as *Compositae.* The Compositae family is the largest family of flowering plants in the world and consists of 14,000 species. All these plants have one special distinguishing feature. Instead of having single flowers, they have numbers of small flowers or " florets " grouped together in large heads, that is, composite heads of florets : hence their name Compositae.

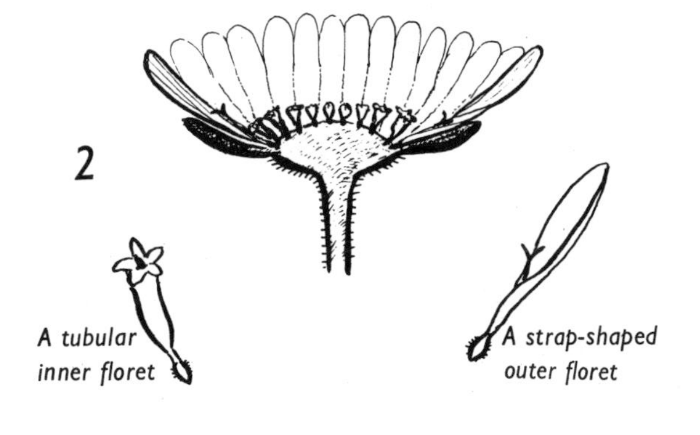

Diagram 2 shows how the florets of the daisy are arranged in a vertical section of the whole head. Make a vertical section of your own daisy " flower " and you will see that it is not really one flower at all but a whole head of flowers or florets. The Latin name of the daisy is *Bella perennis.* The second name " perennis " shows that it is a perennial herb—that is, it remains alive during the winter time and blooms each year. You can therefore classify your daisy plant, or *Bella perennis,* as an *angiosperm* (flowering plant), a *dicotyledon* and a member of *Compositae.*

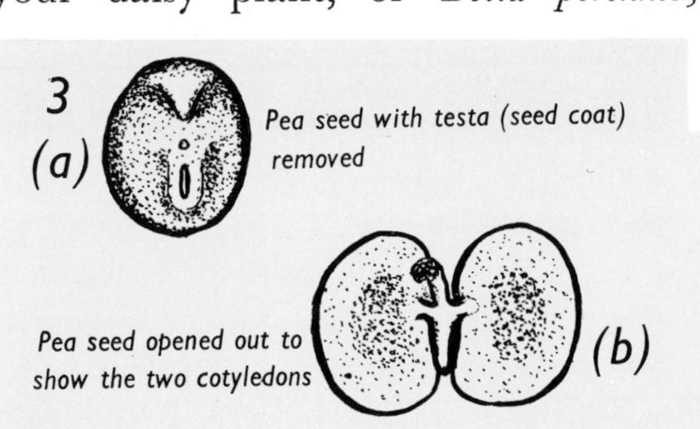

An extra note about dicotyledons : as well as having net-veined leaves, they also have two halves (cotyledons) in their seed. You can see these two cotyledons in the seed of another dicotyledon, the pea.

CLASSIFYING AN ANIMAL—The mouse

Let us try to classify the house mouse in the same manner that we classified the daisy.

We know that the mouse has a backbone and it therefore belongs to the group of backboned or *vertebrate* animals.

There are five groups of vertebrates.

These are :

1. Fishes—which have oily scales
2. Amphibians, such as frogs, toads and newts, which have moist, naked skins
3. Reptiles, which have hard dry scales
4. Birds, which have feathers
5. Mammals, the highest group of vertebrates, which have fur or hair as a skin covering.

To which group would you say that the mouse belongs ? Obviously it is a *mammal* because it has hair or fur.

House mouse

Mammals have other special characters too. One is that they feed their young on milk which they produce in special glands called *mammary glands*. Hence the name : mammals.

There are several different groups of mammals and many of these groups can be identified by their *teeth*. Members of one group, the carnivores, to which cats, lions and tigers belong, have sharp, curved, tearing teeth for eating flesh. The teeth of rats, mice and squirrels are chisel-like in front with flatter, grinding teeth at the back. In between the front and back teeth is a space with no teeth, where the animal has cheek pouches (see diagram).

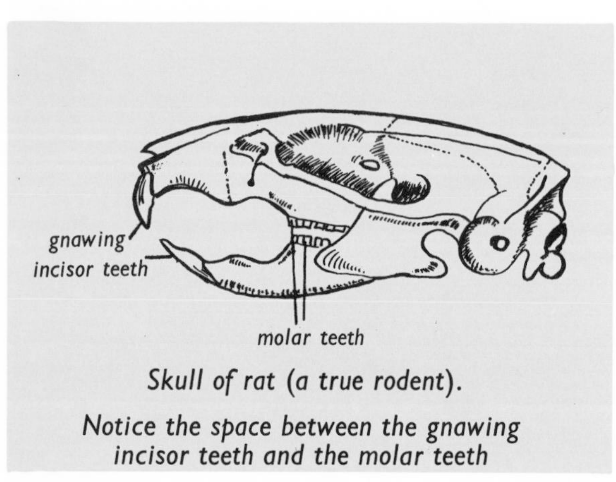

Skull of rat (a true rodent).

Notice the space between the gnawing incisor teeth and the molar teeth

The group of mammals which has these chisel-like teeth is called the group of *rodents* and the name rodent means " gnawing teeth ". Among the rodents you will soon discover the Latin name of the house mouse. It is *Mus musculus*.

ANIMALS AT PLAY AND AT WORK

Young foxes at play

Kitten with ball of string

Young animals which belong to the higher groups, such as mammals (furry animals) and birds, are looked after by their parents for some considerable time. During this time the young animals often play together as you have seen kittens and puppies play.

This play of young animals is often a kind of training or preparation for their adult life. A kitten playing with a ball of string or wool is really going through the performance of catching and playing with a mouse or bird, before settling down to eat it. Young fox cubs, snapping at each other and pretending to bite, are practising for later and more serious skirmishes when they will compete for food or for mates, or when they will be protecting their families from enemies.

Most adult (grown up) animals work extremely hard, rearing families, hunting for food, and building their homes. Some, of course, are better equipped for their hard life than others. As you saw on page 290, moles have special fore-limbs shaped for digging. Below you can see a mole's " fortress " with tunnels and a central chamber where the nest is made and the young are reared. Other examples of industrious animals are the spider, which spins a wonderful web ; the beaver, which builds dams across rivers ; the dormouse, which weaves a most beautiful winter nest ; and, of course, all nest-building birds.

Section through mole hill showing mole in nest with three babies

ANIMAL EQUIPMENT

Leopard

When you saw the kitten playing on the previous page, did you realise that a kitten, and indeed all cats, are really animals of prey? Cats are related to lions, tigers and leopards.

The picture on the left shows you a leopard in a frightening mood, ready to spring on its prey, but the way that a kitten springs on to a ball is very similar except that the kitten is playing and the leopard is deadly serious.

Notice its wide jaws and sharp fangs. The claws of all cats are re-tractile, which means that they can be pulled back into their sheaths.

The carnivorous or flesh-eating animals, such as the leopard above, feed on the gentler vegetable feeders or herbivores. In Africa, for instance, lions and leopards kill and eat many kinds of deer or buck. Most deer-like animals are very swift runners and quite often escape from their enemies in this way. These animals often have protective equipment such as horns or antlers, as seen on the head of the kudu on the right. In the mating (or rutting) season, males often fight, locking their antlers together.

Head of kudu

Head of rhino

The rhinoceros (see drawing on left) can have either one or two horns standing up from the middle of its snout.

Instead of horns or antlers, some animals have tusks, like the wart hog on the right.

Head of wart hog

Not only are certain animals equipped with claws, hooves, antlers and horns for their " battle for life " ; the bodies of some animals also show special adaptations which give them advantages over other animals. For instance, the neck of the giraffe (see picture) is exceptionally long. In fact the giraffe is the longest-necked animal in the world.

Giraffes

To this animal, its long neck is a great advantage. It helps it to look over the tops of trees, and thus to get a good view of the surrounding country and of any approaching enemies. It also enables the giraffe to chew off the tender shoots at the tops of these trees, which are its favourite food.

There is, however, one great disadvantage about having a long neck. It is difficult to get a drink. As the giraffe also has very long legs, the only way it can drink is by straddling its legs wide apart so that its head and tongue can reach the water comfortably. This " straddling " process often takes a considerable time and it is very interesting to watch.

Giraffe drinking

It seems that some animals, like the carnivores (flesh eaters) are equipped for *attack* and some, like the various deer, the giraffes and so on, are equipped more for *defence*. There are many defensive measures to be observed in plants and in animals. One of these is camouflage, which you will read about on the next page.

CAMOUFLAGE AND MIMICRY

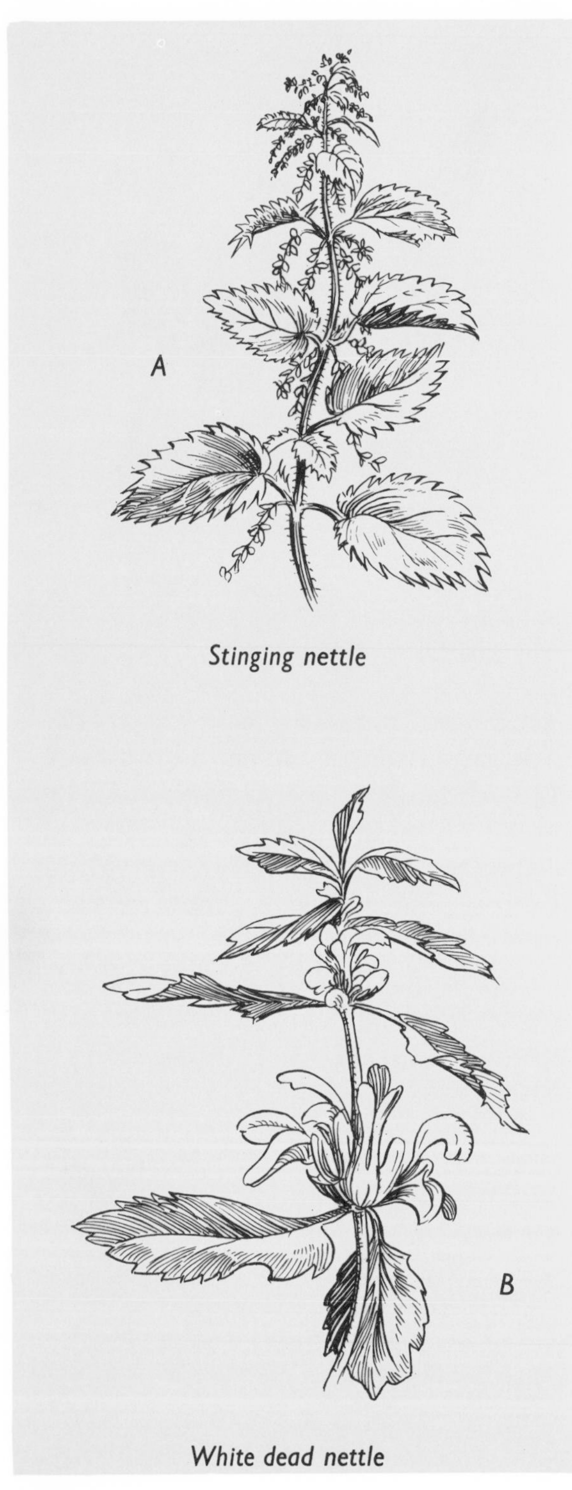

Stinging nettle

White dead nettle

What is camouflage? Perhaps you have heard that, during wartime, buildings of importance were disguised by painting them in a special fashion so that they looked like their surroundings and could not easily be picked out by enemy aircraft. Camouflage (disguising) can be a sort of " defence mechanism "; we frequently find camouflage in the world of plants and animals.

If you study the colour page opposite page 320, you will see that the colours and markings of the animals shown there fit in well with their backgrounds and in this way they may be practically invisible to their enemies. In the same way, the stripes of the zebra make it difficult to distinguish the animal amongst the long grass stalks in its homeland, and the spots of a leopard camouflage it in the patches of sunlight which filter through the bushland.

Mimicry is another of these " defence mechanisms " and is found in plants as well as in animals. It is rather like camouflage in that it deceives the observer. Look at these two nettle plants. A is the true stinging nettle and you know what to expect when you touch it ! B is the white dead nettle which does not sting at all but it looks so much like a stinging nettle that people hesitate before plucking it and so its mimicry is " defensive ".

MIMICRY AND DEFENCE IN INSECTS

Mimicry is found amongst insects. Here are diagrams of a stick insect (A) and a leaf insect (B) which look so much like parts of a plant (twig and leaf) that they may escape being seen for a long time and so remain safe.

The markings on some insects, particularly on the wings and bodies of moths and butterflies, often have a defensive quality. Look, for instance, at the mark like a skull on the body of the death's head moth (C), and the "eyes" on the wings of the emperor moth (D). Do you think that these markings could frighten off enemies? Some entomologists (or insect specialists) think so.

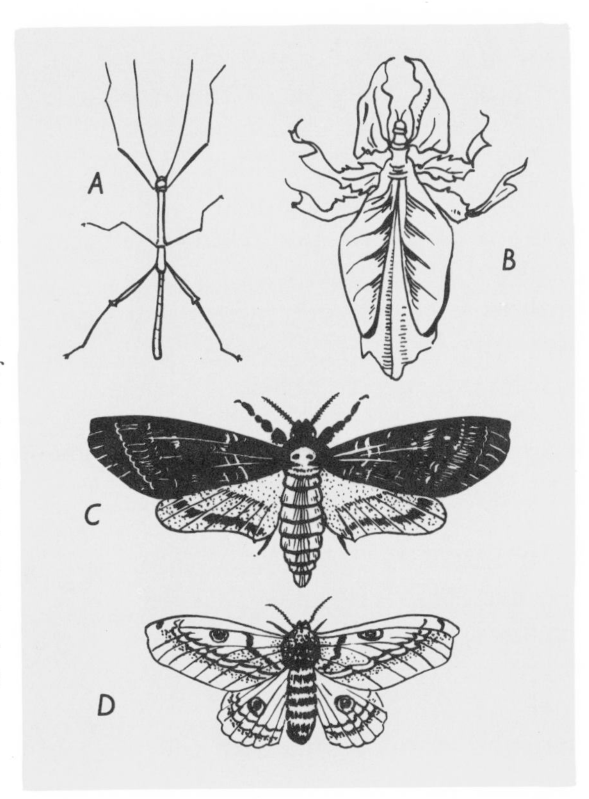

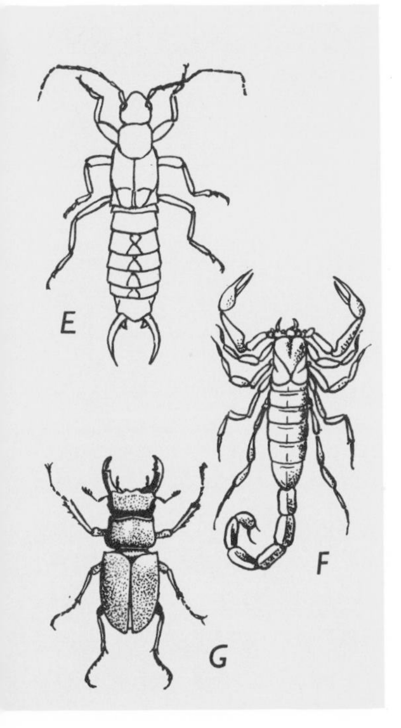

Insects often possess special defence mechanisms with which they are able to frighten off their enemies. Diagram (E) shows an earwig. This insect can inflict a painful wound by nipping with its pincer-like tail appendages.

The scorpion, shown in diagram (F), has large pincers like a crab, which it uses to seize its prey. In its tail is a sting. The scorpion curls its tail up and round to sting its victim and this sting can cause death.

Diagram (G) shows a stag beetle which has antler-like appendages rather like the antlers of a stag on a very small scale. There is no doubt that these weapons look very impressive to the beetle's enemies.

PROTECTING WILD LIFE

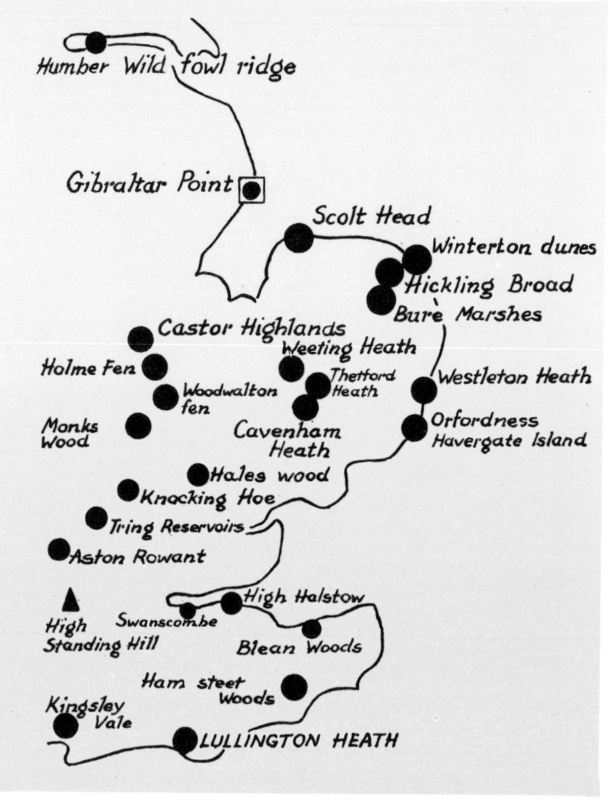

Although animals and plants have various ways of protecting and defending themselves against their enemies, they have very little defence against man. Perhaps you do not think of mankind as the enemy of many kinds of wild life, but when you remember that large areas of our countryside are becoming " built up areas ", you will realise that there is a danger of much of our wild life being crowded out.

Naturalists everywhere realise this danger. The Nature Conservancy, part of the National Environment Research Council, works hard to safeguard our beautiful scenery and wild life.

A place where Nature is specially protected is called a Nature Reserve. There are some large nature reserves in Scotland, including Beinn Eighe in Ross-shire, and another in the Cairngorms. In Wales, too, special areas of beautiful countryside are protected, including Cader Idris, Morfa Harlech and Skomer Island.

The avocet

The map on this page shows you many of the beauty spots in south-eastern England which are protected by " Nature Conservation ".

Look for Havergate Island where a rare bird, the avocet, now breeds. This bird is a wader (note its long legs) and has a slender curved beak. Since Havergate Island has been a protected area, avocets have become more numerous and are now breeding there regularly.

AN ISLAND NATURE RESERVE

This beautiful island, Skomer, has been under the protection of the Nature Conservancy since 1959. It lies off the coast of Pembrokeshire (West Wales). There are thousands of puffins and guillemots on this island. Below right are two guillemots with their chick. On Skomer there is also a special little red vole which is found nowhere else in the world. A picture of Skomer voles is given below. They are bigger and tamer than other voles and have brought much fame to Skomer in the world of science.

PROTECTION OF WILD LIFE IN AFRICA

There are several enormous national parks in Africa where the wild life is protected by law. One famous park in South Africa is the Kruger National Park which you can find on this map. This great sanctuary for wild animals covers an area of more than 19,000 square kilometres and is situated in the northern Transvaal between the Crocodile river in the south and the Limpopo river in the north. It is actually larger than the whole of Wales.

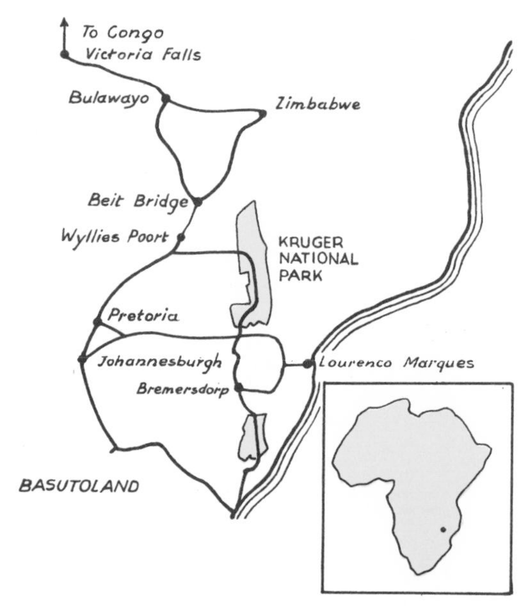

The dot on the small map of Africa shows the position of the Kruger National Park. The large map shows it in more detail

Tourists can travel through the park in the dry season. It is not possible to go during the wet season because much of it is a swampy, malaria-infested area at that time.

A rondavel

Tourists stay at various camps in the park which are fenced off to keep out lions and other dangerous animals, and fires are kept burning there all night. Visitors stay in little round huts called " rondavels ".

Zebra and many forms of deer roam about in the park, as well as giraffes and other gentle creatures, some of which fall as the prey of lions, leopards and other carnivorous flesh-eaters which also live there. You can hear the trumpeting of the large-eared African elephant as you rest in your camp at night. The elephant is a vegetarian.

WILD LIFE IN AUSTRALIA

Australia is a very large island, a land mass which must have become separated off from other land masses many millions of years ago. Probably because of this, the animals of Australia are different from the animals in any other part of the world and seem almost as strange to us as though they were creatures from another planet. For instance, there is the duck-billed platypus, a furry animal which lays eggs. This you read about on page 273. There is also the kangaroo, which you also read about on page 273, with a pocket for its baby.

Koala bear

The koala bear is a fat little Australian bear which spends most of its time in trees. It feeds mainly on the gum tree and eats the tender shoots which it finds at the ends of branches. Baby koala bears are first reared in their mother's pouch and are later carried about on her back. They are gentle and slow in their movements and look rather like children's " teddy bears ". They have soft greyish-brown fur, round ears and a black rubber-like nose.

Even the native cats of Australia are different from our cats. They are slender and rather weasel-like. They are about 45 centimetres long, and very ferocious. They kill birds and steal their eggs.

Australian cat

The native dog of Australia is probably one of the few native dogs left in the world. This " yellow dog dingo ", as it is sometimes called, has a fox-like head and a bushy tail and is very savage.

Australian dog (Dingo)

THE BISON—An animal which may become extinct

Bison (wild cattle) of different types are found in Europe and America, although there are not now many of them. There is only one small herd of European bison left in the world. It is in a private park in England (Woburn Park in Bedfordshire).

The animal in this picture is an American bison or American " buffalo ", as it is often called.

American bison

Notice its great shoulder hump, the shaggy, rust-coloured hair and the tuft on its tail which is held straight up when it charges. At one period, these bison herds roamed about all over North America, particularly in the western plains as far north as Canada and as far south as Mexico. There were literally millions of them. There are now no wild bison in North America, but a flourishing herd of about 14,000 has been saved and may be found in the Wood Buffalo National Park in Canada. This is an enormous wilderness of 44,000 square kilometres in Alberta.

Head of South African buffalo

Another type of buffalo is found in South Africa and is known as the Cape Buffalo. It was almost wiped out in parts of the country by a disease called rinderpest, but there are now some flourishing herds in the National Parks, where they are protected. Like all buffalo, these carry their heads lower than the top of their shoulders. Their horns, which are present in both males and females, are thick at the base, very heavy, and meet together in the middle of the forehead. Full-grown buffalo may weigh up to 680 kilogrammes.

MUSTANGS—THE WILD HORSES OF NORTH AMERICA

A mustang

At one time, large herds of wild horses roamed the plains of Texas in North America. In the " Westerns ", or cowboy films that you see, you may have learned about these wild mustangs which roamed the prairies. They were the descendants of the great wild herds and they originated in the following way.

Horses were brought to America by the Spaniards in the 16th century. Some of these escaped and ran away into the prairies, eventually becoming wild. The Spaniards called these runaway horses *mestengos,* and this name became abbreviated (shortened) to " mustangs ". It was from these runaway *mestengos,* that the mustangs of the prairies originated. A stallion or male horse used to run about with as many as sixty mares and the stallions used to fight savagely for possession of the mares. Thousands of these beautiful horses were trapped and sold by the settlers. Some were put to work in mines. Some were killed and sold for horsemeat. Many were lassoed and killed in a ruthless way.

It looked as though the wild mustang would become extinct in the plains of Texas, Arizona, New Mexico and throughout the West. Animal lovers and humane societies made protests about the cruelties which some of them had witnessed. At last a law was passed in Wyoming which saved hundreds of horses from being killed for horsemeat. All profits for the sale of this horseflesh were banned and so the wild mustangs were allowed to roam free.

In spite of this new law, only a small number of America's wild mustangs have survived. These remain on high land in the rocky western plains and very rarely venture down on to the roads. They are almost impossible to catch on horseback. It is said that they panic and run wild whenever they see planes flying low over their territory.

THE STORY OF THE NORWEGIAN LEMMING

This little animal, only 12 centimetres long, belongs to the same family as rats, mice, squirrels and dormice—the rodent family. It looks rather like a fat little field vole. It has brownish-yellow fur and bright, beady eyes.

Lemming

The lemming likes best to live fairly high up in the mountainous country of Scandinavia (Norway and Sweden). Lemmings feed upon tussocks of grass and also upon lichens and mosses. They also like to chew the bark and leaves of the young silver birch trees and juniper bushes which grow in this kind of country. Lemmings need a great deal of vegetable food of this kind and, as they reproduce (multiply) very rapidly, the food in their territory soon gets used up and then the lemming families have to move farther down into the valleys in search of more food. There are some years when the number of lemmings becomes enormous. These are called "lemming years". Many of the narrow Scandinavian valleys lead down to the sea. The great mass of lemmings in search of food goes on down these river valleys until the animals eventually reach the sea. Those behind push on and the animals in front are pushed into the sea.

It is possible that the lemmings hope to reach another shore by swimming away from the mainland. Unfortunately this does not happen and many hundreds of them are drowned.

The march of the lemmings

Toad

Tiger moth

ANIMAL CAMOUFLAGE

Guillemots

Rabbit

A NEW ANIMAL IN BRITAIN—THE SQUIRREL-DORMOUSE

The squirrel-dormouse is neither a squirrel nor a dormouse, although it looks a little like both animals. It is about 15 centimetres long, has soft grey fur and a bushy tail. It was known to the Romans, who used to catch and eat it, as the edible dormouse. Because it is rather fat, it is also called the fat dormouse. Its Latin name is *Glis glis.* It comes from south-eastern Europe.

This animal has not been living in Britain for very long and is not *indigenous* to this country (not a native of this country). The first of these pretty little creatures arrived about the beginning of the century, having been brought here by Lord Rothschild who let several pairs loose in Tring Park in Hertfordshire. There are now several hundreds of these animals, many of which live in Whipsnade Park in Bedfordshire.

Squirrel-dormouse (Glis glis)

The squirrel-dormouse is a rodent, like the lemming, the squirrel and the dormouse to which it is closely related. With its sharp, chisel-like teeth, it can gnaw bark from trees and chew nuts and fruit. It is particularly fond of apples and has been seen to enter lofts and storehouses where these are stored. During winter time the fat dormouse hibernates, or goes into its winter sleep, although not for so long a time as the bat or the true dormouse. The edible dormouse has a long and rather flat kind of tail which it curls around itself when it goes to sleep.

The squirrel-dormouse eating an apple

This photograph shows the squirrel-dormouse eating an apple. Notice how it holds the piece of apple in its delicately fingered hand. It will hold a nut or a cherry in the same way.

ANIMALS OF COLD COUNTRIES

Polar bear

You may have seen a polar bear with its white coat in a zoological garden. In such surroundings its white colour makes it conspicuous (noticeable) but among the snows of its native land, within the Arctic Circle, it is perfectly camouflaged, or hidden, among its white surroundings. It has little need to fear enemies, however, for it is one of the most savage of animals and would attack man, given the opportunity.

Seal

The polar bear swims beautifully and feeds mainly on fish. It is never found in the Antarctic.

Several animals from the snowy lands of the Arctic Circle have white fur to match their surroundings. The Arctic hare and the Arctic fox are two of these. When the snows melt in the short Arctic summer, the coats of these animals then take on a darker hue.

Stoat

One little animal which you probably know quite well—the stoat—is brown in summer but becomes white with a black tip to its tail in snowy wintertime, even in this country. When it is arrayed in this white covering, it is known as the *ermine*.

Seals are cold weather animals and they are found both in the Arctic and the Antarctic. Seals, although they live in the sea, are not fish but warm-blooded mammals (fur-covered animals).

Another cold weather animal, a swimming bird known as the penguin, is found in the Antarctic, but not in the Arctic.

Cape Penguin

THE PINK-FOOTED GOOSE FROM ARCTIC LANDS

In 1951, Peter Scott and James Fisher re-discovered the main breeding grounds of the pink-footed goose in Iceland. These large and beautiful birds also have nesting sites in Greenland and Spitzbergen. The pink-footed goose is a grey goose and you may have seen it for it is abundant in Britain in the winter. In summer, however, when the ice caps in the Arctic lands are melting and flowers are blooming in the valleys, the pink-footed geese are busy rearing their families there.

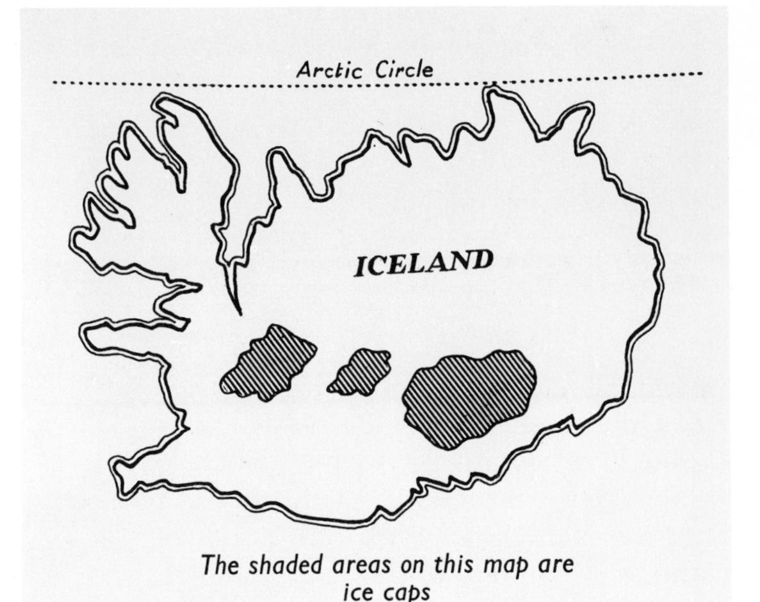

The shaded areas on this map are ice caps

In a book called *A Thousand Geese* you will read how Peter Scott and James Fisher caught and ringed a thousand geese in the summer of 1951. Geese are usually caught in rocket nets which shoot up and catch many birds together. After the scientists have ringed or examined the birds, they are allowed to go free. The rings, which are put on the birds' legs, help scientists to discover more about bird movements, for other scientists discovering the ringed birds can send reports of the flight distances from the place of ringing. For instance, in October 1951, less than three months after returning from Iceland, Peter Scott's party caught 530 pink-footed geese in Scotland which were wearing the rings they had put on them in Iceland.

Pink-footed goose

WHICH ARE THE HIGHEST ANIMALS?

By " highest " animals, we mean the most highly evolved or most advanced, or those best fitted by Nature to survive in this competitive world. Often these are also the most intelligent animals. The group of furry animals known as the Primates (to which man himself belongs) are known to be highly intelligent. The chimpanzee, which in recent years made a successful space flight, had been trained to pull levers and feed himself mechanically.

Young chimpanzee examining its toes

Apes and monkeys can walk upright

Baby chimpanzees are like human babies in many ways. The one in the illustration is examining its toes in the same way that a human baby does. You can see that it has the same number of toes as we have but the big toe and second toe on the foot can be used to pick things up in the same way as we use our thumb and first finger.

Apes and monkeys (including the chimpanzee above) can walk on two legs like their relative, man, although they drop on to all fours occasionally and often *run* on all fours. If you turn back to page 282, however, you will realise how much more advanced man is than even the most intelligent ape.

CAN ANIMALS TALK TO EACH OTHER?

According to legend, King Solomon was supposed to have the gift of speech with animals. An author named Konrad Lorenz has written a book about animal conversation called *King Solomon's Ring* which you would enjoy reading. To what extent, do you think, can animals communicate with us and with each other ?

A dog can understand certain sounds and words which it hears repeated several times and which it associates with certain happenings. A dog usually understands what " going for a walk " means and a well-trained dog will sit down when you say " sit ". Many of our pets can make us understand their wants, and whether they are pleased, by their actions or the sounds they make.

Worker bee

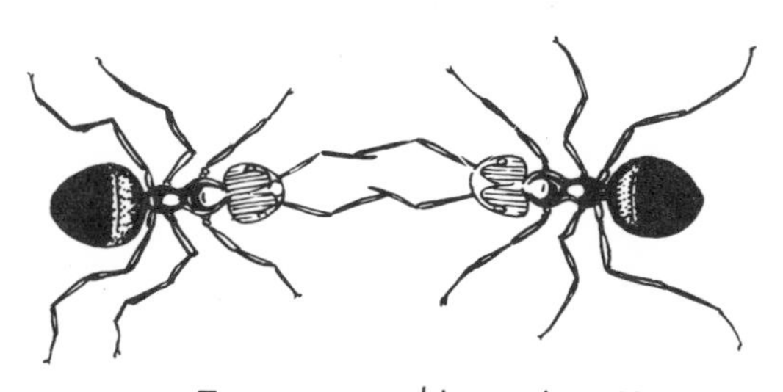

Two ants touching antennae

Although the majority of animals do not appear to use a form of speech to each other, they do have other means of communication. We have heard birds calling to each other, particularly in the mating season. We are told by apiarists (bee-keepers) that bees can tell each other the direction and position of " good honey flowers " by means of a specially symbolic dance which they perform in front of their hive. Male and female moths can be brought together over long distances by the vibration of their antennae (feelers). Ants have been seen communicating with each other by touching antennae as you see in the diagram. Some animals, such as bats, can make and hear sounds which are too highly pitched to be heard by the human ear, and fishes are known to receive and understand vibrations in the water which are made by other fishes.

LIFE UNDER THE SEA

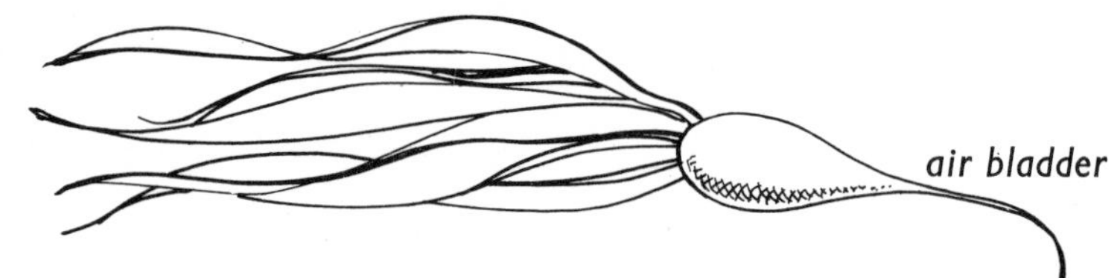

Plants and animals in the sea differ considerably from those living on land. In many cases they are more primitive for, as you read on page 283, scientists state that life must have begun in the sea, and many forms of water-life have not altered to the extent that terrestrial or land forms have, since their environment has remained the same for millions of years.

On parts of the ocean floor and on rocks which stand out from this floor grow beautifully coloured seaweeds. Some seaweeds are floating types or have bladders which help to keep their foliage afloat, like this giant kelp in the picture.

Scientists are finding that seaweeds are valuable. Bacteriologists use a gelatine called agar, which is made from seaweed, on which they rear cultures of bacteria. Seaweeds are often gathered for agricultural manure. A seaweed called Irish Moss has been used by cooks for stiffening blancmange for centuries and now is often put into medical lotions.

Corals, which very often look like beautifully coloured sea plants, are not plants at all. The many branches of the coral structure are made by innumerable small animals called coral polyps, which are relatives of the sea anemone. The coral structure is their home. In tropical seas, corals are often found in the form of reefs—" coral reefs "—which occur in the Pacific, the Indian Ocean and the Red Sea.

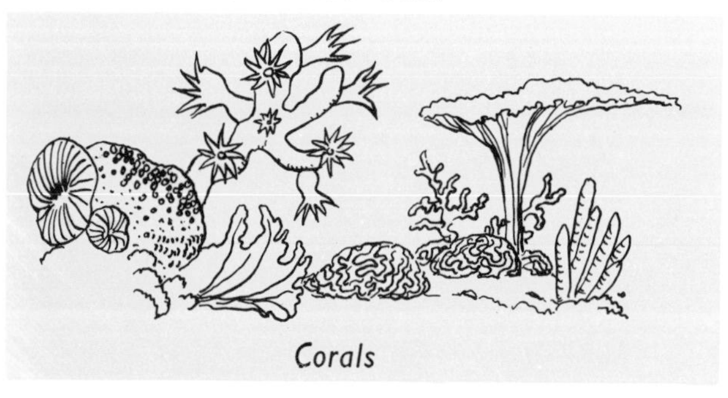

Corals

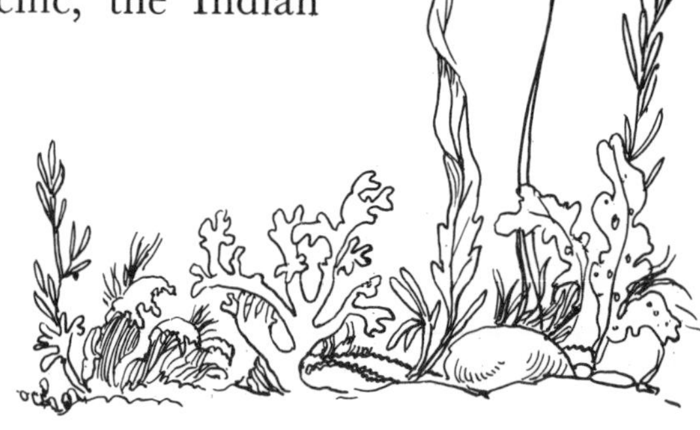

Seaweed

UNDER-SEA EXPLORATION—
Porpoises and dolphins

Diver

Idea for frogman's flippers sketched in the 16th century by Leonardo da Vinci

You must have heard a great deal about under-sea exploration by " frogmen " who wear " aqualungs " (breathing apparatus) and flippers. Your television screen has recorded for you adventures of under-sea explorers such as Hans Hass, so that many of the beautiful creatures which inhabit the deep oceans are no longer such mysteries to you as they were to all of us before new scientific methods made these explorations possible.

Through these investigations, divers have even been able to make friends with some of the large sea creatures, such as whales, dolphins and porpoises. These creatures, although aquatic (water dwellers) are not fish, but mammals. They are intelligent and are related to the land mammals, which are our most advanced animal group, and which include the Primates (ape and monkey group) which you read about on page 324.

Porpoises and dolphins move about in herds or schools and often race alongside ships, leaping and tumbling in the water. One of the most interesting and friendly creatures of the sea is the bottle-nosed dolphin. These animals inhabit great seas in all parts of the world and are great favourites in marine aquaria. They can be trained to feed from the hand and will play like puppies with their keepers who often descend into the large sea tanks to feed them.

Porpoise

Bottle-nosed dolphin

FISHES

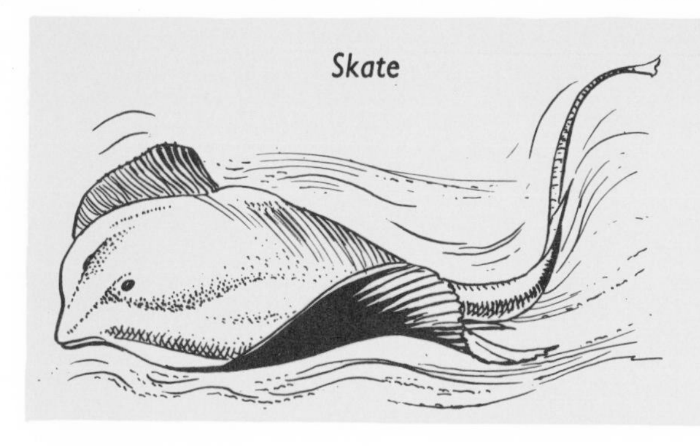

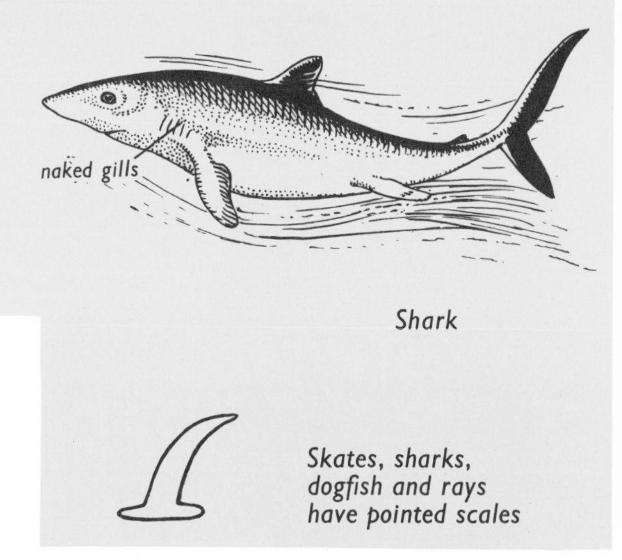

Remember that porpoises, dolphins, whales and seals are mammals and are related to warm furry land mammals such as dogs, cats and tigers. They must not be confused with fishes. Sharks, on the other hand, are fishes. They are related to the dogfish, skates and rays and have pointed scales like the ones shown here.

Sharks, dogfish, skates and rays belong to a special family of fishes called the *Cartilaginous* fishes because they all have gristly bones and pointed scales.

Most other fishes belong to the family of *Teleost* fishes which all have hard or " bony " bones and rounded scales. There are other differences too. Look at the mouth of the shark (above). You will see that it is *underneath* the head while that of the bony fish (such as the trout, salmon, mackerel or herring) is at the end of the head (see below). Look at the tail of both. The bony fish has an evenly divided tail, while the cartilaginous fish has the upper lobe of its tail larger than the lower one. You will see also that the bony fish has a cover over its gills, while the gills (behind the eye) of the shark, a cartilaginous fish, are just naked slits. You will remember that gills are the breathing apparatus of a fish.

Bony fish such as trout, salmon, mackerel and herring have rounded scales

A " bony " fish

SOME STUDIES IN NATURAL SCIENCE

This book helps you to study life in many of its different forms. Now that you are coming nearer to the end of the book, you will have realised what a great number of different sides there are to this study of life. The name given to the complete study of life is *biology*. This comes from a Greek word *bios* meaning " life ".

Biology is made up of two main studies which, in themselves, contain many others. The two main studies are *botany*, which deals with plants, and *zoology*, which deals with animals.

Within botany and zoology, or associated with these, are many branch studies. Here are a few of them :

Entomology— the study of insects
Ornithology—the study of birds
Piscatology—the study of fish
Mycology—the study of fungi (mushrooms, toadstools, etc.)
Conchology—the study of shells.

These studies are also sciences and the " -ology " part of each word means " a science ". Thus in " piscatology " the first part of the word means fishes (Latin—*pisces*) and the second half means " a science ", so that piscatology means the science of the fishes.

Now see whether you can name the sciences which deal with each of these five things :

ORNITHOLOGY

Ornithology is the study (or science) of birds. Although all birds have common features such as feathers and beaks, they differ from each other in many ways, such as in the colours of their plumage and in their beaks and feet. It is by these characters that we recognise them. We recognise a robin by its red breast, a wren by its small size and cocked-up tail, a chaffinch by its pink breast, slate blue head and white flashes on the wings. Try to remember how you recognise a thrush and a blackbird.

The birds mentioned above are common types found in gardens, hedgerows and woodlands in our own country. There are, however, many birds which are not so common and which have many special features. Think for instance of the swan, the heron, the eagle and the various ducks you see on park lakes. They are different from the other birds we mentioned. For instance, the swan has a long neck and can glide beautifully on the surface of the water. Ducks are also water birds and, like swans, have webbed feet, but their necks are not so long. Herons have long legs like storks and can wade out into lakes and streams, fishing with their long beaks. Eagles are birds of prey with large, strong bodies and wings, curved beaks and long talon claws.

In spite of the differences between bird types, most birds have features in common. One of the first things a good ornithologist must learn is the general shape and structure of a bird's body. Study this picture of a typical bird and make sure that you remember the names of the different parts of its body.

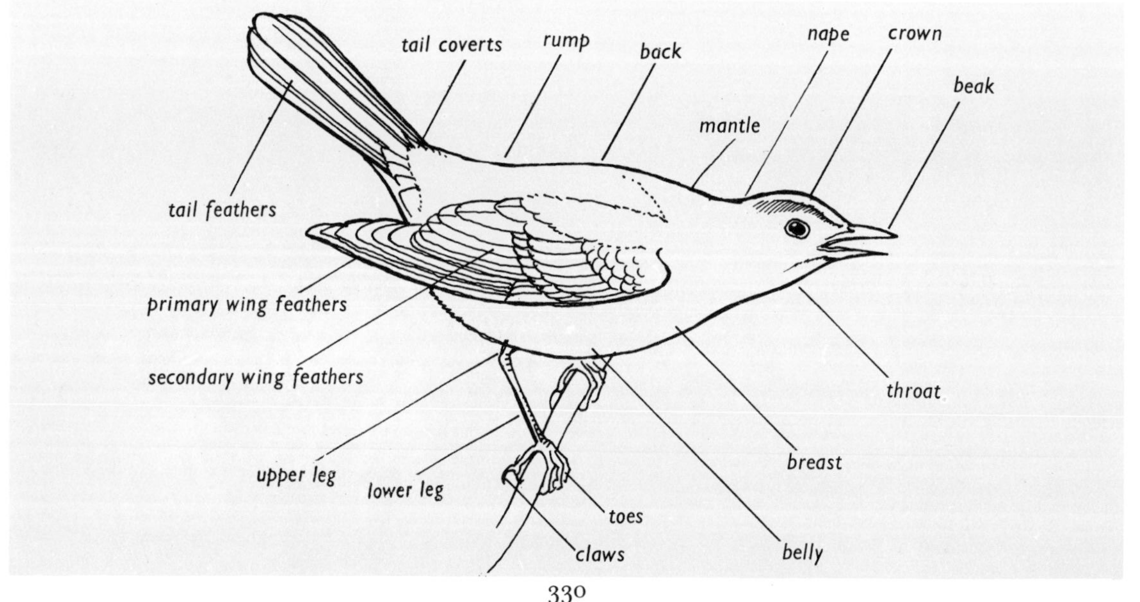

ENTOMOLOGY

Entomology is the study or science of insects. There are thousands of different insects and they are found in all parts of the world. Similar types are grouped into " orders ". For instance, there is one order called the *Hymenoptera* which includes bees, ants and wasps. Another order, the *Lepidoptera*, includes all the moths and butterflies. The *Coleoptera* includes all the beetles.

Look at these words again. They are Greek words and each one ends in *ptera*, meaning " wing ". *Lepido* means " scales ", so you would expect that Lepidopterans (moths and butterflies) would have little scales on their wings, which, in fact, they do. *Coleos* means " sheath " and beetles have hard wing covers or *elytra* which protect the flimsy wings underneath—thus their name *Coleoptera* or " sheath winged ". You can see one below.

Male great diving beetle (Dytiscus) *showing flimsy wings under the hard wing covers*

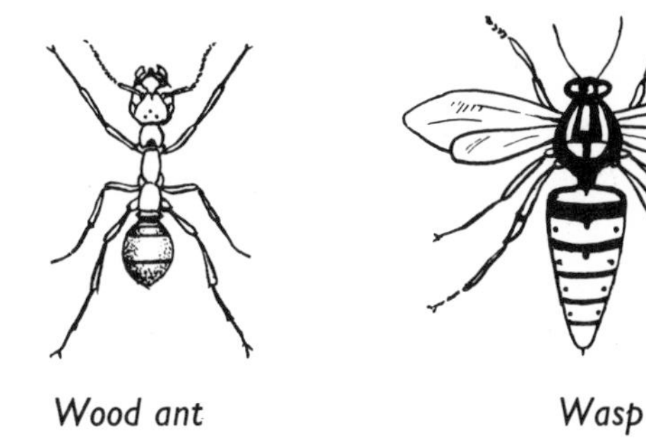

Wood ant *Wasp*

Some other insect orders are :

Orthoptera (straight winged insects), which includes grasshoppers and crickets. These have long, straight wings.

Hemiptera (half winged) includes all those insects we call " bugs ", such as plant lice and froghoppers.

Diptera includes the two-winged flies such as house flies and bluebottles.

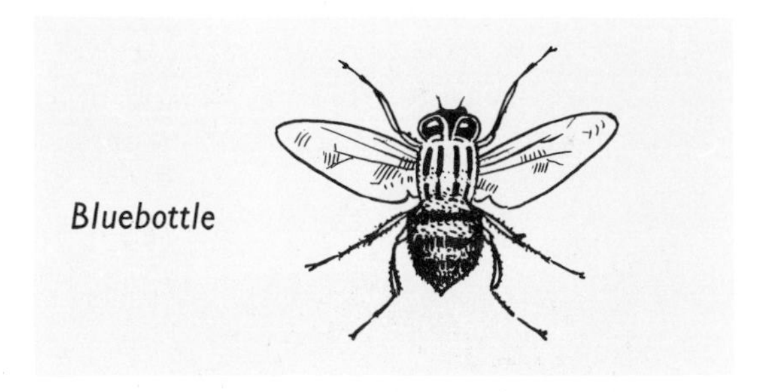

Bluebottle

One of the greatest entomologists of all time was a Frenchman, Jean Henri Fabre, who lived from 1823 to 1915. During the whole of his long life, he studied insects and wrote books about them. His books were translated into other languages, including English. You would enjoy these books by Fabre :—*Insect Adventures* and *The Life of the Fly*.

You have been introduced to the study of fishes (piscatology) on page 328, the study of birds (ornithology) on page 330 and the study of insects (entomology) on page 331. Among the other " ologies " mentioned on page 329 were *mycology*, the study of fungi, and *conchology*, the study of shells. All these " ologies " are special branches of the very large science, Natural Science.

Looking underneath the cap of two toadstools

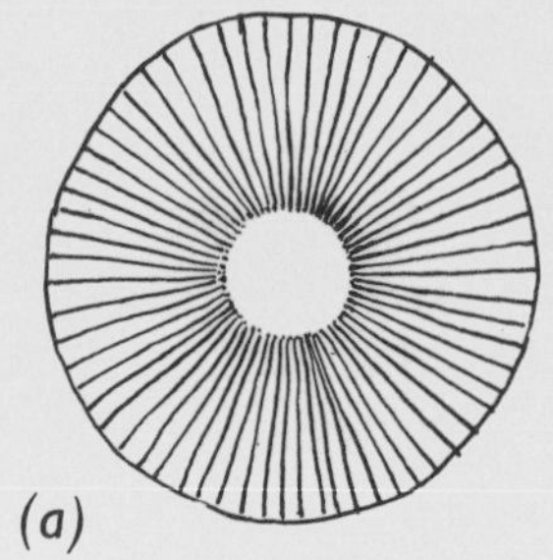

(a)

Toadstool with gills

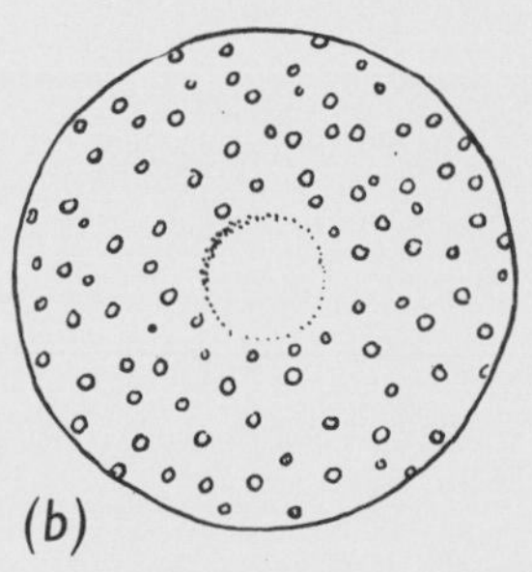

(b)

Toadstool with pores

MYCOLOGY, THE STUDY OF FUNGI (see also pages 300–303)

Fungi form a very large group of plants, none of which has any chlorophyll and all of which are therefore non-green. Some fungi such as *moulds* (see page 300) are very simple. Mushrooms and toadstools are higher types of fungi and have special types of body structure. They usually have fruiting bodies which often have the shape of a cap on a stalk, just like the mushroom that you pick or buy to eat. Underneath the cap are gills like diagram (a) or pores as in diagram (b).

Giant tuft

Some of the " pore " types of fungi do not have a " cap " and stalk, but are arranged rather like brackets growing on tree stumps. The one you see in the picture on the left is called the giant tuft and often measures 60 to 90 centimetres across.

There are some beautiful and dainty fungi called " fairy clubs ". This one on the right, whose proper name is *calocera*, is bright orange and is found on fir trunks in October.

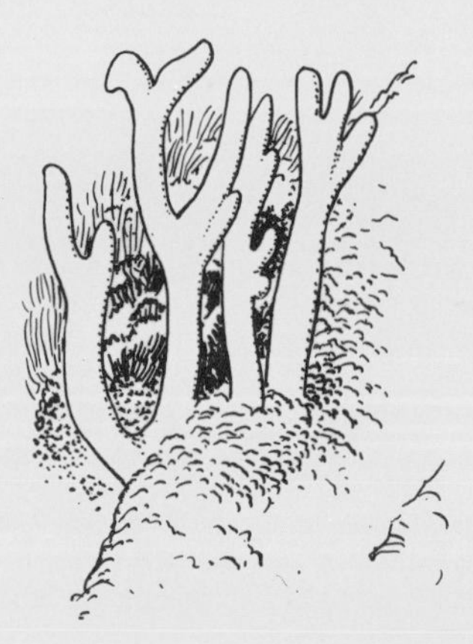

Calocera
(" fairy club ")

Conchology, the Study of Shells

If you collect shells, identify them and find out a great deal about them, you may eventually become a *conchologist,* a specialist in shells. In this book you are only introduced to these various studies. You must find other books which tell you more about them before you are really proficient, because often a whole book is written about just one of these lesser sciences.

Shells are the protective outer skeletons of the soft-bodied animals known as *molluscs.* Not all molluscs live in water—the common garden snail, for example, is a mollusc. A large proportion of molluscs do live in water and many of these live in the sea or in salt water pools (marine molluscs).

Most of the shells you pick up on the beach are empty. The soft bodies of the tenants of these shells have decayed, or they may have been eaten by sea birds. The oyster catcher, for instance, prizes open the double shells of mussels and other bivalves (two shelled molluscs) and eats the inhabitant. Look for the two shells in the mussel picture on the right.

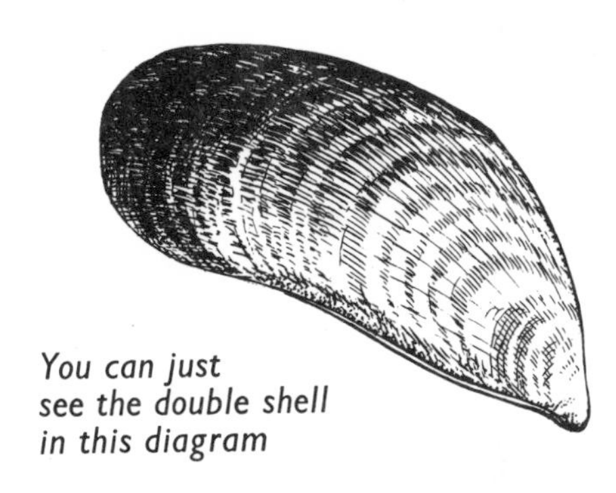

You can just see the double shell in this diagram

The shells of mussel — a bivalve

Molluscs which have only one shell are called univalves. The whelk in the picture below is a univalve.

The whelk is a univalve. You can see its shell in colour opposite page 337

Univalves are often considered to be a more advanced type of mollusc because the inhabitants of these shells have distinct heads with tentacles. Sometimes the creature's eyes are on the ends of the tentacles. Univalves have a rough, rasp-like tongue to help with their feeding.

Molluscs may be found in fresh and salt water. You will see some marine or salt water types in the colour page opposite page 337.

THE GREAT VARIETY OF LIVING THINGS

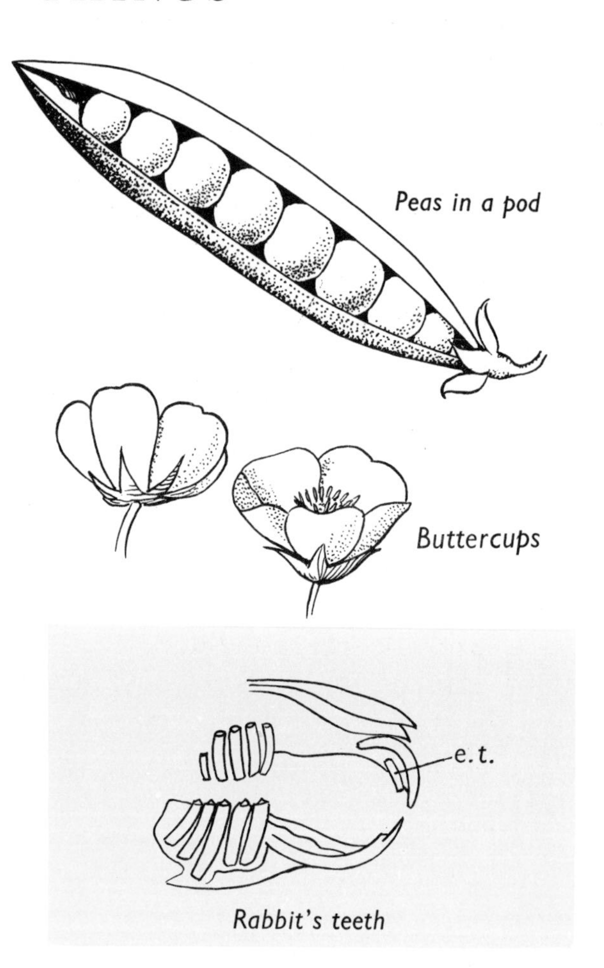

Peas in a pod

Buttercups

Rabbit's teeth

While you have been reading this book, you must have become aware of the great variety of form and structure which exists in the living world. No two things are really alike. Even " identical " twins have slight differences and we are told by scientists that no two peas in a pod are alike. All buttercup plants look alike, but there are many different kinds or *species* of buttercup, although all these " buttercup types " belong to one big family called *Ranunculaceae*.

When living things of different types have certain " likenesses " or qualities in common, we consider that they are related and put them together in the same family groups. Rats, mice and hamsters are alike in several ways, particularly in their teeth, so we put them in the same family of RODENTS.

Rabbits and hares also have gnawing teeth but, since they have two extra teeth at the back of their front ones, they are now put in a separate group called *Lagomorpha*. You can study these extra teeth (e.t.) in the diagram above.

Compare this diagram with the one of a true rodent's skull on page 308. Notice particularly the top front teeth.

Plants or animals of the same kind or species usually mate within that species. Where there is mating or a " cross " between plants or animals of *different* species, the result is called a " hybrid ". You have heard of hybrid roses, where two species of garden roses have been cross fertilised. There are hybrid animals too, such as " mule " canaries, and the mule itself is a hybrid—part horse, part ass.

FAMILY CHARACTERISTICS—Heredity

There are certain distinguishing features by which we recognise each plant or animal as a *species*, that is a type which differs from all other plant and animal types in the world. Species belong to families known as *genera* and members of one genus (singular of "genera") show family likenesses. Genera are themselves grouped into larger categories known as *orders* and similar orders are grouped together into *classes*.

Hare

Rabbit

Now let us look at two animals which are related to each other. We can see that the rabbit and the hare shown here are similar in body form and yet there are also certain differences. In both animals the backbone is curved and the hind legs are long, but we can see that the rabbit has a more rounded back than the hare which is a longer, leaner animal. The hare has distinguishing black tips to its prominent ears.

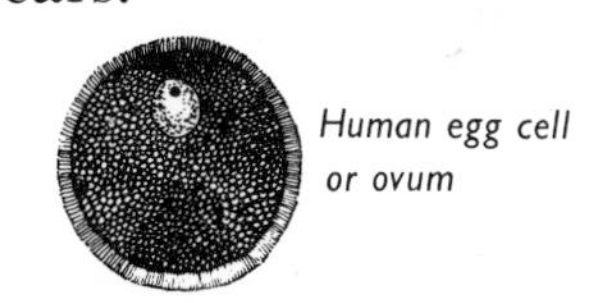

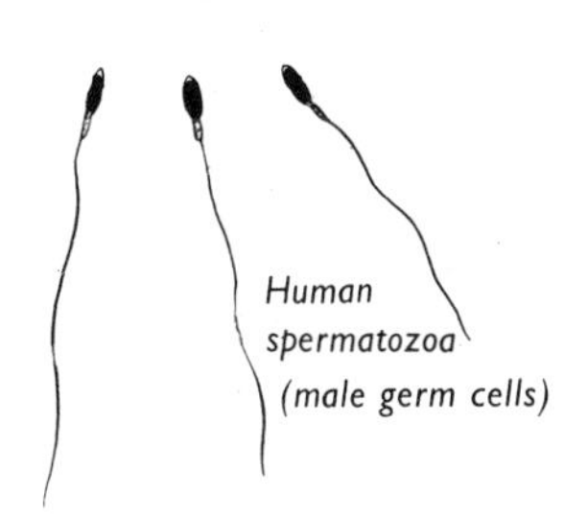

The union of one sperm with the egg will fertilise the egg

Look carefully and see what other differences you can find, and what similarities too.

Family likenesses are inherited from parents and reappear in successive generations in definite mathematical ratios, as Mendel discovered (see page 293).

It was many years after Mendel's death, however, before scientists could tell us how hereditary characters were passed from one generation to the next. We now know that the fertilised egg contains hereditary or genetic material from both parents—a double dose, in fact.

HOW HEREDITARY CHARACTERS ARE PASSED ON*

A new individual is formed when a sperm unites with an egg. Sperms and eggs are produced by male and female parents respectively. These are the " sex cells " and their correct name is *gametes.* Like other living cells, each gamete possesses a nucleus and within the nucleus is the important hereditary material which is known as *chromatin.* This sometimes appears as a network within the nucleus, but during the active division of a cell, it appears in the form of little bodies called *chromosomes.*

Cell division happens in living cells whenever new tissue is forming, as in plant shoots or in the elongating parts of roots or in every growing part of a young animal. This type of cell division is called *mitosis.* You will see later how the chromosomes behave in this kind of division.

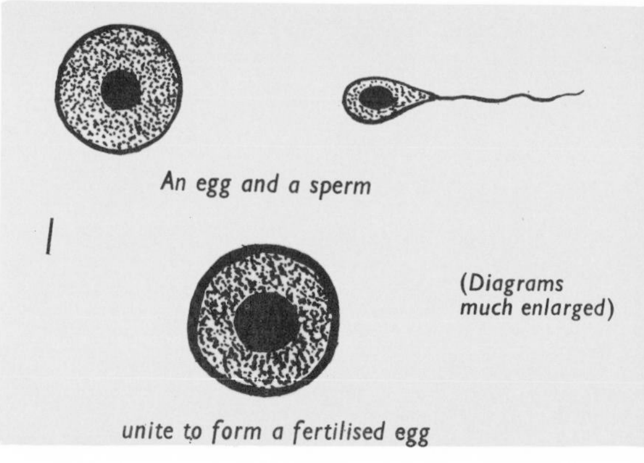

Gametes (eggs and sperms) have a special kind of cell divison in which chromosomes again play an active part, as you will see if you study the diagrams on the following pages. The main thing to remember now is that the number of chromosomes is halved in the gametes so that eggs and sperms each have half the number of chromosomes which are in the cells of the parents. When an egg and a sperm unite to form a new individual, the correct number of chromosomes is restored. The special division of gametes in which the number of chromosomes is halved is known as *meiosis.* You will see diagrams of this later.

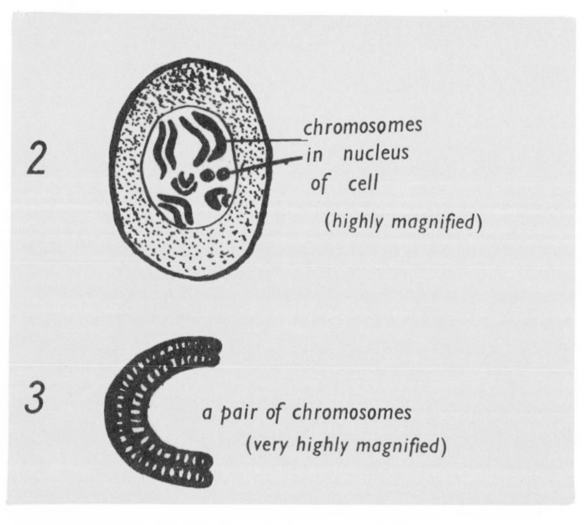

Chromosomes of various shapes are shown in the nucleus of a cell in Fig. 2. You will notice that the chromosomes are arranged together in *like* pairs. The markings on the chromosomes in Fig. 3 are *genes.* Each gene controls some special hereditary character, such as eye colour.

* Pages 336 to 340 give you more advanced information on cells and particularly on the behaviour of germ cells. You may find them difficult now but this knowledge will be most helpful to you in later studies in biology.

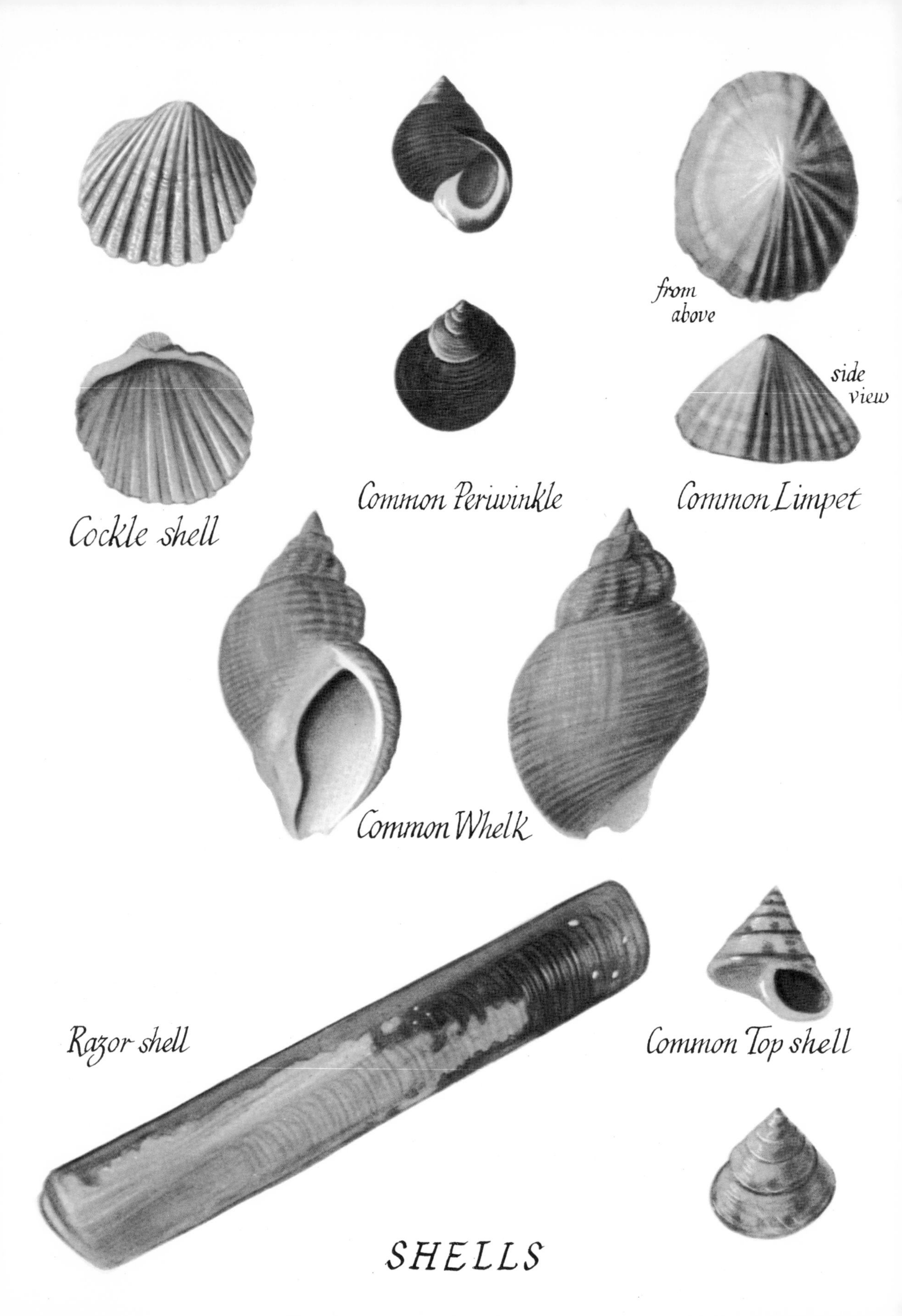
Cockle shell
Common Periwinkle
from above
side view
Common Limpet
Common Whelk
Razor shell
Common Top shell
SHELLS

THE BEHAVIOUR OF CHROMOSOMES—CELL DIVISION

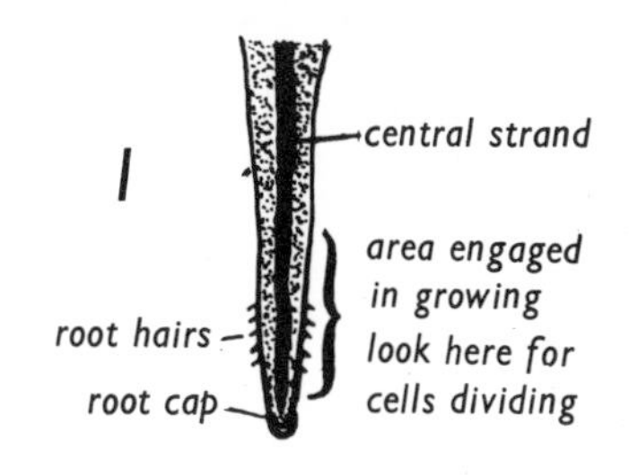

Young root tip

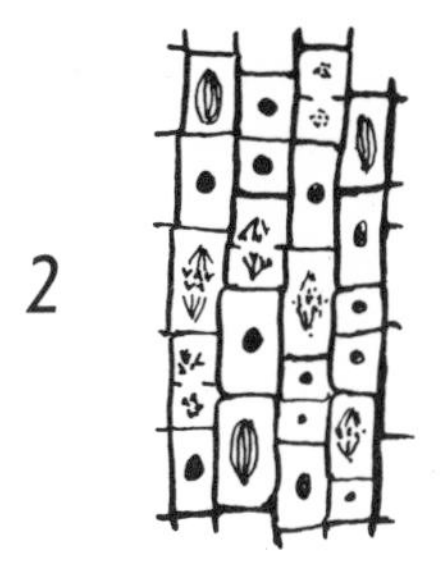

Part of a root tissue showing dividing cells

Each chromosome splits into two

Chromosomes can be seen clearly (with a powerful microscope) when young and active cells, like those behind a root tip, are dividing. In the growing areas of roots and shoots of plants, and wherever new tissues are forming in plants or in animals, microscope sections, such as those in Fig. 2, will show cells which are in the process of dividing.

The five diagrams across the bottom of the page show five stages in the behaviour of cells when they are dividing into two. The chromosomes appear at stage (b). You can see in stages (c) and (d) the separating chromosomes and the formation of two new cells, each with its own nucleus (e). The words with the diagrams explain the stages of development in more detail.

Since each chromosome splits along its length into two similar chromosomes (see Fig. 3), the new nuclei (plural of " nucleus ") will have the same number of chromosomes as the original cell. This " ordinary " kind of cell division is called *mitosis* or *mitotic cell division,* and is the one shown and described below. You will notice that at one stage, (b), a " spindle " appears. This is connected with the withdrawing of the chromosome " halves " to opposite ends of the original cell.

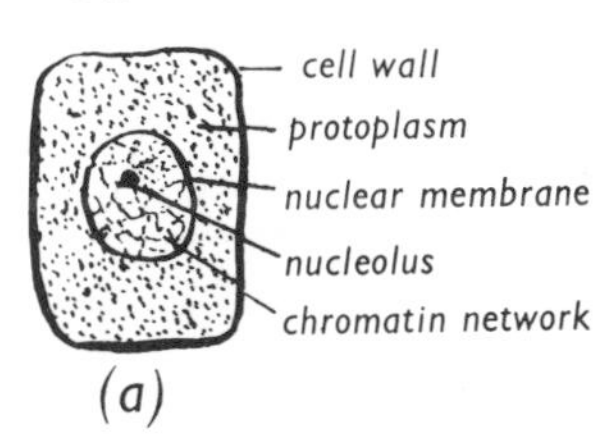

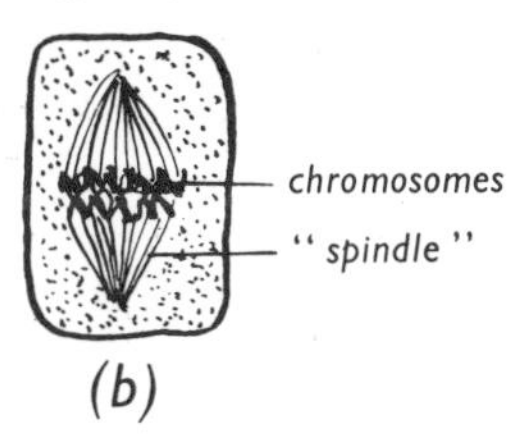

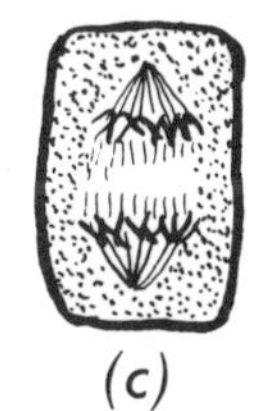

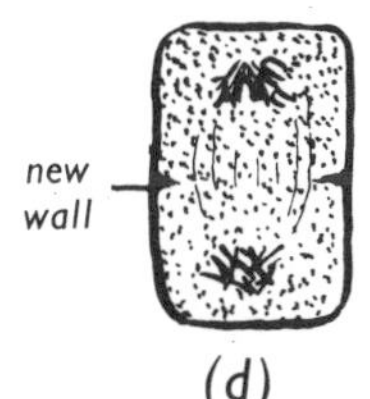

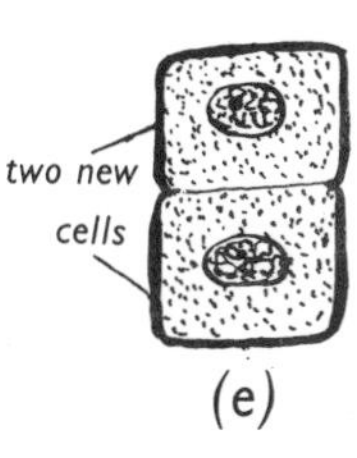

(a) Cell before division. Notice the network of chromatin within the nuclear membrane

(b) Nuclear membrane has disappeared. Chromosomes appear along " equator " of cell. A nuclear " spindle " forms

(c) Chromosomes split and move apart

(d) A new cell wall is forming. The chromosome halves are widely separated

(e) Two new cells are formed. The chromosome halves form the new nuclei and disappear in new chromatin networks

CELL DIVISION IN GERM CELLS

Those cells (germ cells) which produce the gametes (eggs and sperms) are like other cells in many respects. They each have a cell wall, protoplasm and a nucleus. The germ cells have special work to do, however ; they have to help to produce a new individual.

You have just learnt that in " ordinary " cell division, or mitosis, the chromosomes are halved so that each of the two new cells has the same number of chromosomes as the original cell. The number of chromosomes, therefore, remains constant throughout the cells of the new tissues.

In germ cells, however, the situation is different for when gametes (eggs and sperms) unite, they each give their chromosomes to the new fertilised egg, which develops into the new individual. You might think, then, that when an egg and a sperm unite, the fertilised egg and all the cells of the new individual will have twice as many chromosomes, and when *that* new individual produces offspring, their cells will each have twice as many again. But, of course, this does *not* happen. The number of chromosomes remains the same in new individuals because there is a halving of their number in the germ cells.

Germ cells which give rise to gametes (eggs and sperms) undergo a special kind of division known as *meiosis,* described on the next page. Meiosis is often called " reduction division " because it reduces the number of chromosomes in the gametes by half. When the gametes unite, the correct number of chromosomes is restored.

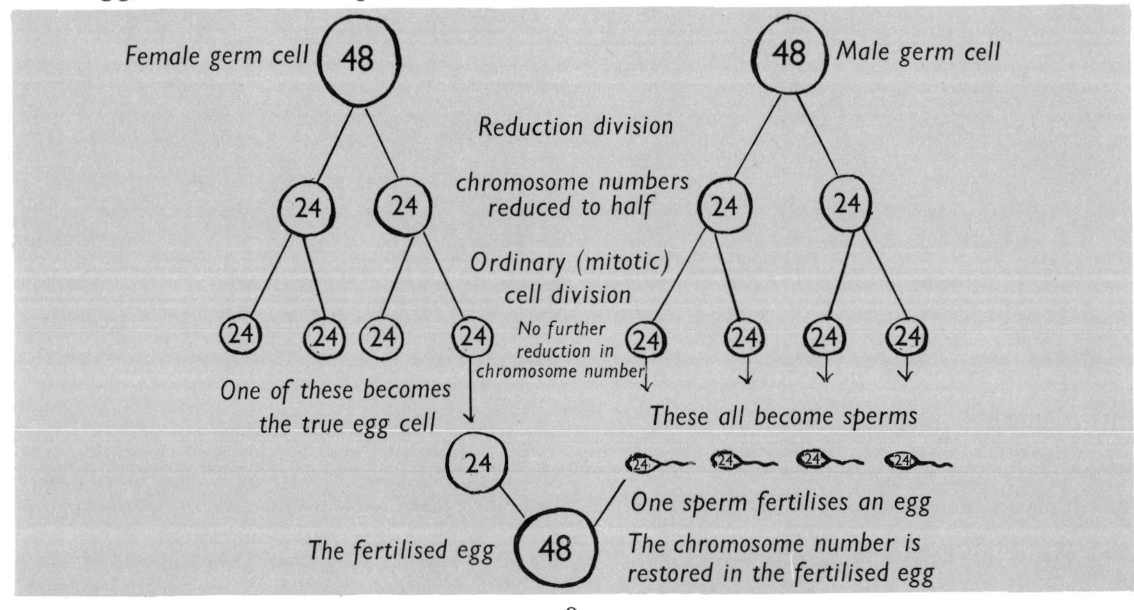

REDUCTION DIVISION (MEIOSIS)

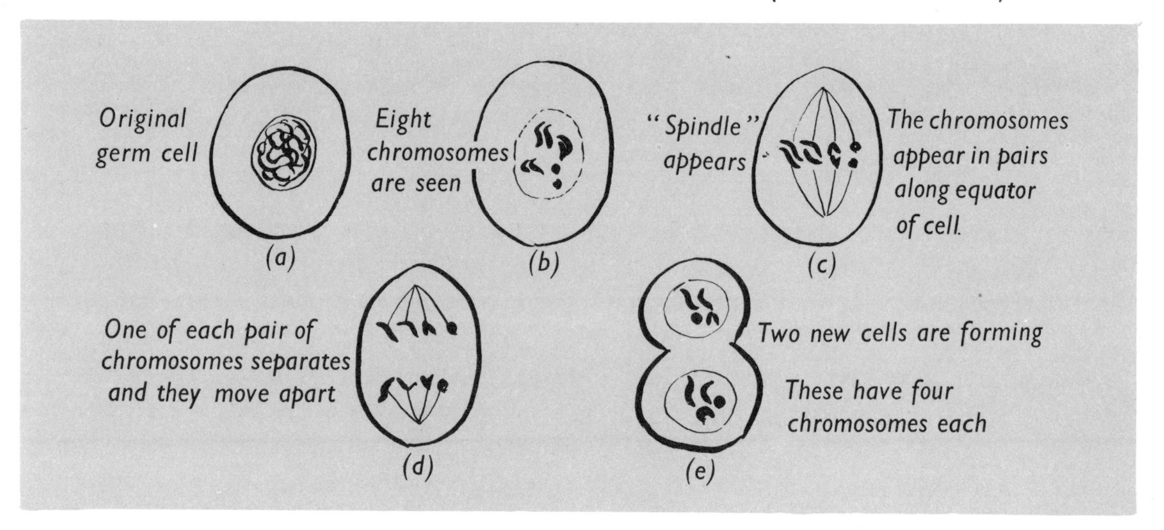

The diagram on the opposite page illustrates the reduction by half of the number of chromosomes in the gametes, so that when a sperm unites with an egg, the original number of chromosomes is restored. The diagrams on this page illustrate more clearly what happens to the chromosomes of the original germ cell in order that this " halved " number should occur. (In this germ cell there were originally eight chromosomes.)

Instead of splitting into two as in mitotic division (described on page   you will see from these diagrams that the chromosomes come together in like pairs and that one whole chromosome of each pair goes into each of the two new cells. These two new cells then divide again by ordinary mitotic division and there is no further reduction in the number of chromosomes. The first (reduction) division is called meiosis to distinguish it from the normal cell division —mitosis.

The germ cell which is undergoing meiosis (reduction division) in this particular set of diagrams contained eight chromosomes originally, which as you can see, are reduced to four in each of the two new cells. It is from these two cells that the gametes (either eggs or sperms) are produced and, as you know, when an egg and sperm unite, the original number of chromosomes is restored.

Original germ cells—male or female —each produce four cells. In the male, *all* these become sperms which *could* fertilise an egg. Only one of the four female cells becomes an egg. There are always more sperms than eggs, probably because sperms are often unlucky in reaching an egg. " Spare ones " are necessary.

THE "BLUEPRINT" FOR A NEW INDIVIDUAL

To sum up, our family characteristics or hereditary characters are passed down from parent to offspring by means of a wonderful mechanism in the germ cells.

These germ cells, like all other living cells in a plant or animal body, contain a *nucleus* which can resolve itself into *chromosomes*, and these chromosomes contain the *genes* which are the bearers of hereditary characters. You will have noticed with what care and exactness this chromatin material is halved when cells divide and the chromosomes split in two, half of each going to one new cell and half to the other (mitosis), and how the germ cells only retain half the original number of chromosomes when they divide (meiosis) so that the new individual will not have too many chromosomes in each cell.

Baby animals resemble their parents because the "blueprint" or plan of their structure is already in the genes on the chromosomes of the parents' germ cells.

SUMMARY OF PART FOUR

All possible worlds are open to the naturalist. We have asked you to look at Life from the simplest plants and animals to the highest forms of life.

We have discussed where living things get their energy for life. We found that green plants are the food-makers of the world, and that they absorb energy for their food-making from the sunlight.

In the cells of all living things is a specially active substance called protoplasm.

Living things prove that they are alive by growing and feeding and reproducing their kind. Some living things such as animals have various kinds of movement (locomotion) and even plants can make certain movements.

Animals and plants all begin life as a single cell. This cell goes on dividing until a new young plant or animal is formed. In the early stages, this little organism is called an embryo. Some of these animal embryos look very much alike.

Animals look after their young with varying degrees of parental care. The higher animals such as mammals or fur-covered animals usually make the best parents and look after their offspring for a considerable time and even train them. Birds are good parents too. Fishes do not usually look after their young. An exception is the stickleback.

On each of the five pages of the summary you will see illustrations of many of the things you have read about in this book. See how many of them you can name.

Summary

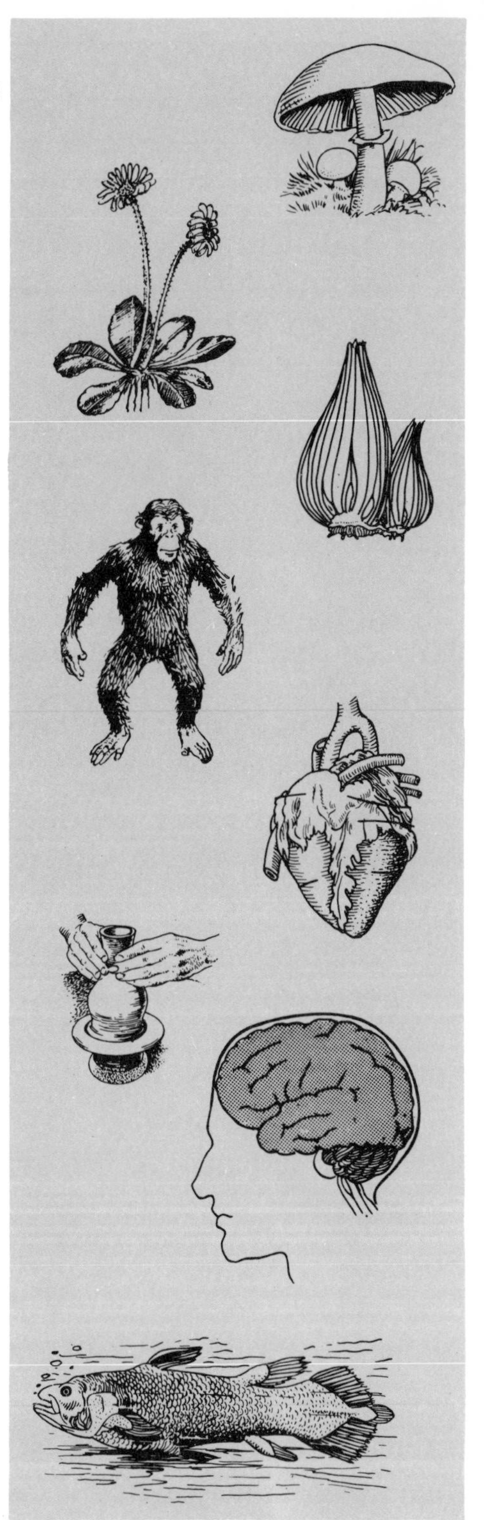

Lower plants have spores and higher plants have seeds. Spores and seeds are rather like animal eggs in that they are produced by the union of male and female cells—except in the case of certain kinds of spores.

Sometimes plants reproduce without using male and female cells. This kind of reproduction is called vegetative reproduction. Many of the vegetables we know are associated with it, such as bulbs (onions) and tubers (potatoes).

Cells form tissues in plants and animals, and tissues are often combined together in the form of organs, such as the lungs, heart, liver, etc. Each organ has some special function or work to perform. An organ of a plant would be its leaf or root.

One of the most important animal organs is the brain, the centre of the nervous system. True brains are found only in higher animals, although even worms have little nervous knots or ganglia which serve them very well.

The brain of man is better than that of any animal. The apes come nearest to man in their type of brain. Man has used his brain and hands to make use of his environment and has developed his superior intelligence to a remarkable degree.

Since life on this planet began, it has changed from simple to very complicated forms. Where did life first begin on earth ?

We have reason to believe that life started in the sea and that, as time went on, many animals and plants emerged from the water and became adapted to life on land. Not all animals and plants developed in this way. There are plants and animals today which still cannot live out of water (seaweeds and fishes), and there are some (amphibians) which live partly in water and part of their time on land, while still others, such as whales, have gone back to living in water although their ancestors must have lived on land.

The kind of creatures which lived in the past are revealed to us in fossil evidence, as in the rocks of the Grand Canyon of Colorado. Fossil evidence also shows how present-day creatures have developed and changed through the ages, e.g., the horse.

Comparing animal and plant structure and comparing embryos shows us that certain plants or animals are related in groups and that these groups could have had common ancestors.

Animals and plants transmit many of their own characters to their offspring. These are called hereditary characters. These characters often form a sort of equipment which enables the organism to live, obtain food, and defend itself against enemies.

Animal and plant defence often takes curious forms. Sometimes animals are disguised or " camouflaged " so that they fit in with their environment and are not easily detected by enemies, and we see examples of mimicry in both plants and animals for a similar reason.

Summary

Some of our beautiful plants and animals are in danger of becoming extinct unless we do something to preserve them. There are national societies such as the Nature Conservancy in this country and similar bodies in other countries or continents, as in Africa, which aim at the preservation or conservation of wild life. Two animals which came near to extinction are the mustang and the bison.

The wild life of other countries should be studied alongside our own for *comparison*. Animals in Australia show remarkable differences from our own animals, probably because the land mass of Australia was cut off from that of Eurasia a very long time ago and the animals in these two land masses developed along different lines.

Sometimes animals, and plants too, become overcrowded because they increase at an alarming rate. This can result in tragedy, as in the case of the Norwegian Lemming.

Animals and plants which are not indigenous, or native, to a country but which have been introduced into it are called " aliens ". Two such aliens, one animal and one plant, are the squirrel-dormouse and the Oxford ragwort.

Animals have their own methods of avoiding the disaster of cold and lack of food which occurs seasonally. Some hibernate and some migrate. Although winters in our own country are cold, we still have winter visitors, such as the Snow Goose, which comes from even colder climates.

How does a knowledge of Nature help man ?

Summary

In many ways it has helped him to understand and control disease, both in himself and in his crops and stock. Certain great scientists such as Pasteur, Fleming and Lister did valuable work which helped our understanding of bacteria and the way in which these may be either harmful or helpful. They were helped in their studies by the invention by Robert Hooke of an efficient type of compound microscope.

The study of Natural Science has many branches and the great numbers and varieties of plants and animals have been, and are still in the process of being, identified and classified by naturalists.

This great variety of living things strikes us more forcibly when we realise that no two living things in the world (even identical twins or peas in a pod), are really alike.

Animals and plants which are most alike belong to the same group or family, or are of the same species. Plant and animal families have many different species. Cross breeding of species produces hybrid types and there are many hybrid types of cattle and plants which have been produced by the intervention of man. Cross breeding can take place in Nature, although on the whole animals and plants tend to select mates from their own species.

Parents pass on to their offspring their own hereditary characters. The mechanism by which these hereditary characters is passed on is a wonderful one and you have read about the chromosomes and genes which bear the blueprints of the new individual.

TEST YOUR KNOWLEDGE

See how many of the following questions you can answer. Check your answers and give yourself a mark for each correct one.

1. What properties of living things distinguish them from non-living things?
2. What is the name of the living substance found in living cells?
3. What name is given to the pores on the back of a leaf?
4. What is "soil indicator" used for?
5. What is the name of the green colouring matter in plants?
6. What energy do green plants use in food-making?
7. What gases do we breathe out?
8. What is the commonest colour in Nature?
9. What is odd about *Mimosa pudica*?
10. Name a bird that has "gliding flight".
11. Give another name for a male gamete.
12. What is an immature baby animal or plant called?
13. What male fish builds a nest?
14. Name the brown spots on the back of a fern frond.
15. Name the "saddle" on an earthworm.
16. What is the "thinking area" of a brain called?
17. Where do you think life first began?
18. Name those animals which can live in water or on land.
19. Name the first "flying reptile" type of bird.
20. Give the name of a prehistoric type of horse.
21. Name an edible fungus.
22. Name the group of animals with gnawing teeth.
23. Name the protected bird on Havergate Island.
24. Name the protected mammal on Skomer Island.
25. Name the animal which commits mass suicide.

ANSWERS

1. Feeding, growing, reproducing, being sensitive. 2. Protoplasm. 3. Stomates. 4. Testing the acidity of soil. 5. Chlorophyll. 6. Solar energy (energy from sunlight). 7. Carbon dioxide and water vapour. 8. Green. 9. Its leaflets close up and droop when they are touched. 10. The gannet. 11. Sperm. 12. An embryo. 13. The stickleback. 14. Sori. 15. Clitellum. 16. The cerebrum. 17. In the water. 18. Amphibians. 19. Archaeopterix. 20. Eohippus, Mesohippus or Merychippus. 21. Morel or mushroom. 22. Rodents. 23. Avocet. 24. Skomer vole. 25. Norwegian lemming.

GLOSSARY

(Explanation of words and terms)

AMPHIBIOUS	living partly in water—partially AQUATIC
ANGIOSPERM	flowering plant
AQUATIC	living in water
BLASTULA	the partially-formed body of an animal EMBRYO
CARBON DIOXIDE	a gas in the atmosphere which is used by plants in food-making
CARNIVORES	animals which feed on flesh
CARPEL	the case which contains the SEEDS of a flower
CELLS	units of living PROTOPLASM
CHEMICAL FORMULAE	symbols which show arrangements of the elements in chemical substances—for example, H_2O stands for water which is made up of hydrogen and oxygen in the proportions of 2 to 1
CHLOROPHYLL	the green colouring matter in plants
CHLOROPLASTS	PLASTIDS which carry CHLOROPHYLL
CHROMOSOMES	bodies formed by the cell NUCLEUS during cell division
COLOUR INDICATOR FOR SOILS	a chemical substance which can be used to test the degree of acidity in soils
CYTOPLASM	the general PROTOPLASM of a cell (compare with NUCLEAR PLASM)
EGG	the unfertilised female productive cell of an animal
EMBRYO	a very young plant or animal—almost complete and at the stage when it is ready to be born or hatched, or is ready to grow into a fully developed ORGANISM
ENERGY	a power or force which enables living things to be active—other forms of energy are seen in light and heat
EPIDERMIS	TISSUES which form skin
EXTINCT	having died out—no longer in existence
FERTILE EGG	the fertilised egg cell or female reproductive cell of an animal
FERTILISATION	the union of male with female cells to make a SPORE or a SEED or a FERTILE EGG
FOSSILS	remains or imprints of once-living things found in rocks

Glossary

FRUIT	the ripe OVARY of a flower (there are ovaries in female animals too)
FUNCTION	a special work or task
GANGLIA	little knobs of nervous tissue
GASTRULA	the embryo of an animal a little further developed than the BLASTULA
GENES	the bearers of hereditary characters in the CHROMOSOMES
GEOLOGY	the science of rocks
GEOTROPISM	growing towards the earth's centre or responding to the force of gravity (as in the growth of plant roots)
HELIOTROPISM or PHOTOTROPISM	growing towards light as seen in plant shoots
HERBIVORES	animals which feed on vegetable matter
HUMUS	decaying material found in soil
INVERTEBRATE	having no backbone
MAMMOTHS	early (very large) land animals, now EXTINCT
NUCLEAR PLASM	the special PROTOPLASM in the NUCLEUS
NUCLEUS	the centre of control in a cell
ORGANISM	this word is useful—it can be used to mean either a living plant or animal
ORGANS	parts of the body made of certain TISSUES and devoted to special FUNCTIONS
OVARY (of a flower)	CARPEL or carpels which contain SEED and, when fertilised, will form a FRUIT
OVULES	unfertilised SEEDS
OXYGEN	a gas in the atmosphere—plants and animals use this gas in breathing
PARASITE	a plant which feeds entirely on another living plant
PIGMENT	colouring matter
PLASTIDS	small bodies within the cells which carry special substances —some carry CHLOROPHYLL (CHLOROPLASTS)
PLEUROCOCCUS	a simple one-celled plant
POLLEN GRAINS	special kinds of male or "sperm" cells which are found in flowering plants in the form of a yellow dust or powder
POLLEN TUBE	the pollen grain grows down to its OVULE by making a pollen tube

Glossary

PROTEIN — a body-building substance found in meat, fish, white of egg, etc.

PROTOPLASM — the substance of *life*

REPRODUCTION — the methods by which plants and animals multiply in number

SAPROPHYTE — a plant which feeds on the dead or decaying remains of other plants—and sometimes of animals

SEEDS — the reproductive bodies of higher plants

SEGMENTATION — a repetition of parts like the segments of an earthworm; also stages in the early cell division of an egg

SETAE — walking bristles of an earthworm

SPERMS — male reproductive cells in some plants and in animals—it is the union of a SPERM with an EGG or with an OVULE which results in a new animal or a new plant

SPIROGYRA — a green water weed made of strings of CELLS, each one with a spiral CHLOROPLAST

SPORANGIA — containers for SPORES

SPORES — the reproductive bodies of lower plants

STIGMA — the part of a flower which receives the pollen—often it has a sticky surface

STOMATES — the pore openings on the underside of leaves through which gases from the atmosphere can enter the plant

TENTACLES — " feelers "—or long outgrowths from the bodies of plants or animals which are sensitive to outside conditions—they are found in the hydra, in such animals as snails and in octopi, etc.

TISSUES — body layers made up of CELLS

TROPISMS — growth movements in plants

VEGETATIVE PROPAGATION / VEGETATIVE REPRODUCTION — reproducing or multiplying by other methods than by seeds (in plants)—usually by some form of budding—tubers and bulbs are organs of vegetative reproduction

VERTEBRATES — having a backbone

YOLK — food substance found in an animal egg and on which the EMBRYO can feed

YOLK SAC — often attached to the stomach of a young fish to provide it with food until it is able to obtain nourishment by mouth

BOOKS OF IDENTIFICATION AND REFERENCE

FLOWERS

British Wild Flowers C. A. Hall Black
Flowers of the Coast Ian Hepburne Collins
Garden Flowers H. R. Wehrhahn Burke (Young Specialist)
Wild Flowers Alois Kosch Burke (Young Specialist)
Wild Flowers W. J. Stokoe Warne (Observer Books)
Flowers and their Visitors Janet Davidson Black (Picture Information Books)

TREES

Trees and Shrubs J. Bretaudeau and J. G. Barton Hamlyn (Little Guides in Colour)
Trees and Shrubs W. J. Stokoe Warne (Observer Books)
Trees Clare Williams Black (Picture Information Books)

GRASSES, FERNS, FUNGI

Common Fungi E. M. Wakefield Warne (Observer Books)
Ferns Francis Rose Warne (Observer Books)
Grasses, Sedges and Rushes Francis Rose (Observer Books)
Mushrooms Pierre Montarnal and J. G. Barton Hamlyn (Little Guides in Colour)
Fungi George Parkinson Black (Picture Information Books)

FOSSILS

Fossils Frank H. T. Rhodes, Herbert Zim and Paul R. Shaffer Hamlyn (Little Guides in Colour)
Life Before Man: The Story of Fossils Duncan Forbes Black (Junior Reference Books)

POND LIFE

Pond and Marsh James Whinray Black (Picture Information Books)
Pond Life John Clegg Warne (Observer Books)
Pond Life Wolfgang Engelhardt Burke (Young Specialist)
Pond Life R. L. E. Ford Black (Young Naturalist's Series)

SEASHORE LIFE

Common British Sea Shells William S. Forsyth Black
Sea Shells of the World Tucker Abbott and Herbert Zim Hamlyn (Little Guides in Colour)
Seashore A. Kosch, H. Frieling and H. Janus Burke (Young Specialist)
The Naturalist on the Seashore E. M. Stephenson Black (Young Naturalist's Series)
Seashore Ian Murray Black (Picture Information Books)

FISHES

Fishes Tony Burnaud and Peter Whitehead Hamlyn (Little Guides in Colour)
Freshwater Fishes A. Laurence Wells Warne (Observer Books)
Know Your Fish Leonard Brown Black

REPTILES

Reptiles Alfred Leutscher Burke (Young Specialist)
British Snakes Leonard G. Appleby Baker

INSECTS

British Insects George E. Hyde Black (Young Naturalist's Series)
Know Your Butterflies C. A. Hall Black
Butterflies W. J. Stokoe Warne (Observer Books)
Butterflies and Moths Georg Warnecke Burke (Young Specialist)
Butterflies of the Wood S. and E. M. Beaufoy Collins
Insects Herbert Zim and Clarence Cottam Hamlyn (Little Guides in Colour)
Insects Matthew Prior Black (Picture Information Books)

BIRDS

A Bird is Born E. Bosiger and J. M. Guilcher Oliver and Boyd
Birds S. Vere Benson Warne (Observer Books)
Birds Bertel Bruun and Philippe Degrave Hamlyn (Little Guides in Colour)
Bird Watching Peter Clarke Newnes
Common British Birds W. Willett and C. A. Hall Black
Lesser Known British Birds W. Willett and C. A. Hall Black
More About British Wild Birds Eric Pochin Brockhampton

WILD ANIMALS

The Small Water Mammals Maxwell Knight Bodley Head
Wild Animals Keith Shackleton Nelson
Wild Animals of the British Isles W. J. Stokoe Warne (Observer Books)

ASTRONOMY

Stars Herbert Zim and R. H. Baker Hamlyn (Little Guides in Colour)

GENERAL

How to Use the Microscope C. A. Hall and E. F. Linssen Black
The Junior Naturalist's Handbook Geoffrey Watson Black
The Penguin Dictionary of British Natural History Penguin and Black (hard cover edition)
Conservation James Whinray Black (Picture Information Books)

OTHER BOOKS OF INTEREST

Animals after Dark Maxwell Knight Routledge and Kegan Paul
Born Free (The story of a lioness) Joy Adamson Collins
King Solomon's Ring Konrad Lorenz Methuen
My Family and Other Animals Gerald Durrell Hart Davis and Penguin
Out of the Wild Fred Speakman Bell
Ring of Bright Water Gavin Maxwell Longmans
Seal Summer Nina Warner Hooke A. Barker and Pan
The Book of Experiments Leonard de Vries Murray
Zoo-Man Stories T. H. Gillespie Oliver and Boyd

INDEX

(This index will help you to look up things quickly. The numbers refer to the pages of the book. When some of the page numbers of a subject are more important than others, they are printed in **heavy type**.)

D

E

F

G

H

T

V

W

Y

Z